An Analysis of Radiographic Quality

THIRD EDITION

Lab Manual and Workbook

Daniel P. Donohue, BS

AN ASPEN PUBLICATION ®
Aspen Publishers, Inc.
Gaithersburg, Maryland
1995

Library of Congress Cataloging-in-Publication Data

Donohue, Daniel P.
An analysis of radiographic quality / Daniel P. Donohue.—
3rd ed.
p. cm.
Includes bibliographical references and index.
ISBN 0-8342-0678-1
1. Radiography, Medical—Quality control. I. Title.
[DNLM: 1. Radiography—laboratory manuals. 2. Technology,
Radiographic—laboratory manuals. WN 25 D687a 1995]
RC78.D69 1995
616.07'572—dc20
DNLM/DLC
for Library of Congress
95-171
CIP

About Aspen Publishers • For more than 35 years, Aspen has been a leading professional publisher in a variety of disciplines. Aspen's vast information resources are available in both print and electronic formats. We are committed to providing the highest quality information available in the most appropriate format for our customers. Visit Aspen's Internet site for more information resources, directories, articles, and a searchable version of Aspen's full catalog, including the most recent publications:
http://www.aspenpublishers.com
Aspen Publishers, Inc. • The hallmark of quality in publishing
Member of the worldwide Wolters Kluwer group.

Editorial Services: Ruth Bloom
Library of Congress Catalog Card Number: 95-171
ISBN: 0-8342-0678-1

Printed in the United States of America

3 4 5

Table of Contents

Preface

The first edition of *An Analysis of Radiographic Quality* introduced the theory and principles related to the production of a recorded image of radiographic quality in a detailed, practical format that included text, workbook, and laboratory assignments. The student was introduced to the concepts, controls, formulas, and factors and learned to analyze the radiographic image by independently examining and assessing the influence of each of the multiple factors that contribute to the production of the radiographic image. The material was introduced logically and in a sequential format so that the student learned and profited by the previous lessons. Completing the review worksheets and performing the experiments helped to build the students' understanding and confidence so that by the completion of the textbook, they had achieved a practical, working knowledge and possessed the basic skills necessary to produce recorded images of radiographic quality.

The second edition in 1984 reflected changes in the terminology within the profession, introduced new technology, and expanded the format in several of the chapters.

The third edition presents the subject of radiographic quality in an entirely new direction based on the adoption of quality controls and the use of levels of radiation that are "as low as reasonably achievable"—the principles of ALARA. It is more than just a minor revision. Every chapter has been revised and expanded to provide greater explanations and examples of the application of the principles, formulas, and concepts that go into the recording of the radiographic image. The principles of ALARA have been introduced related to the elimination of motion, the reduction of repeat examinations, and the avoidance of unnecessary exposure to the patient, as well as emphasis on the accountability and responsibility of technologists in the performance of their duties. The significance and importance of the adoption of quality controls is introduced throughout the entire text.

The latest terminology has been used throughout the text; many figures have been revised; and new figures, diagrams, and tables have been added. The basic format of what to do, how to do it, and why it has to be done has been expanded with many additional examples. The workbook format has been retained and the chapter review worksheets expanded to include an additional 110 questions, going from 175 questions to 285, including 19 questions covering quality control in the new chapter introducing this important function. All of the laboratory experiments have been revised and reworked to reflect the current state of the art, and an additional nine experiments have been added, including experiments related to quality assurance, the influence of the central ray angle on radiographic density, and the proper application of radiographic grids, to name but a few.

The basic design of the film and intensifying screens related to the operation of film-screen imaging systems has been completely changed to reflect the current applications of these materials. An expanded section on the patient factors related to the thickness and opacity of the part has been introduced in the chapters on radiographic density and radiographic contrast. New sections dealing with macroradiography; milliamperage output; timer accuracy; automated exposure controls; central ray angle/source-image distance relationship; and the darkroom and film processing, including a new section on radiographic artifacts, have been added to the text. The entire section on radiographic grids has been revised to make the presentation more understandable with expanded directions and examples and new experiments and tables.

An entirely new chapter, "Quality Control," introduces three major elements associated with quality control: (1) the principles and elements of a quality assurance program and the accountability of radiologic technologists in the performance of their duties; (2) the introduction and application of the concepts of standardized exposure technique charts with examples of an optimum kilovoltage (kVp) variable mAs technique system; and (3) the application of a radiation safety program for radiology personnel, the patient, and other assisting personnel through use of the principles of ALARA in the radiology department.

In the final chapter, "A Summary of Radiographic Quality," a revised and expanded diagram identifying all of the multiple factors and influences affecting the radiographic quality of the recorded image and their position within the total production of a film of radiographic quality is provided.

In the preparation of this third edition, I thank a number of close friends within the profession as well as those instructors and program directors who were kind enough to provide suggestions and critiques of the previous edition.

1

Introduction

The production of an image of **radiographic quality** should be the goal of the radiologic technologist in every radiographic procedure.

The radiographic image represents the culmination of a complex procedure that involves many considerations and choices by the technologist. In order to perform radiographic procedures effectively and to minimize repeat exposures, the technologist must be familiar with the equipment, materials, and multiple exposure factors that are utilized in the production of the radiographic image. More importantly, the technologist must also recognize and understand the influences associated with the interrelationships and interdependencies of these technical factors in order to identify the quantitative and qualitative values related to their application in the production of an image of radiographic quality.

Of equal importance to the proper application of the technical factors utilized in the production of an image of radiographic quality are the ethical and moral directives associated with the application of x-radiation on human beings. The technologist must also be able to analyze the application of these principles in a responsible manner related to the radiation protection and safety of both the patient and the personnel performing the radiographic procedure.

All radiographic procedures should be performed by applying the principles of **ALARA**. That means that the radiographic image should be produced utilizing the least amount of radiation necessary to achieve an image of radiographic quality—a level of radiation that is **as low as reasonably achievable (ALARA).**

What is meant by an image of radiographic quality? An image of radiographic quality is one that possesses a sharpness of the recorded structural details and a clear visualization of the structural details recorded. Within this definition, there are actually two separate concepts to appreciate.

Sharpness of the Recorded Image

Examine the recorded image on the radiograph closely. Are the individual lines that actually form the image keen, distinct, thin-edged, and detailed as **recorded** on the film? Many of the technical factors contributing to the sharpness of the recorded details of the image are geometric in nature. That is, they are measurable factors related to radiographic quality. The amount of unsharpness that contributes to the loss of structural details within the recorded image can be accurately determined and objectively evaluated. Therefore, we can refer to the sharpness of the recorded details as a **geometric property** of the radiographic image.

Visibility of the Recorded Image

The visibility of the recorded structural details refers to the image's **photographic properties** and is controlled by the technical factors that contribute to the radiographic density and contrast of the radiographic image. Although a quantitative value for the production and control of radiographic density exists, the photographic properties of the image are dependent on the overall relationship between the radiographic density and contrast of the recorded image and would more appropriately be referred to as the **quality component of the radiographic image**. As such, the evaluation of the photographic properties of the image and therefore the visibility of the recorded details can frequently be more subjective than the assessment of the sharpness of the recorded details associated with the geometric properties of the radiographic image. To evaluate the visibility of the recorded details and, thus, the photographic properties of the recorded image, you must investigate the radiographic image for two distinct elements:

1. **Radiographic density** represents the quantity factor when applied to the visibility of the recorded details. The evaluation must determine whether a sufficient quantity of radiation has affected the radiographic film so that all of the structures of interest are recorded and visible.
2. **Radiographic contrast** refers to the quality factor when applied to the visibility of the recorded details. For this evaluation, you must determine whether acceptable differences have been demonstrated between adjacent recorded structures so that each structure of interest can be clearly identified and visualized as a separate, individual image.

Radiographic Quality

Utilizing the appropriate materials and exposure factors to produce a radiographic image that possesses excellent recorded details is of little value if the radiographic image possesses poor photographic properties and the structures of interest cannot be visualized. It is impossible to produce an image of radiographic quality if the recorded structures possess insufficient or excessive radiographic density or if inadequate radiographic contrast between the structures of interest has been demonstrated. In such an example, the technologist has not adequately visualized the structures of interest, even though they may have been recorded with geometric accuracy and possess excellent possible recorded detail. (**Perform Experiment 1.**)

It can also be concluded that a radiographic image that possesses excellent photographic properties demonstrating that the proper radiographic density and appropriate radiographic contrast have been achieved is no guarantee of the production of an image of radiographic quality. A radiographic image that possesses excellent photographic properties is of little value if the actual recorded details of the image are unsharp, indistinct, or blurred. (**Perform Experiment 2.**)

In order to produce a recorded image that can be properly diagnosed and accurately interpreted and one that possesses radiographic quality, an appropriate balance between the geometric and photographic properties of the recorded image is necessary. Images of radiographic quality must therefore achieve the proper balance of the best characteristics of these equally important properties of the recorded image (Figure 1-1).

The technologist has at his or her disposal a variety of equipment and materials, together with a number of exposure factors, formulas, and techniques, that can be employed and manipulated in order to produce a desired effect. It is important to recognize that the principles that govern the sharpness of the recorded details (geometric properties) are **different** from those that govern the visibility of the recorded details (photographic properties). Having a comprehensive knowledge and a clear understanding of these principles; the application of equipment or materials; or the selection and adjustment of the exposure factors, formulas, and techniques does not have to be a technical guessing game. An unsure technologist who has to guess about the application of the appropriate materials or the selection of the required technical factors for a given radiographic procedure is performing an injustice and disservice to the profession and an indefensible act on the patient. Such a technologist is unnecessarily producing inappropriate numbers of repeat exposures and examinations, thereby significantly increasing the patient's overall radiation exposure, contrary to the principles of ALARA.

The technologist must be fully knowledgeable of all of the principles and factors influencing the production of the radiographic image and have a thorough understanding of how each of the selected materials or exposure factors affects the recording of the radiographic image. Secure in this knowledge and understanding, the technologist can adjust or manipulate these materials or factors as required by a given set of circumstances and do so with little appreciable loss of radiographic quality. In those instances where adjustments in the exposure factors are necessary and may considerably affect the overall radiographic quality of the image, the technologist can analyze the needs of the examination and the patient; weigh the

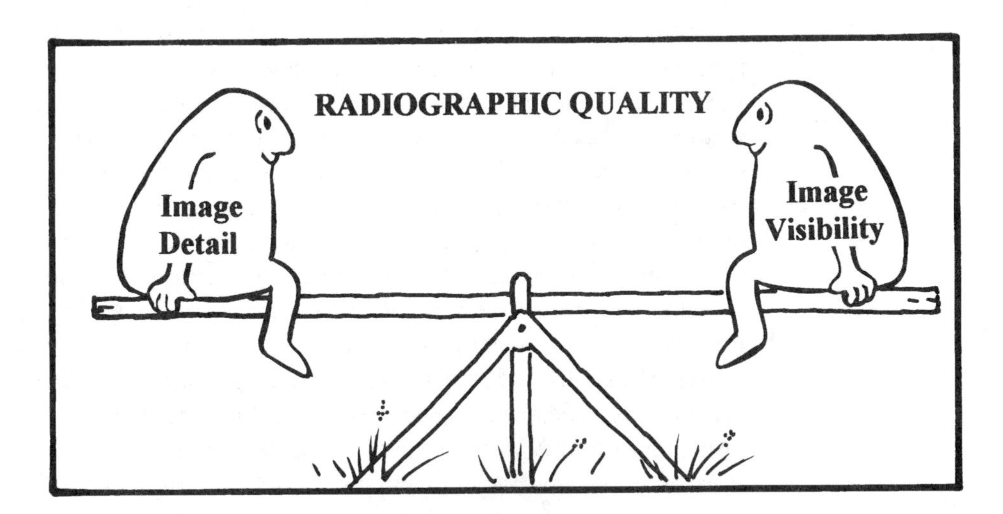

Figure 1-1 Balance between the Geographic and Photographic Properties of the Radiographic Image

various choices available; and, with an understanding of the principles involved, select those adjustments that are acceptable and will produce the least amount of radiographic quality loss.

An Analysis of Radiographic Quality

To analyze radiographic quality is to learn to recognize and understand the effects of the different equipment and materials and the multiple exposure factors that influence the recorded details and their visibility within the radiographic image. To accomplish this, the technologist must identify and examine each of the major properties of the radiographic image. After identifying the major properties, it is necessary to be able to distinguish the principles that influence each property, regardless of whether or not the influence is favorable to radiographic image quality.

As these image properties and the technical factors that influence them are examined, it will be observed that the manipulation of a single exposure factor frequently results in a change of more than one property of the recorded image. It is important for the technologist to be aware of these multiple influences as well as the interrelationship and interdependency of these factors on one another. By recognizing this, the technologist will be able to consider the proposed adjustment in light of the multiplicity of influences that the change may produce and determine whether the change is acceptable or not.

It is not uncommon for the technologist to have to select materials and/or exposure factors that may not produce a radiographic image of the highest quality. As an example, when performing an examination on a small child, the technologist may find it advantageous to reduce the required time of exposure in order to eliminate the loss of recorded detail associated with patient motion. To reduce the time of exposure for the given examination and thus the overall exposure, it may be necessary to select alternative radiographic materials, such as a higher speed recording system. However, the higher speed recording system may not be capable of recording the radiographic image with the same level of recorded details as the original, but slower, system. Nevertheless, it could be an appropriate alternative. In this instance, a calculated loss of recorded detail may be acceptable in order to eliminate the greater loss of recorded detail associated with the unsharpness produced by the movement of a child during the examination. A decision such as this would be a good example of the application of the principles of ALARA and would potentially prevent additional unnecessary exposure of the patient.

There are many instances when it may be necessary to accept a less than ideal trade-off related to the radiographic quality of the image as a result of the circumstances of the patient's condition, the specifics of the examination being performed, or the limitations of the equipment or materials available to the technologist. By analyzing the influence of the change(s) on the major properties of the radiographic image, the technologist can determine the impact of the change and determine whether the adjustment would be acceptable. Only by analyzing the principles underlying the production of the radiographic image and having an understanding of the influences and effects of the various materials and exposure factors can the technologist select those materials and factors that will minimize the loss of radiographic quality. A technologist cannot be expected to perform radiographic examinations with accuracy and repeatability without a comprehensive knowledge and understanding of the principles contributing to the quality of the radiographic image. The ability to analyze radiographic quality is essential to the elimination of repeat exposures and examinations, minimizing patient exposure; to the application of the principles of ALARA; and to the elimination of the pitfalls inherent in the technical guessing game. **An analysis of radiographic quality** is the only way to ensure the production of an image of radiographic quality.

Major Radiographic Properties

The major properties associated with the production of an image of radiographic quality are reviewed below.

Major Properties of the Radiographic Image

Recorded detail
Distortion
Density
Contrast

- **Recorded detail.** The actual formation of the structural lines of the image recorded. The sharpness of the recorded details of the individual lines contained within the radiographic image produced.
- **Distortion.** The misrepresentation of the size and/or shape within the radiographic image produced compared with the actual size and shape of the structures being recorded.

Recorded detail and distortion refer to the geometric properties of the recorded image. The materials and exposure factors involved with this property of the radiographic image can be measured against known values and can be expressed in terms of image unsharpness (loss of image detail) and image misrepresentation (distortion). In an analysis of the radiographic quality of an image, recorded detail and distortion can be loosely identified as opposing values or at least as inversely related to one another. That is, as the distortion of the radiographic image increases, the recorded details of the image decrease.

- **Radiographic density.** Demonstrates the completeness of the recording of the image and the ability to visualize that image, indicating that sufficient quantities of radiation have traversed the body part being examined and have been recorded on the radiographic film. Radiographic density is represented by the overall blackening of film emulsion in different amounts dependent on the quantity of radiation employed and the size and opacity of the body structures being examined.
- **Radiographic contrast.** Describes the ability to distinguish each recorded structure separate from adjacent structures, indicating that proper penetration of the body part being examined and the control of radiographic fog were achieved. When we are discussing radiographic contrast, we are actually describing the relationship between the different radiographic densities recorded. Of all the major properties of the radiographic image, radiographic

contrast can prove to be more difficult to achieve or assess. Proper radiographic contrast for a given examination may differ as a result of the purpose for the examination and may be considered somewhat subjective in its evaluation. As an example, the radiographic contrast required when performing an examination of the chest for the investigation of a possible rib fracture would be totally inappropriate when performing an examination of the chest for potential lung pathology.

Radiographic density and contrast refer to the photographic properties of the recorded image. The completeness of the recording of the image and the proper visibility of that image are dependent on achieving a proper relationship between these photographic properties. Figure 1-2 summarizes the relationships among the four major factors involved in the production of an image of radiographic quality.

In the following chapters, you will investigate each of the four major properties of the radiographic image. You will examine the multiple technical factors that control and influence these properties and identify the interrelationships and interdependencies among them. You will quickly learn how to apply these factors in order to produce an image of radiographic quality.

You will discover that there is a difference between a diagnostic radiographic image and an image of radiographic quality. All radiographic images **must** be of diagnostic quality, but not all images are capable of measuring up to the standards of radiographic quality. You will soon be able to examine a radiographic image and determine whether it meets the criteria required for a diagnostic quality image. This is an admirable goal, even though the technologist should always strive to produce images of radiographic quality. You will also learn to analyze the image to determine exactly what properties of the image require improvement and how you can utilize your equipment, choose your materials, and manipulate your exposure factors in order to produce the changes necessary to achieve the highest quality radiographic image possible. This is the purpose of *An Analysis of Radiographic Quality.*

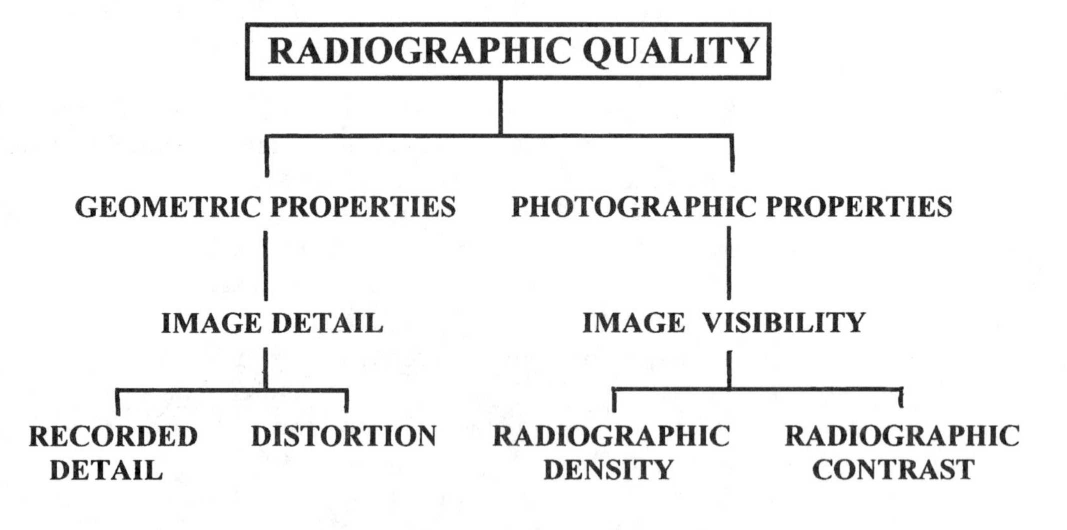

Figure 1-2 A Basic Outline of the Radiographic Quality Relationship

Introduction Review Questions

Name ____ Date____________________

1. Define **ALARA** as applied to the production of a radiographic image.

2. Why should the application of the principles of ALARA govern the conduct of every radiographic examination?

3. Describe the proper relationship between the geometric and photographic properties of an image of radiographic quality.

4. The recorded details of the radiographic image refer to:
   a. the geometric properties of the recorded image
   b. the photographic properties of the recorded image
   c. the ability to interpret the recorded image properly
   d. the distortion of the size or shape of the recorded image

5. The visibility of the radiographic image refers to:
   a. the geometric properties of the recorded image
   b. the photographic properties of the recorded image
   c. the ability to interpret the recorded image properly
   d. the distortion of the size or shape of the recorded image

6. Of the four major properties contributing to the radiographic quality of the recorded image, identify the geometric properties and provide reasons for your choice.

7. Of the four major properties contributing to the radiographic quality of the recorded image, identify the photographic properties and provide reasons for your choice.

8. Is it possible to produce a radiographic image possessing excellent recorded details and still classify it as lacking in radiographic quality? Explain your answer.

9. Will producing a radiographic image that possesses excellent image visibility assure you that the resultant image will be of radiographic quality? Explain your answer.

10. Discuss the significance of the relationship between the image details and the image visibility related to radiographic quality.

11. Discuss the interrelationship between major properties of the radiographic image as it relates to the selection of equipment, materials, or exposure factors.

12. Describe the contributions of the four major properties of the radiographic image to the overall radiographic quality of the image.

13. Describe how you could avoid the mistakes inherent in the technical guessing game, and relate your answer to the application of the principles of ALARA.

2 Recorded Detail

The recorded details of the radiographic image are analyzed by examining the multiple factors that affect the resolution capabilities and the quality of the linear patterns that contribute to the recording of the image. Image resolution refers to the actual number of structural lines that define the recorded image. The quality of the image resolution refers to the distinctness or sharpness of each individual line that contributes to the recording of the radiographic image.

The technologist must recognize the fact that a geometrically sharp image is impossible to achieve. One or more of the multiple factors that contribute to the formation of the recorded image will always produce some unsharpness. Herein lies the key to an understanding of the principles related to the radiographic property of recorded detail. It is far more practical to identify the **degree of unsharpness** that can be tolerated within the recorded image than to identify the amount of recorded detail that has been achieved. Image unsharpness is present to some degree in every radiographic image; however, it is not considered detrimental to radiographic quality until it reaches a level where it is actually visible and begins to detract from the appearance of the recorded image.

Geometry is that branch of mathematics dealing with the properties, measurement, and mutual relationships of points, lines, surfaces, and solids. The recorded image is actually created by various amounts of minute strands of silver deposited within the emulsion of the film. These strands of silver have a measurable value of length and width. Additionally, the overall radiographic image is produced by the specific relationship of the source of radiation, the object being examined, and the recording media. As such, the level of unsharpness associated with the materials selected and the geometric arrangement of the equipment organized by the technologist can be controlled. The level of unsharpness produced by these relationships can be accurately measured and evaluated utilizing the mathematical principles of geometry.

The average visual acuity of the human eye when inspecting a radiographic image at the **normal viewing distance of 18″ to 24″ (46–61 cm)** enables the viewer to visualize and to discriminate up to a total of 5 line pairs within the space of a millimeter. A **line pair** (LP) refers to the width of a line and the width of the space between it and an adjacent line. An evaluation of a radiographic image that has recorded 5 line pairs/mm would reveal a total of 5 lines and 5 spaces visible in each millimeter. Measuring the 5 LP/mm, we would determine that each of the 5 lines and spaces would measure 0.1 mm. Therefore, the recording of each line pair requires a total space of 0.2 mm. Since up to 5 LP/mm are visible with average visual acuity, it is obvious that an intolerable loss of image detail exists when the width of a recorded line pair measures more than 0.2 mm. A radiographic image that records a line pair in a space wider than 0.2 mm has already begun to demonstrate unsharpness.

LEVEL OF UNSHARPNESS
Visibility = 5 LP/mm
>0.2 mm line pair width = intolerable level

This does not mean that a radiographic image with a loss of recorded detail exceeding 0.2 mm is automatically identified as a poorly defined image or one that possesses intolerable levels of unsharpness. What it does indicate is that the visible level of unsharpness begins at this point and becomes progressively worse until the image details are visualized at a level that is unacceptable.

The following factors influence the degree of unsharpness recorded in the image and thus affect the radiographic quality of the image.

Motion
Materials } Measurable
Geometry

The level of unsharpness attributed to the materials selected and the geometric arrangement of the equipment utilized for any given examination are easily identified and measured. The loss of recorded detail attributed to motion, although a greater influence on the recorded image, is far more difficult to control, measure, and quantify.

Motion

Motion is the single most detrimental factor contributing to a loss of recorded detail. In diagnostic radiography, certain examinations, such as those performed on infants and children, emergency and operating room procedures, and mobile (portable) radiographic examinations, are especially susceptible to problems associated with patient motion. All possible methods of eliminating motion should be employed. Control of motion must be the technologist's first consideration in the selection of equipment, materials, and technical exposure factors for every radiographic procedure.

The loss of image detail associated with motion can be classified into two types: (1) voluntary and (2) involuntary. **Voluntary motion** is generally referred to as motion that is under the conscious control of the will. The majority of patients can and will "hold still" and maintain the specific position required by the technologist during the radiographic procedure. However, the problem of voluntary motion control becomes obvious when you are confronted with an injured patient who cannot hold still, an elderly patient who is unable to maintain an uncomfortable position, a patient who is unable to understand the directions provided by the technologist, or the infant or child who refuses to hold still for the examination. The control of problems associated with voluntary motion depends on the knowledge and skills of the technologist. It includes a suitable explanation of the procedure; the utilization of materials to aid in the comfort of the patient and to assist him or her in achieving and maintaining a difficult or uncomfortable position; and, when necessary, the proper use of immobilization devices to eliminate patient motion. Obviously, the selection of exposure factors that permit the most rapid time of exposure are among the primary considerations when attempting to control patient motion. However, the most rapid exposure time is of little value if the patient cannot or will not hold still for the procedure. Therefore, the control of voluntary motion requires a multifaceted approach.

Involuntary motion, on the other hand, presents an altogether different consideration for controlling recorded detail loss. Involuntary motion is the motion attributed to the physiological activity of the systems of the body. Since the **rate of motion** (number of motions occurring per second) and the **amplitude of motion** (distance traveled or total amount of measurable motion) differ from system to system and frequently from organ to organ within the same body system, the technologist must have a thorough knowledge of human physiology in order to recognize and understand the problems that involuntary motion present in a given radiographic examination.

An example of involuntary motion is the movement of the heart associated with the cardiac cycle of the circulatory system. An explanation of the procedure to patients in order to gain their confidence and assistance, although necessary with every examination, is of little value in controlling involuntary motion. Immobilization of the internal organs is also impossible. Therefore, in this instance, as with most problems associated with involuntary motion, the selection of materials and exposure factors that enable the technologist to perform the procedure by utilizing the most rapid exposure time is fundamental to considerations involving motion.

With each contraction of the heart, an amplitude of motion approximating 5 mm is produced. In a cardiac cycle beating an average of 60 times a minute, each cycle represents 1 sec. The total involuntary motion associated with the cardiac cycle would be 5 mm/sec. The actual motion of the cardiac cycle will be recorded within a 5-mm space on the radiographic film. Recognizing that our visual acuity enables us to discriminate up to 5 LP/mm, for all practical purposes, the 5-mm/sec motion associated with the rate and amplitude of the cardiac cycle represents the potential loss of recorded detail of 25 LP/mm/sec (5 LP/mm × 5 mm = 25 LP/mm/sec). An exposure time of 1 sec would produce a total blurring of the recorded image within the 5-mm space associated with the motion of the cardiac cycle. Considering this, the time of exposure required to eliminate the loss of image detail associated with the involuntary motion of the cardiac cycle can easily be determined. By dividing the total recorded detail potentially visible in the 5 mm represented by the recorded image of the cardiac cycle (25 LP/mm/sec) by the time required for the total involuntary motion (1 sec), you will determine that the maximum time of exposure to prevent recorded detail loss would be 0.04 (1/25) sec. Exposure times greater than 0.04 (1/25) sec will begin to demonstrate involuntary motion within the recorded image of the heart (Figure 2-1).

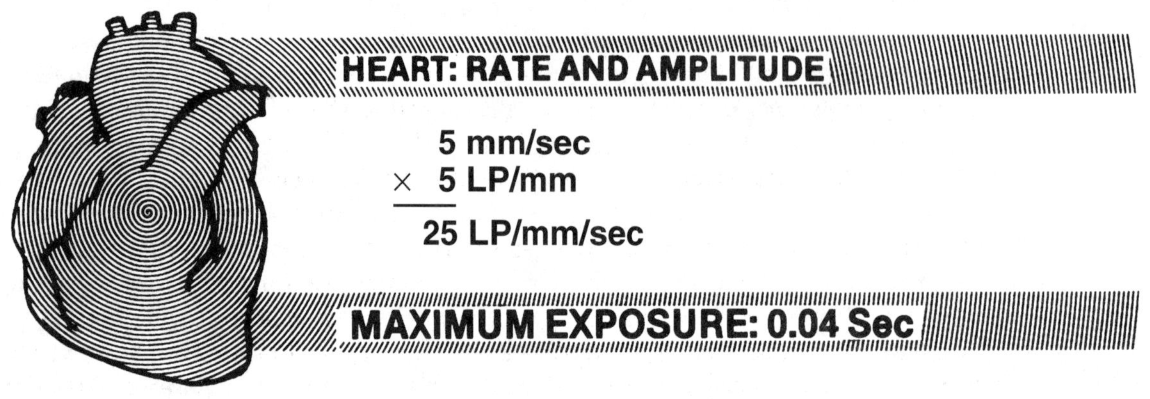

Figure 2-1
Calculation of Exposure Time To Eliminate Motion Unsharpness Resulting from Heart Contraction

It is important to recognize that the motion attributed to the circulatory system differs from the motion associated with other systems. The rate of motion and thus the involuntary motion that occurs in the gastrointestinal system is less than that of the cardiac cycle, although the amplitude of the motion is greater. In the gastrointestinal system, the peristaltic motion of the stomach, small intestines, and colon is more associated with a tonic spasm and is considerably less frequent than the motions of the cardiac cycle. The movements of the stomach associated with the digestive process are more intense and represent a greater rate and amplitude of motion than the slower, more controlled motions associated with the movement of solid wastes through the colon. In the genitourinary system, the movements associated with the physiological functions of the kidneys and the flow of urine through the ureters are more rhythmic and controlled and less spasmodic than the motion associated with the gastrointestinal system. Overall, the time of exposure required to eliminate involuntary motion associated with the gastrointestinal and genitourinary systems is less critical than the time of exposure required with the cardiac cycle of the circulatory system. Exposure times up to 0.5 (1/2) sec may be selected without demonstrating a visible loss of recorded detail. For each organ system, the rate and amplitude of involuntary motion must be considered before the time of exposure necessary to eliminate loss of image details can be properly selected.

Control of Motion

The control of motion, and therefore the elimination of unnecessary repeat exposures, is a practical application of the principles of ALARA. Examples of methods utilized to control motion are listed below:

FACTORS IN THE CONTROL OF MOTION

1. **Explanation of the procedure**
2. **Position achievers and maintainers**
3. **Immobilization of the part**
4. **Reduction of exposure time**

Explanation of the Procedure

A proper explanation of the procedure is necessary in all radiographic examinations regardless of whether the problem is voluntary or involuntary motion. How comprehensive and detailed an explanation will be is guided by considerations such as the patient's age, condition, and ability to understand. Talk with your patients, not at them. Meeting patients in a positive manner and talking with them to ascertain their level of understanding and their ability to cooperate is the first step in the explanation process. Never assume that patients have ever had a previous radiographic examination. Gaining patients' trust and placing them at ease by discussing and explaining what is going to happen will help to ensure their understanding and cooperation. When patients understand what the procedure involves and exactly what is expected or required and why, they will frequently assume the most difficult of positions and maintain them for lengthy periods of time in order to ensure a successful examination. Most patients want to cooperate. Providing them with an explanation of the procedure will help you to establish a supportive atmosphere and ultimately enable you to gain their cooperation.

Never assume that, as a result of the patient's age or condition, an explanation is unnecessary. Even patients in a semicomatose state may have a level of hearing of which you are unaware. With infants and children, be sure to explain the procedure to the parent or family member present and to adjust the explanation to the children's level of understanding. Be sure to involve the children in the examination and talk with them about what you are doing in addition to explaining to and reassuring the parent. Although a detailed explanation would be inappropriate with infants, a pleasant, well-modulated voice and reassuring gestures and methods of handling infants can do much to relax and calm them and make your job easier. With elderly patients, do not assume they cannot hear or will not understand you. Above all, adjust your explanation without assuming a tone of condescension or adopting a patronizing attitude. All patients are entitled to be treated with dignity and respect. For patients who are deaf or speak a foreign language, request assistance from family members or other staff members who use American sign language or speak the foreign language. If this is not possible, use gestures, use demonstrations, and, if necessary, draw simple instructional pictures in an effort to gain patients' understanding and cooperation.

Position Achievers and Maintainers

Frequently, the problems associated with motion is not the patient's lack of understanding or desire to cooperate, but an inability to achieve or maintain a difficult or uncomfortable position required for the radiographic examination. The inability to achieve or maintain a required position may occur as a result of the patient's age, physical condition, or injury. An elderly patient may be unable to straighten a limb completely or, due to a condition such as kyphosis or scoliosis, may find it difficult to lie down upon the hard surface of the examination table. A normally thin patient or one who has suffered severe weight loss as a result of a disease may find it uncomfortable or even painful to have to lie on the bony processes of the spine or pelvis. An obese patient may find it difficult to breathe in a recumbent position or may have difficulty achieving or maintaining the required position due to excessive weight and size. In the case of an injured patient, the pain or discomfort associated with placing the injured part into a specific position upon the examination table may prevent him or her from achieving or maintaining the required position. In each of these examples, the patients want to cooperate and assist you in the performance of the examination, but circumstances beyond their control make it difficult or impossible for them to achieve the required position or to maintain it during the examination.

In these situations, specific materials and methods must be employed in order to assist the patient to achieve the required position and to maintain it until the exposure can be taken. Materials such as table pads and pillows, sandbags, radiolucent positioning sponges of various sizes and shapes, compression bands, and masking or adhesive tape can be used effectively to assist the patient during the examination.

Table pads and pillows should be employed in all examinations where the opacity of the material will not interfere with the requirements of the examination. Their use will greatly enhance the comfort of the patient and the likelihood of maintaining the required positions for longer periods of time.

Sandbags can be used to achieve an internal rotation of the lower extremity required for certain positions of an ankle examination and for many other examinations of the upper and lower extremities. The added weight of the sandbag will assist the patient in maintaining the position once it is achieved.

Radiolucent positioning sponges are available in a variety of sizes and shapes, many designed for specific purposes related to difficult positions. As an example, a large, 45°-angle sponge can be employed to achieve and maintain the proper position of the patient for an oblique examination of the lumbar spine. Similar sponges of varying degrees of angle can be utilized to achieve and maintain the required positions for oblique examinations of multiple body parts, including most of the upper and lower extremities.

Compression bands, frequently used in conjunction with positioning sponges, can be placed across the body part being examined to assist the patient in maintaining a difficult position once the position is achieved. Compression bands are most frequently used with large body parts and are placed across the torso, pelvic region, or thorax.

Although compression bands can also be employed with small body parts, masking tape and/or adhesive tape is more frequently utilized for that purpose. Masking tape is less expensive, easier to apply and remove, and more readily adapted for achieving and maintaining a desired position than is adhesive tape. As an example, masking tape, used in conjunction with positioning sponges, can be used to maintain the head in the appropriate position for the examination of the skull or to achieve and maintain the required oblique position of a foot or ankle.

Technologists must be familiar with all of these materials and utilize them to advantage in the performance of their examinations. To help eliminate motion, materials such as these should be employed in all examinations, even those performed on the alert, cooperative patient. The consistent, well-planned application of these devices will help to lessen the potential for unnecessary, repeat exposures; to reduce motion; and to improve the recorded details of the radiographic image.

Immobilization of the Part

The primary means of eliminating image detail attributed to voluntary motion is patient immobilization. Infants and small children; patients with acute injuries;

adults in pain or suffering from conditions that produce spasms or tremors; or adults under the influence of medication, drugs, or alcohol may require immobilization of the body or body part in order to achieve a successful examination. Considerable care must be exercised in the application of immobilization devices to prevent patient discomfort or injury. A general rule when applying immobilization devices is to immobilize only when necessary, only those structures necessary, and only to the degree necessary to suspend the motion of the part. Many excellent commercial devices as well as simple materials located within the radiology department or physician's office can be adapted and applied to help immobilize the patient. A variety of methods of immobilization have been developed that are often quickly arranged and very simply applied to meet the requirements of a given situation or examination. Using these devices effectively and safely does require some forethought and planning on the part of the technologist. Immobilization devices can be classified into two major categories: (1) whole body immobilizers and (2) selected part immobilizers.

Whole Body Immobilizers

Whole body immobilization is frequently needed on the infant or small child. Devices such as brat boards, mummy wraps, or compression bands can be effectively employed and adapted to meet the requirements of the examination. On adults, compression bands and, to a lesser extent, mummy wraps are the devices most frequently employed. Preparing the infant for an examination of the abdomen, the application of a mummy wrap used in conjunction with a compression band can be an effective method of whole body immobilization. A mummy wrap together with the application of a compression band and the use of positioning sponges, sandbags, and masking tape can enable the technologist to effectively immobilize the infant or child for a successful skull examination. When longer periods of immobilization are required such as in genitourinary examinations that require multiple exposures over a specific sequence of time, the use of a brat board together with positioning sponges and a compression band will enable the technologist to perform the procedure and adjust the patient into different positions required by the examination quickly and effectively. Similarly, the application of a compression band together with the effective use of positioning sponges will enable the technologist to immobilize the adult for examinations of the spine and pelvic girdle as well as for longer periods of time such as those associated with gastrointestinal or genitourinary examinations. The mummy wrap, although effective in whole body immobilization, is more difficult and time consuming when applied to the adult patient.

Selected Part Immobilizers

For the immobilization of individual body parts such as arms, hands, legs, and feet, materials such as masking or adhesive tape, sandbags, long-handled radiolucent (Lucite) paddles, and compression bands often provide quick, effective, and safe means of immobilization. Placing the part to be examined upon the film holder, supporting the part with positioning sponges, and placing a series of masking tape strips across the part will effectively immobilize the part during the examination. Placing sandbags on each side of the head and employing a strip of masking tape across the forehead of the patient when performing a cervical spine examination will effectively immobilize the structures of interest. Two- and three-foot lengths of clear plastic having a wide surface at one end can be effectively employed in the examination of extremities of infants and small children. The plastic paddle can be placed over the part to be examined, and with sufficient pressure by the person assisting with the examination, it can effectively immobilize the part as well as help to maintain the position. Masking tape is perhaps the most effective device to immobilize a small body part, achieve a desired position, and maintain the position during the examination. The proper utilization of position achievers and maintainers together with the use of whole body or selected part immobilization when required by the circumstances of the examination represent positive examples of the application of the principles of ALARA.

Reduction of Exposure Time

Although a reduction in the time of exposure is the major factor to consider when the recorded detail loss is attributed to involuntary motion, it is also an appropriate solution to problems related to voluntary motion. Using a rapid time of exposure provides the technologist with a "margin of safety" related to some of the other factors affecting radiographic quality over which he or she has less control.

A reduction in the time of exposure can be achieved by the substitution of different imaging materials or by the manipulation of specific exposure factor relationships. Reducing the time of exposure whether by the selection of different imaging materials or the adjustment of specific exposure factors will frequently enable the technologist to eliminate the loss of image details associated with motion. However, this adjustment would not only influence the recorded details of the image, it would also affect other major properties associated with the production of an image of radiographic quality. Reducing the time of exposure will also affect the overall radiographic density of the image and thus the visibility of the recorded image. The radiographic density of the image will be decreased, and the photographic properties of the image will be reduced. It is essential to the radiographic quality of the image that the overall radiographic density be maintained. Therefore, the influence on the radiographic density of the image must be evaluated prior to the selection of alternative imaging materials or different exposure factors. By analyzing the influences and selecting the appropriate adjustments, you will be able to maintain the radiographic density of the image and achieve the desired reduction in the time of exposure. The correct application of these principles will enable you to eliminate the loss of recorded details associated with motion.

It is possible that the selection of alternative recording materials or the manipulation of exposure factors may produce an adverse effect on properties of the radiographic image other than radiographic density. Therefore, a careful analysis of any planned adjustment is necessary to prevent any appreciable change in the properties that contribute to the radiographic quality of the image. There are many occasions when your analysis will suggest a specific adjustment, but in understanding the interrelationships between the major properties of the radiographic image, you may find it necessary to choose an alternative, less ideal change—a less desirable trade-off.

The following factors can be adjusted in order to reduce the time of exposure while maintaining the radiographic density of the image. In the example provided below, each change is performed progressively.

1. **Milliamperage/Time (mAs) Relationship** The radiographic density can be maintained with an increase of the milliamperage together with a proportional

decrease in the time of exposure. The overall mAs remains the same (see Chapter 4). For example:

If the original technique required 100 mAs:

100 mA × 1 sec = 100 mAs

you can change to:

400 mA × 0.25 (1/4) sec = 100 mAs

2. **Film/Intensifying Screen Systems** The radiographic density can be maintained with an increase in the speed of the film/intensifying screen recording system and an appropriate reduction in the overall mAs (see Chapter 4). For example:

   If the original technique required 100 mAs:

   400 mA × 0.25 (1/4) sec = 100 mAs using a system speed of 100

   you can change to:

   400 mA × 0.0625 (1/16) sec = 25 mAs using a system speed of 400

3. **Source–Image Distance (SID)** The radiographic density can be maintained with a decrease in the SID and an appropriate reduction in the required mAs by application of the inverse square law formula (see Chapter 4). For example:

   If the original technique required 25 mAs:

   400 mA × 0.0625 (1/16) sec = 25 mAs using a 40" (102 cm) SID

   you can change to:

   400 mA × 0.05 (1/20) sec = 20 mAs using a 36" (91 cm) SID

4. **kVp/mAs Relationship** The radiographic density can be maintained by an increase in the kilovoltage by 15 percent and an appropriate reduction in the mAs by 50 percent, using the kVp/mAs relationship formula (see Chapter 4). For example:

   If the original technique required 20 mAs:

   400 mA × 0.05 (1/20) sec = 20 mAs using 60 kVp

   you can change to:

   400 mA × 0.025 (1/40) sec = 10 mAs using 69 kVp

An adjustment in the time of exposure is an important factor to consider in order to eliminate motion and improve recorded detail. Four different methods to maintain the radiographic density while reducing the time of exposure have been introduced. In these examples, the exposure time has been reduced from 1 sec to as little as 0.025 (1/40) sec.

REDUCTION OF EXPOSURE TIME
mA × sec = mAs
Original: 100 × 1 = 100
New: 400 × 0.025 = 10

Although the previous discussion identified four methods of reducing the time of exposure, it should not be implied that all of the methods are equally recommended, nor should they be used simultaneously or in conjunction with one another. However, they were presented in the order that one might anticipate using them when attempting to eliminate recorded detail loss by reducing the time of exposure. Let's review the application of each and analyze its influence on the radiographic quality of the image.

The manipulation of the **milliamperage/time relationship** should be considered first when attempting to control motion by reducing the time of exposure. Increasing the milliamperage enables you to reduce your time of exposure and maintain the radiographic density. It produces no appreciable loss to the other principles contributing to the radiographic quality of the recorded image. However, if an increase in the milliamperage requires you to select a larger size tube focus due to equipment output limitations, there will be a measurable loss in the geometrically recorded details, and this loss must be considered.

The use of a faster **film/intensifying screen recording system** is recommended as a second choice after you have exhausted your milliamperage/time manipulation possibilities. Faster recording systems enable you to reduce your time of exposure and maintain radiographic density but will increase the loss of recorded detail resulting from the use of higher speed materials. More rapid recording systems are frequently associated with a reduced image detail recording capability. The amount of recorded detail loss depends on the resolution properties of the intensifying screen speeds you have selected.

A reduction of the **SID** is not routinely recommended. However, if you have exhausted your choices of intensifying screen speeds, it would be the next recommended change. The application of the inverse square law will enable you to maintain radiographic density when you reduce the SID; however, not only does this increase the loss of geometrically recorded detail, it also records the image with increased size and shape distortion.

The manipulation of the **kVp/mAs relationship** is another method that can be employed in an effort to reduce the time of exposure. Among the four listed changes, the manipulation of the kVp/mAs relationship was provided as the last consideration. Although the overall radiographic density of the image can be maintained utilizing an increase in the kVp, the scale of radiographic contrast and thus the visibility of the details are altered. The recording of the structural details may not be affected, but the visibility of those details may be compromised. Remember, an image of radiographic quality provides a proper balance between the recorded details and the visibility of the details.

However, if your evaluation of the radiographic image shows that motion is the factor contributing to the reduction in radiographic quality, any and all means at your disposal should and must be employed to eliminate this unacceptable influence. (**Perform Experiment 3.**)

In investigating recorded detail loss due to motion, it can be seen that a thorough knowledge and understanding of the multiple factors that contribute to the recording of the radiographic image is essential if the manipulations required to eliminate the motion are to be successful and produce a minimal loss of radiographic quality.

Motion Summary

1. Motion is the single most detrimental factor affecting the radiographic quality of the image by producing the greatest loss of recorded detail.
2. Determine the type of motion you are dealing with in order to choose the most effective method(s) to eliminate it.

3. When utilizing methods of immobilization or employing devices to achieve or maintain a position, apply these materials with caution and care. The student technologist should practice and demonstrate proficiency in the application of these materials and devices within a structured program of simulated procedures and competency-based evaluations.
4. Analyze the effect of your proposed changes on each of the four major properties of the radiographic image in order to select adjustments that will eliminate the influence of motion while producing a minimal loss of radiographic quality.
5. Consider each adjustment individually. Select more than one adjustment only when the circumstances require multiple changes.
6. Remember, when necessary, any and all means to eliminate the loss of recorded detail associated with motion should be applied.
7. Review the motion produced in Experiment 2 and its significant effect on the recorded details of the radiographic image.

Material Unsharpness

An analysis of radiographic quality continues with an investigation of the imaging media and materials selected to record the image. They must be chosen carefully, keeping in mind the requirements of the examination and the limitations presented by the radiographic equipment. The influences that different combinations of imaging materials have on the radiographic image are included in an investigation of material unsharpness. The film-screen recording system has three major factors influencing material unsharpness, including (1) the type of radiographic film; (2) the type and speed of the intensifying screens; and (3) the contact achieved between the film and the screen (see Figure 2-2). In order to begin our investigation of material unsharpness, we must start with a definition of the modulation transfer function of the image recording system.

Modulation Transfer Function

The **modulation transfer function** (MTF) defines and measures the capabilities and limitations of radiographic imaging systems and is therefore intrinsic to an investigation of material unsharpness. MTF employs a complex mathematical concept to evaluate the recording capabilities of the materials utilized in the production of an image. The MTF objectively measures the inevitable loss of imaging capability between the image-forming radiation emerging from the body part and the recorded image of the imaging system used. The MTF defines the imaging capability of the system in terms of sharpness and resolution.

Sharpness as it relates to the imaging system and material unsharpness describes the ability of the x-ray film, intensifying screens, and the film-screen combination to record a sharp-edged image. An evaluation of the sharpness of the image requires a measurement of the **line spread function** (LSF). The LSF evaluates and measures the imaging system's ability to record an extremely narrow beam of x-rays passing through a slit 10 microns wide. The x-ray beam should be recorded as an extremely sharp line 10 microns wide. To determine the LSF, one measures the actual width of

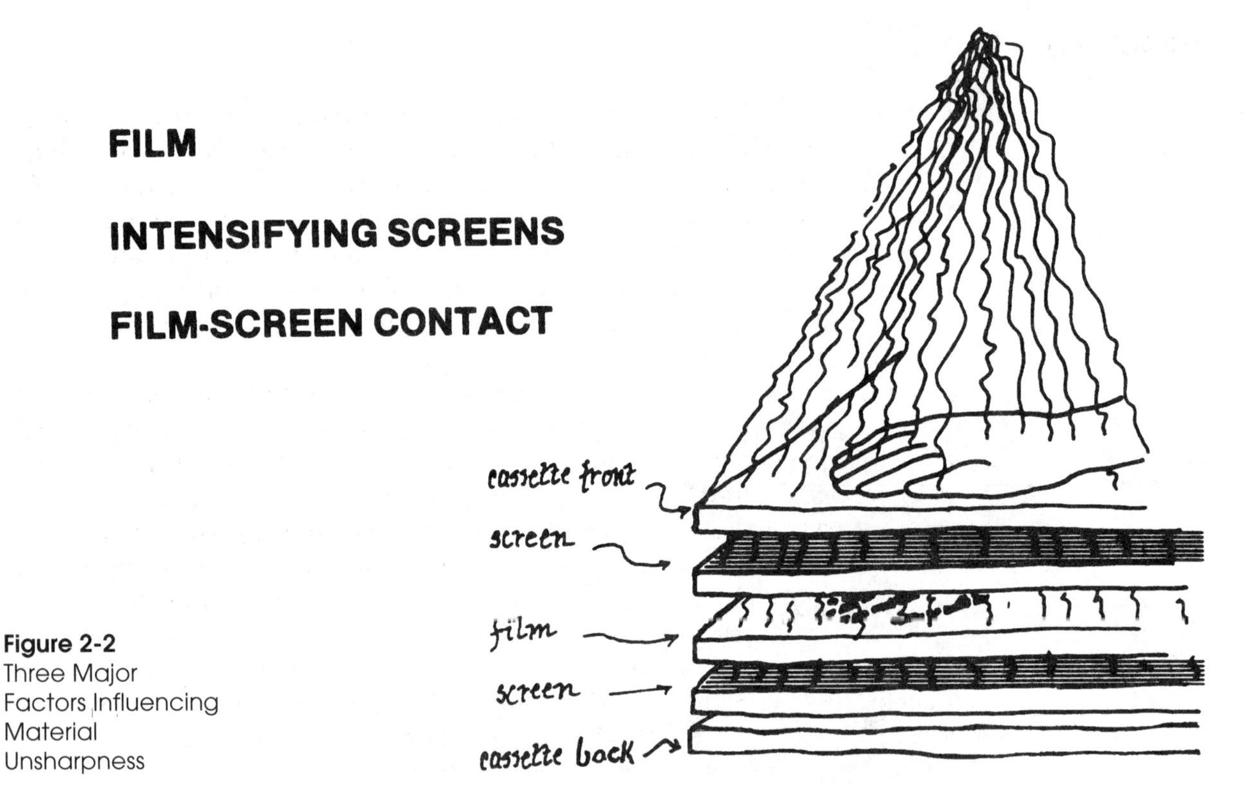

Figure 2-2
Three Major Factors Influencing Material Unsharpness

the recorded line at various points along its length and plots the relative density of the line at these points from the center of the line to its edge with a microdensitometer. How close to the 10-micron-wide measurement and how equal the radiographic density is throughout the recorded line are reflective of the image system's ability to record a line with sharpness. A system that can record a line with sharp, keen edges and with minimal line spread function would be identified as a system with excellent recording capabilities.

Resolution as it relates to the imaging system and material unsharpness is the ability to record a specific number of separate images within a limited, defined space. Resolution is usually identified as the number of line pairs per millimeter that the imaging system is capable of recording.

A combination of the line spread function (sharpness) and the line pairs per millimeter (resolution) related to the application of a given image recording system determines the quality of the recorded details of an image. An actual determination of the MTF for general radiography presents a number of practical difficulties. Image resolution can easily be identified by testing the system with a resolution grid capable of resolving at least 10 LP/mm. Image sharpness identified through an analysis of the LSF of the recording system requires expensive and accurate instrumentation not normally found in the average radiology department or physician's office. Therefore, for all practical purposes, our investigation of material unsharpness will be limited to an examination of the image resolution properties of the image recording system.

Radiographic Film

The radiographic film provides the ideal medium for the recording of the image. The two major components of the radiographic film are (1) the film base and (2) the light-sensitive emulsion. Most films used in radiography have a light-sensitive emulsion on both sides of the film base and are referred to as **double-emulsion films.** Single-emulsion films are still manufactured and used in limited applications such as mammography examinations. Recent advances in film-manufacturing technology allow double-emulsion films also to be employed in the performance of mammography examinations.

A problem known as film "crossover" exists with double-emulsion film. Film crossover is caused by the light emission of one intensifying screen passing through the emulsion layer on one side of the film base and essentially producing an image upon the adjacent emulsion layer. Since the x-ray film within the cassette is sandwiched between two adjacent intensifying screens, the problem is multiplied twice (see Figure 2-2). This double crossover occurs from the light emission of each intensifying screen. The crossover effect produces unsharpness as the light emission from each screen is further scattered and diffused by the film emulsion grains as it passes over from one side of the film base to the other. To combat this problem, film manufacturers have added sensitive dyes to the film emulsion, which has reduced this crossover effect. The added dyes have helped to reduce image unsharpness by adsorbing much of the light that would normally pass through the film base and affect the adjacent emulsion. (See Figure 2-3.)

Film Base

The film base is most often manufactured as a polyester substance that is both flexible and unbreakable and has greater dimensional stability and storage life than

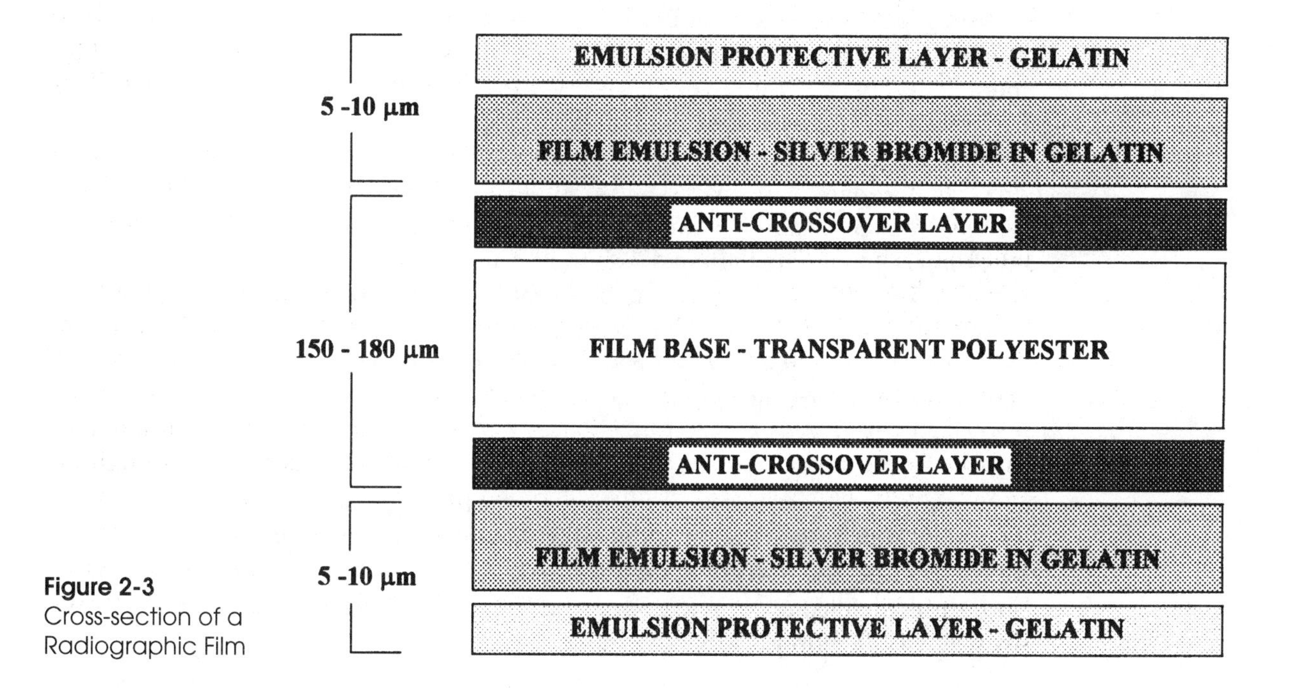

Figure 2-3
Cross-section of a Radiographic Film

films utilizing a cellulose triacetate base. It is manufactured to be nearly transparent to light to avoid any unwanted pattern on the film that would interfere with or detract from the recording of the image. Frequently, various dyes are added to the film base to reduce eye strain or glare when viewing the radiographic image. The film base is approximately 180 microns thick.

The surface of the base is coated with an adhesive subcoating to ensure an excellent adherent bonding between the base and the emulsion. Recently, to further prevent problems associated with crossover exposure, additional dyes in the form of an anticrossover layer have been added between the film base and the emulsion. This anticrossover layer together with the dyes contained within the emulsion have virtually eliminated the material unsharpness associated with the crossover problem.

Film Emulsion

The film emulsion consists of a mixture of gelatin and silver halide crystals. The silver halide crystals are made up of silver bromide and silver iodide and comprise the light-sensitive component of the emulsion. The gelatin provides an excellent medium for the suspension of the silver halide crystals. The emulsion is coated on each side of the film base to a thickness of approximately 5 to 10 microns each. Added to the surface of the light-sensitive film emulsion is an additional layer of gelatin sometimes referred to as the **supercoat.** This additional layer of gelatin helps to protect the light-sensitive emulsion from scratches and other problems associated with the handling of the film during cassette loading and unloading and during film processing.

Film employed in radiography is available in a wide range of speed (sensitivity to a given quantity of radiation) and latitude (capability of recording density over a wide spectrum of contrast values). The film's speed, latitude, and imaging resolution capabilities are determined by the process in which the silver halide crystals are manufactured and mixed into the gelatin. The number of sensitivity specks on the lattice of the crystal that aids in the formation of the latent image, the crystal's size and shape, and the concentration and distribution of crystals within the emulsion all affect the performance characteristics of the film. Recent advances in the development of the shape of the silver halide crystals have increased their image resolution properties. In the past, the silver halide crystals were three-dimensional, pebble-shaped structures. However, recent advances in the development of the silver halide crystal have enabled the manufacturer to produce crystals that are flatter, more two dimensional, and tubular. This enables the crystal to be better oriented within the emulsion, allowing its flattened side to lie parallel to the film base. This allows the silver grains of the latent image to be formed in two dimensions rather than three, which increases the imaging resolution properties of the film. The subject of film speed and latitude is discussed in greater depth in Chapter 5. There are two major types of film utilized in radiography: (1) nonscreen film and (2) screen-type film.

Nonscreen Film

Nonscreen film is rarely employed in radiography today; however, we will provide a brief description of it for historical purposes. Its current utilization would be contraindicated to the responsible application of x-radiation on human beings and the principles of ALARA. Nonscreen film is manufactured to be especially sensitive to the energy of x-radiation and for use in direct exposure radiography without the application of intensifying screens. It was manufactured to have a thicker film emulsion, and the greater quantities of silver halide crystals in the thicker film emulsion made the speed of nonscreen film considerably faster than screen-type film. Depending on the manufacturer, the quantity and concentration of silver halide crystals, and the thickness of the emulsion layer, the nonscreen film could be up to eight times faster than the screen-type film. However, the thicker emulsion layer made the nonscreen film unsuitable for use in a cassette with intensifying screens. Its main application was in procedures that demanded extremely detailed images such as those required in mammography, the detection of early bone pathology, and the investigation and localization of foreign bodies within the extremities. However, advances in the manufacture of screen-type film and intensifying screens together with reductions in material unsharpness associated with their use have enabled screen-type film used with intensifying screens to replace nonscreen film in recent years.

Screen-Type Film

Screen-type film is manufactured to be especially sensitive to the specific spectrum of light emitted by intensifying screens and is used to advantage in procedures employing intensifying screens. Two types of screen-type film are manufactured. Panchromatic film is designed to be sensitive to all colors of the light spectrum. Orthochromatic film is designed to be sensitive to all colors except red. Adjustments in the various dyes employed in the manufacture of screen-type film enable the manufacturer to limit the film to have greater reaction to the specific spectrum of light emitted from the intensifying screens. Most intensifying screens emit light within the blue or green spectrum. Although screen-type film, being sensitive to x-radiation, can be employed with any intensifying screen, it is used to its greatest advantage when employed with an intensifying screen emitting the spectrum of light to which it is especially responsive. The actual speed of the screen-type film when compared with nonscreen film is slower. (**Perform Experiment 4.**)

The use of intensifying screens with screen-type film significantly increases the overall speed of the procedure, enabling a substantial reduction in the quantity of radiation required to produce the desired radiographic density. The increase in the speed of the procedure results from an inherent property of the intensifying screen referred to as the **intensification factor.** The intensification factor is affected by a number of influences, including film speed, screen speed, and kilovoltage. The intensification factor of screens is examined in Chapter 4. The reduction in the required exposure with the application of intensifying screens employed with screen-type film can enable the technologist to eliminate problems attributed to motion. The application of various film-screen recording systems represents a positive example of the principles of ALARA.

Film Application in Radiography

Since screen-type film is responsive to x-radiation, radiography can be performed as either a direct exposure procedure relying on the actual exposure to produce an image on the film or as a procedure employing a film-screen recording system. However, the application of direct exposure radiography is no longer advocated and is seldom employed in radiography because of the significant increase in exposure required compared to a similar procedure performed utilizing a film-screen recording system. The responsible application of x-radiation in the production of radiographic images will no longer permit the quantities of radiation needed for direct exposure radiographic procedures.

Of historical note, direct exposure radiography was advocated for use with either nonscreen film or screen-type film used in a cardboard holder device for small body parts, soft tissue structures, and the investigation and localization of embedded foreign bodies. There were certain advantages of direct exposure radiography compared to procedures utilizing intensifying screens, and these were often employed as justification for performing direct exposure radiography. Some of the advantages included increased recorded detail, increased exposure latitude, and elimination of artifacts often attributed to the use of intensifying screens. However, with the increased concerns related to the biological hazards of radiation and the emphasis on performing minimal-dose radiography according to the principles of ALARA, direct exposure radiographic procedures are no longer advocated. Improvements in the manufacture of screen-type film and intensifying screens with greater speeds and image-resolving capabilities enable film-screen recording systems to be effectively used in all radiographic applications. It can be employed for both small and large body parts, procedures that can be performed as tabletop examinations, and those that require the use of a grid and Bucky apparatus. The main advantage of film-screen recording system radiography is the significant reduction of patient exposure when compared to direct exposure radiography. Depending upon the kilovoltage selected and the speed of the film-screen recording system, procedures employing screen-type film with intensifying screens can reduce the required exposure up to several hundred times.

For example, utilizing screen-type film in a cardboard film holder for a knee examination employing 60 kVp (intensification factor of 80 ×) could require an exposure of 200 mAs as a direct exposure procedure. Using a calcium tungstate film-screen recording system with a speed of 100, the same procedure could be performed using 2.5 mAs (200 mAs divided by the intensification factor of 80 × = 2.5 mAs). Utilizing more rapid recording systems, the exposure could be reduced even further. In addition to the significant reduction of exposure to the patient, a reduction in the mAs from 200 to 2.5 also enables the technologist to reduce the time of exposure significantly, thereby eliminating the problems associated with motion.

However, with the application of intensifying screens, there are some disadvantages that must also be considered:

- A loss in the recorded detail resulting from the decreased resolution capabilities is associated with the application of intensifying screens. This loss can be somewhat offset by the selection of slower speed systems that won't have as pronounced an effect on the recorded detail. In general, as the speed of the system increases, the imaging capability of the system decreases.
- A significant contrast enhancement is associated with the use of intensifying screens. Intensifying screens inherently produce a higher contrast (short-scale contrast) than does direct exposure radiography. This can be somewhat offset by the use of intensifying screens designed to provide an increase in the latitude of contrast that they can record and by the use of radiographic films possessing contrast extended sensitometric properties.

- Exposure latitude is reduced. That is, the margin for exposure selection between the minimum and maximum exposure that will enable you to produce a diagnostic image is less when employing intensifying screens. For procedures employing the Bucky apparatus, the application of automated exposure control can significantly reduce errors associated with exposure factor selection (the technical guessing game).
- Greater potential for film artifacts are associated with the use of old, cracked, or dirty intensifying screens. However, the implementation of a planned quality assurance program within the department or office will help to reduce this problem.

Thus, even though the application of intensifying screens in radiography does present some potential disadvantages, many of these problems can be eliminated or reduced to an acceptable level, or an appropriate technical trade-off can be applied that will more than compensate for problems associated with the use of film-screen recording systems.

Recording Properties of Radiographic Film

As far as the resolution properties of radiographic film are concerned, recent advances in the manufacture of screen-type film, such as the near elimination of the crossover effect and the production of silver halide crystals with a newer, two-dimensional shape (flatter and tubular in design), have added significantly to the recording capabilities of the film. Even without these recent advances, all films employed in radiography are capable of resolving far greater numbers of LP/mm than the human eye is capable of visualizing (Figure 2-4). The increased structural details visualized within a radiographic image are more the result of how the film was exposed (direct exposure versus intensifying screen exposure) than any inherent property of the film itself.

The LSF of x-ray film is almost impossible to measure. The grain size of the silver halide crystals within the film's emulsion are frequently smaller than 2.0 microns and considerably smaller than the 10-micron width the LSF is capable of measuring.

X-RAY FILM

CAMERA
100 LP/mm
2500 LP/In

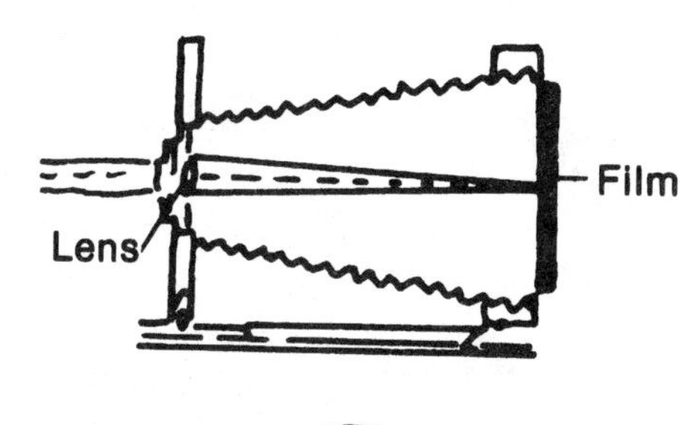

HUMAN EYE
5 LP/mm =
125 LP/In

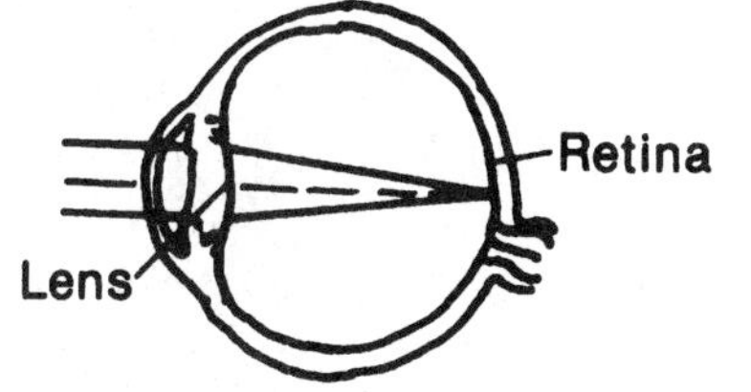

Figure 2-4
Resolution Power of X-ray Film Versus Visualizing Ability of the Eye (LP, Line Pairs)

Therefore, the so-called "radiographic noise or mottle" (which results from a nonuniform arrangement of the silver actually making up the image) that may be seen when the film is examined with a magnification lens of ×10 power is not a practical consideration related to the material unsharpness of the film itself. For all practical purposes, the film's contribution to "image noise" or to the material unsharpness of the image is nonexistent. (**Perform Experiment 5.**)

Intensifying Screens

Intensifying screens, as the name implies, intensify the action of the x-radiation. Less than 5% of the exposure produced actually contributes to the formation of the latent image on the film. Film-screen combination recording systems have far greater sensitivity than film used alone in direct exposure radiography. Use of intensifying screens result in a considerably lower patient exposure, but with a significant increase in material unsharpness compared with direct exposure radiographic procedures. Direct exposure radiography contributes very little to material unsharpness. However, with recent improvements in the manufacture of screens, the resolution loss measured by the MTF of the system is not as serious as it was in the past.

With film-screen recording systems, the high-energy, invisible x-ray exiting from the patient's body is converted into a lower energy pattern of visible light. The visible light image interacts with the film to produce the latent image. The screen acts as an amplifier to the radiation exiting the body part in the production of the latent image. Screen-type film, which is manufactured to be specifically sensitive to the visible light spectrum emitted by the screens, is used to its greatest efficiency in a film-screen recording system. The visible light emitted by the intensifying screens accounts for more than 95% of the radiographic density of the recorded image. Intensifying screens resemble thin, flat sheets of plastic and are composed of at least four layers. (See Figure 2-5).

Screen Base

The base of the intensifying screen is made of a polyester material approximately 1 mm thick. The base provides the support for the active phosphor layer and should

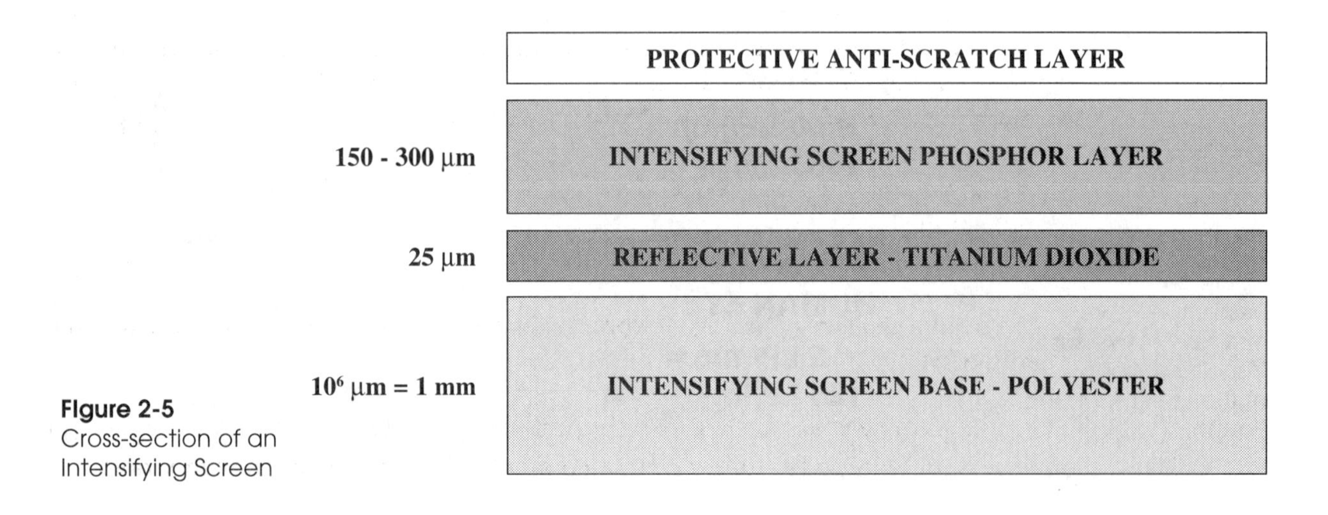

Figure 2-5
Cross-section of an Intensifying Screen

be flexible, moisture resistant, chemically inert, capable of long life (will not discolor with age), and unresponsive to x-radiation.

Reflective Layer

To the surface of the screen base may be added a thin layer (approximately 25 microns) of titanium dioxide, which enhances the amount of light emission directed toward the film. Visible light emitted by the phosphor crystal travels in all directions from its original source. Some of the emitted light travels in a direction away from the film and is therefore lost related to any potential image production. The reflective layer intercepts this light and redirects it back toward the film, increasing the efficiency of the screen and nearly doubling the number of light photons available to produce the latent image. Although increasing the speed of the screen, it also contributes to increased material unsharpness due to a greater diffusion of the light pattern and the added distance traveled when reflected back toward the film. With some screens, a light-absorbing layer is added to the screen base in order to reduce the crossover effect of the light emission. The light emitted from the phosphor crystal is not uniform in its intensity and, as it spreads, produces what is referred to as a **halo effect,** which adds to the material unsharpness attributed to intensifying screens. A light-absorbing layer can reduce this halo effect and improve image resolution. Depending on the purpose of the procedure, the technologist can choose a film-screen recording system related to screen speed, which is translated into decreased patient exposure or screen resolution, which is translated into increased image details.

Phosphor Layer

Intensifying screens depend on the luminescence of the crystals of different elemental compounds embedded in a polymer matrix and coated with an additional thin, scratch-resistant layer to protect the active phosphor layer. Thousands of chemical compounds have the property of luminescence to some degree or another. However, not all are responsive to the energy levels associated with the production of x-rays or have the useful characteristics necessary for screens used in radiography. Intensifying screen phosphors must be capable of **fluorescence;** that is, the ability to emit visible light immediately on the absorption of x-radiation. Additionally, the light emission must stop as soon as the source of radiation is removed. Some phosphor materials continue to emit light or glow even after the source of radiation is removed. This property is known as **phosphorescence.** Phosphorescence, also referred to as **afterglow** or **screen lag,** is an unwanted property related to phosphor materials utilized in the production of intensifying screens. It is, however, a useful property in the production of fluoroscopic screens allowing for the intermittent energizing of the radiation and reduced patient exposure during fluoroscopic procedures. Therefore, the elemental compounds of the phosphors chosen for intensifying screens differ from those employed in the manufacture of fluoroscopic screens.

Positive characteristics of phosphors utilized in intensifying screens include the following:

- High atomic number, so that the probability of interactions with x-rays is high
- Capability of emitting a large amount of light per interaction with x-ray

- Capability of emitting the proper wavelength of light (color spectrum) to match the sensitivity of the film employed with the film-screen recording system
- Production of minimal phosphorescence (afterglow or screen lag)

Conventional Phosphors. Calcium tungstate was discovered in 1896 to be an excellent phosphor material for use in intensifying screens. It possessed the positive characteristics needed and was the main phosphor employed up to the early 1980s. Barium lead sulfate was introduced in 1948 and gained some prominence as it responded faster to exposures in the 70- to 100-kVp ranges. Conventional phosphors emit light over a fairly broad spectrum in the blue, blue/violet, and ultraviolet color spectrum peaking in the 430-nm range. Each of these phosphors are still in general use in radiography today, although they are being rapidly replaced by a group of screens referred to as **rare earth phosphor screens.** Examples of conventional phosphors used for intensifying screens are listed below:

- Calcium Tungstate ($CaWO_4$)—Atomic # 74 Tungsten
- Barium Lead Sulfate ($BaPbSO_4$)—Atomic # 82 Lead

Table 2-1 provides a comparison of the characteristics of direct exposure radiography to the use of different speed calcium tungstate intensifying screens.

TABLE 2-1 Comparison of Characteristics of Direct Exposure and Various Types of Calcium Tungstate ($CaWO_4$) Intensifying Screens Used with Blue Sensitive Film

Type of radiography	Characteristics
Direct exposure radiography	More than 100 line pairs/mm resolving power No practical material unsharpness Long scale contrast Increase exposure latitude Unacceptable radiation exposure required
Detail screen 50 speed	0.25-mm unsharpness Short scale contrast Decreased exposure latitude Significant radiation exposure reduction
General purpose screen 100 speed	0.35-mm unsharpness Short scale contrast Decreased exposure latitude 2 × faster than detail screen
Fast screen 200 speed	0.45-mm unsharpness Short scale contrast Decreased exposure latitude 2 × faster than general purpose screen
Ultrafast 300 speed	0.55-mm unsharpness Short scale contrast Decreased exposure latitude 3 × faster than general purpose screen

Rare Earth Phosphors. In the mid-1970s, new phosphor materials for use in intensifying screens were developed and introduced. These materials have come to be identified as a group as **rare earth phosphors.** Rare earth phosphors encompass a number of elemental compounds that are generally found in less abundance and are more difficult to isolate and purify. The primary advantage of rare earth phosphors over conventional phosphors is their (speed) reaction to x-radiation. Initially, the cost of these new screens compared to conventional phosphor screens was prohibitive, and, although introduced in the mid-1970s, they did not gain a foothold in conventional radiography until the 1980s. As additional rare earth phosphors were developed and competition increased, the cost of replacing a conventional screen system within a radiology department or physician's office became more acceptable. Rare earth phosphors can emit light in the blue, blue/violet, and ultraviolet color spectrum peaking in the 430-nm range or in the green color spectrum peaking in the 540-nm range. In general, all rare earth phosphors produce a more limited, discrete color spectrum range than do conventional phosphors. Some of the more prominent rare earth phosphors are listed below:

- Terbium-Activated Gadolinium Oxysulfide ($Gd_2O_2S:Tb$)—Atomic # 64 Gadolinium
- Terbium-Activated Lanthanum Oxysulfide ($LaO_2S:Tb$)—Atomic # 57 Lanthanum
- Lanthanum Oxybromide (LaO_2Br)—Atomic # 57 Lanthanum
- Terbium-Activated Yttrium Oxysulfide ($Y_2O_2S:Tb$)—Atomic # 39 Yttrium
- Yttrium Tantalate (Y_2TaO_4)—Atomic # 39 Yttrium
- Barium Fluorochloride (BaFlCl)—Atomic # 56 Barium

The two major intensifying screen characteristics of importance to the technologist are **screen speed** and **screen resolution**.

Intensifying Screen Speed

Screen speed indicates the ability of the screen to react and give off light when exposed to a defined amount of x-radiation. Screen speeds are evaluated by a determination of the intensification factor (IF) of the screen. The intensification factor of a screen is defined as the amount of exposure required without screens divided by the amount of exposure required with screens to produce the same radiographic density:

$$\text{IF} = \frac{\text{exposure required without screens}}{\text{exposure required with screens}}$$

If a nonscreen procedure requires an exposure of 400 mAs utilizing 60 kVp to produce a 2.0 radiographic density, and a similar procedure employing a conventional film-screen recording system requires an exposure of 5 mAs to produce a similar radiographic density, then we would conclude that the intensification factor of the screen at this kVp is 80 × (400 divided by 5 = 80 ×).

Screen speeds are influenced by a number of factors, some that are inherent in the manufacture of the screen and others that are influenced by technical factors.

Screen Speed Inherent Factors.

- **Phosphor composition.** The actual conversion ratio of x-ray to light differs for different phosphor materials. As the amount of light per photon of x-ray energy increases, the speed of the screen increases. Rare earth screen phosphors are considerably more efficient in the conversion of x-rays to visible light.
- **Phosphor crystal.** Of significance is the size and concentration of the phosphor crystal in the active layer. The larger the size of the phosphor crystal and the greater the concentration of the phosphor within the active layer, the greater the speed of the screen. A 200 speed $CaWO_4$ conventional phosphor measures approximately 8 microns.
- **Phosphor layer.** As the thickness of the active layer increases, the speed of the screen increases (greater quantities of phosphor). Phosphor layers usually range from between 150 and 300 microns thick and allow for the manufacture of many different screen speeds.
- **Phosphor spectrum.** Different phosphor materials produce different patterns of light emission (different color spectrums). $CaWO_4$ screens emit a blue to blue/violet light. $LaO_2S:Tb$ screens emit a green light. Intensifying screens have their greatest efficiency and speed when employed in a film-screen recording system that takes advantage of the specific, discrete color spectrum of light emitted by the phosphor. The intensification factor of the screen would be reduced if the wrong film sensitivity were employed.

Screen Speed Technical Factors.

- **Radiation quantity.** As the quantity of the exposure increases, the speed of the screen increases as a result of a greater number of crystals being exposed to x-radiation and therefore emitting light.
- **Kilovoltage.** As the kilovoltage increases, a larger amount of exit radiation as well as a higher percentage of higher energy radiation will be emitted from the patient capable of interacting with the screen crystals. Kilovoltage is one of the factors affecting the screen intensification factor.
- **Temperature.** Screens emit more light per x-ray interaction at lower temperatures. This factor is no longer a significant influence in practice since almost all radiographic procedures are produced indoors and in a temperature-controlled environment.

Intensifying screens are manufactured to produce different speeds. Conventional screens used to be classified as either slow, par, high, and ultrahigh speed screens. With the introduction of rare earth screens and the different reactions of their phosphors compared to conventional screens, a new system of speed classification was required. The need to match the screen to a film that responds to the specific color spectrum of light emission from the screen has produced speed values attributed to the combined film-screen recording system rather than to the identification of the screen speed alone. The speed of the screen is no longer the final determinant of the speed of the recording system. There is somewhat of a carryover with terminology, and you will often hear a system referred to as a **detail, general purpose, fast,** or **ultrafast** system. A more correct method of determining the speed of the materials utilized for a given radiographic procedure would be to refer to the designated speed of the film-screen recording system.

All film-screen recording system speeds are based on and compared with the speed of a conventional phosphor, $CaWO_4$ (par speed), screen used with a blue sensitive film. The speed value of 100 has been assigned to this system. All other film-screen recording system combinations are evaluated and given a speed value based on this known 100 speed system. The advantage of rare earth film-screen recording systems is that they begin at a 200 speed, 2 times faster than the 100 speed $CaWO_4$ system, and progress to speed levels of up to 1200 speed, 12 times faster. Table 2-2 provides a comparison of a typical series of exposure factors from a direct exposure procedure requiring 4800 mAs to a rare earth, 1200 speed film-screen recording system requiring 5 mAs to produce the same radiographic density.

Why are rare earth film-screen recording systems so much faster than conventional film-screen systems? It is a combination of higher conversion efficiency and increased x-ray absorption ratios of the phosphors.

- **Higher conversion efficiency**. The ratio of x-ray absorbed and converted to visible light emitted with conventional $CaWO_4$ screens is only 4% to 5%. This means that only 4% to 5% of the exit radiation reaching the screen will be converted to visible light. The rare earth screen ratio of x-ray absorbed and converted to visible light is 15% to 20% (4 to 5 times more efficient).
- **Increased x-ray absorption**. Conventional screens such as $CaWO_4$ absorb only about 20% of the x-ray energy because of the high atomic number of the tungsten employed in its phosphor. The energy required to eject the K-shell electron from the tungsten atom is high. The atomic numbers of the rare earth phosphor materials are all less than tungsten; therefore, the binding energies of their K-shell electrons are less. As a result, rare earth screens absorb three to five times more exit radiation than a conventional $CaWO_4$ screen. Therefore, for each x-ray photon absorbed by the rare earth screen, more light is emitted.

TABLE 2-2 Comparison of Typical Exposure Factors for an Anteroposterior Projection of the Knee (12:1 Ratio Moving Grid). Intensification Factor of 60 kVp = 80 ×

Type of radiography	Exposure factor
Direct (nonscreen) exposure	4800 mAs (100 mA × 12.00 sec at 60 kVp)
50 speed film-screen recording system	120 mAs (100 mA × 1.200 sec at 60 kVp)
100 speed film-screen recording system	60 mAs (100 mA × 0.600 sec at 60 kVp)
200 speed film-screen recording system	30 mAs (100 mA × 0.300 sec at 60 kVp)
300 speed film-screen recording system	20 mAs (100 mA × 0.200 sec at 60 kVp)
400 speed film-screen recording system	15 mAs (100 mA × 0.150 sec at 60 kVp)
600 speed film-screen recording system	10 mAs (100 mA × 0.100 sec at 60 kVp)
800 speed film-screen recording system	7.5 mAs (100 mA × 0.075 sec at 60 kVp)
1000 speed film-screen recording system	6 mAs (100 mA × 0.060 sec at 60 kVp)
1200 speed film-screen recording system	5 mAs (100 mA × 0.050 sec at 60 kVp)

Intensifying Screen Resolution

In any transfer or duplication system, the copy or duplicate loses some of the properties of the original. In radiography, the film-screen recording system is ostensibly a transfer process, and, as with all transfer processes, there is a loss of image detail and an increase in unsharpness attributed to the materials employed to record the image. We have learned that the film is capable of recording an image with far greater resolution properties than we are able to visualize. However, the use of intensifying screens does add to the material unsharpness of the image and a loss of resolution when a measurement of the MTF of the system is employed. With film-screen recording systems, the light diffusion (spreading) emitted by the screen before it is recorded by the film causes blurring (unsharpness). Some of the factors contributing to the material unsharpness attributed to intensifying screens are the following:

- **Phosphor crystal.** The larger the size of the phosphor crystal, the greater the material unsharpness associated with its use (see Figure 2-6). The actual size of the x-ray photon is considerably smaller when compared with the phosphor crystal size, so there is a measurable loss of image resolution capability as the x-ray energy is absorbed and converted to visible light by a much larger size crystal. $CaWO_4$ conventional phosphor measures approximately 8 microns. Rare earth phosphors are frequently smaller.
- **Phosphor layer.** As the thickness of the active layer increases, the material unsharpness attributed to the screen increases. Phosphor layers usually range from between 150 and 300 microns thick. Phosphors within the active layer that are embedded further from the film emulsion have to travel further before interacting with the silver halide crystals and producing a latent image. As light

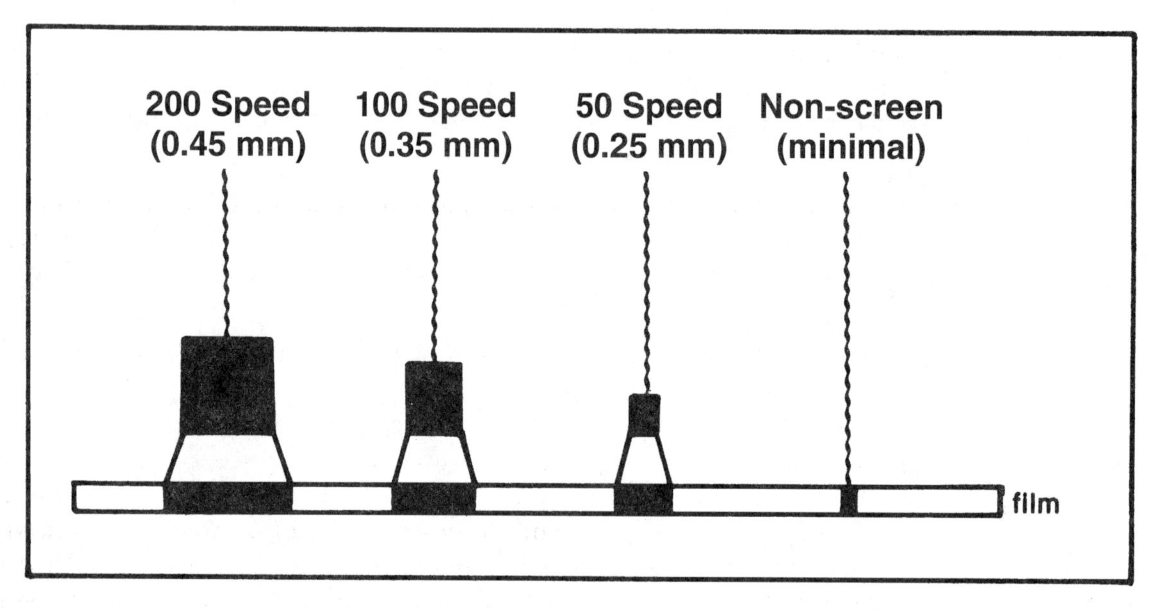

Figure 2-6
Comparison of $CaWO_4$ Phosphor Crystal Sizes to Image Details (a Measure of Unsharpness)

travels further from its source, it spreads in all directions. Additionally, as the light travels through the layers of phosphors within the active layer, it diffuses and begins to create a halo effect, which further diminishes its image resolution capabilities.

Comparing a 200 speed $CaWO_4$ recording system with a 200 speed Gd_2O_2S:Tb recording system, the rare earth system will produce a considerably sharper image. This higher resolution of the rare earth system comes as a result of both its higher conversion efficiency and increased x-ray absorption enabling the active layer of the phosphor to be considerably thinner. The speed of the screen can be maintained with a thinner active layer while improving the resolution properties of the screen. A 400 speed Gd_2O_2S:Tb system will produce images with similar resolution properties as a 200 speed $CaWO_4$ system, although the speed of the system is twice as much. Higher speed systems allow for reduced patient exposure and lower exposures with the potential elimination of motion unsharpness, which is the single factor most detrimental to the recording of image details. (**Perform Experiment 6.**)

In general, rare earth screens employed with proper sensitized film having a recording system speed of 2 to 5 times faster than the 100 speed $CaWO_4$ system will have an image resolution quality comparable to the $CaWO_4$ system. Rare earth systems having a speed of 6 to 12 times faster than the 100 speed $CaWO_4$ system will have some loss of image resolution compared with the $CaWO_4$ system. Most of this image detail loss is attributed to quantum mottle as the system speed increases and the amount of exposure required to produce a given radiographic density is significantly reduced. In an analysis of radiographic quality, the material unsharpness attributed to high-speed recording systems may be acceptable if its application allows for the elimination of motion.

Radiographic Mottle (Noise)

Radiographic noise, or **mottle,** is the term given to a nonuniform or nonhomogeneous appearance of the recorded image more commonly associated with high-speed recording systems. Major sources of radiographic noise include the following:

- **Film graininess mottle** is attributed to the random distribution of the silver halide grains in the emulsion. As the speed of the system increases, less exposure is employed and less silver halide crystals will be affected. The actual observation of film grain mottle is rare.
- **Screen structure mottle** is attributed to the phosphor crystal size and the concentration and thickness of the active layer. The radiographic noise attributed to screen structure mottle is almost unobservable.
- **Quantum mottle** is attributed to random absorption of the x-ray quanta absorbed by the image receptor. Quantum mottle exists to some degree in all film-screen recording system procedures. It becomes more pronounced and observable with faster speed systems. The faster the system, the less mAs required to produce the image. The higher the kilovoltage used, the less mAs required to produce the image. As the quantity of radiation decreases, the random spatial variation in radiographic density produced increases. The

image begins to appear mottled and incomplete. Imaging factors such as radiographic contrast and thus the visibility of the image are important to the perception of quantum mottle.

Once the latent image is formed, the film must be properly processed to demonstrate the ultimate radiographic contrast and visibility of the image. In the film processing room, it is important that the proper safelight filters be employed. With conventional screen radiography, a Wratten 6b, amber filter was employed with safety. However, with the different color spectrum sensitivities of the films employed with rare earth screens, it is necessary to adjust the safelight filters to accommodate the spectral sensitivities associated with the new recording systems. Safelight filters designed more toward the red portion of the light spectrum are employed with rare earth recording systems. An evaluation of the use of a proper safelight is presented in Chapter 4 later in this analysis.

The selection of the film-screen recording system to be employed in a radiology department or physician's office should be chosen carefully with an understanding of all the principles involved in applying these systems. Speed is an important factor in the reduction of patient and operator exposure and the reduction of image detail loss due to motion. Some procedures such as mammography, early detection of osseous system disorders, and extremity radiography looking for small, nondisplaced fractures may require a slower speed system that enables greater resolution of the recorded image. Systems that enable the recording of greater detail at higher speeds are available. Systems with speeds up to 1200 are available for limited procedures such as the abdominal or pelvic examination of a pregnant patient or the examination of infants, small children, or adults where motion is a major factor related to the success of the procedure. The availability of different film-screen recording systems enables the technologist to select the appropriate materials for a given examination keeping in mind the principles of ALARA and the desire to produce an image of radiographic quality while minimizing the exposure to the patient.

Film-Screen Contact

When using intensifying screens, it is vital that the film be in close, uniform contact with the intensifying screens in all areas of the cassette. Any lack of proper (i.e., absolute) film-screen contact will result in an increased level of image unsharpness in those areas exhibiting this poor contact (Figure 2-7).

Poor film-screen contact can occur in any film cassette and may be caused by a number of factors. The most common cause of poor film-screen contact is the presence of artifacts such as dirt, dust, and foreign objects such as bits of paper or other material that enter or accumulate within the cassette during normal operational procedures of opening and closing and replacing film within the film holder. In many cases, the presence of dirt, dust, or foreign objects will be demonstrable and evident as film artifacts visible when the radiographic film is processed. Another cause of poor film-screen contact is the abuse or mishandling of cassettes, including the storage of cassettes often of different sizes on top of one another. This creates uneven pressures upon the cassette frames leading to misalignment and poor film-screen contact. Cassettes should be stored upright on their sides when not in use.

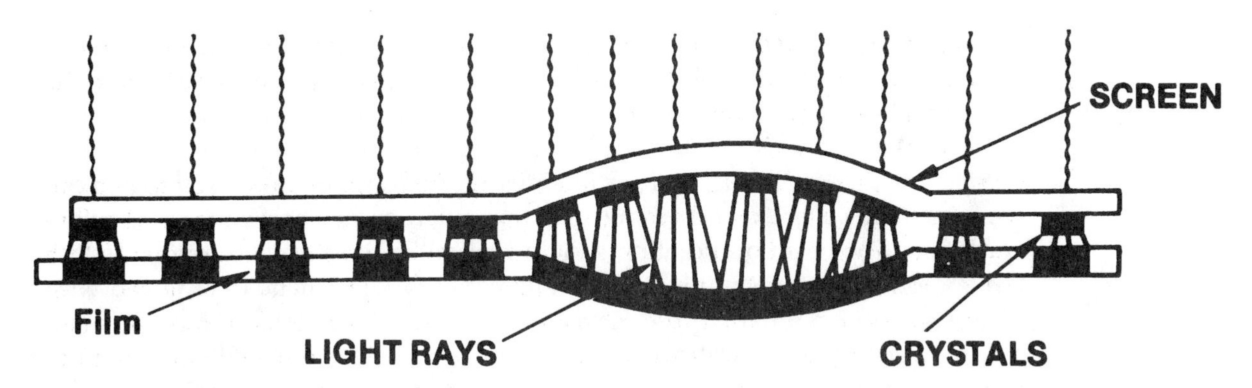

Figure 2-7
Example of Good and Poor Film-Screen Contact. As the Distance Between the Screen and the Film Increases, the Light Rays Spread, Causing an Overlapping of the Rays from One Crystal onto Those from Another. The Resultant Image of Each Crystal Blurs into Those of the Adjacent Crystals.

Dropping cassettes can cause a misalignment of the frame, the hinges, or the locking device. Over a period of time, abused or mishandled cassettes can cause considerable problems associated with poor film-screen contact. Cassettes with misaligned frames or improperly operating hinges or locks can readily allow dust and foreign matter to enter the cassette and lodge between the screen surfaces, creating poor film-screen contact. Additional problems associated with improper cassette handling or maintenance include radiographic fog from light leaks and image artifacts associated with uneven pressure within the cassette or the presence of foreign matter.

To minimize the unsharpness associated with poor film-screen contact, a dedicated program of quality assurance should be implemented within the department that includes a routine check and maintenance of cassettes as well as routine tests to check for foreign matter, scratches upon the surface of the screens, dirt build-up upon the screens and within the cassette, and film-screen contact. Screens should be routinely cleaned using an approved cleaning agent to remove dirt build-up and the surfaces of the screen should be examined for cracks, blemishes, or scratches. Internally, cassettes should be marked or numbered so that when images are produced demonstrating film artifacts or unsharp areas that are suspected to be caused by poor film-screen contact, the cassette in question can be taken out of use and checked. A number of commercial marking systems are available to identify your cassettes.

Film-Screen Contact Testing

A special screen contact test grid specifically designed for the testing of film-screen contact is used to evaluate this factor. The screen contact test grid is a rigid flat sheet of plastic usually the dimensions of a 14" × 17" (36 × 43 cm) cassette. A wire grid of 1/8" squares is embedded within the plastic. Commercial screen contact test grids are precision devices and should be handled with care during use and stored on their side when not in use. They should not be stored flat, and nothing should be allowed to be placed on top of the test grid either during its use or while in storage to prevent the grid from being damaged or misaligned due to pressure or inadvertent abuse. The test grid can be used for testing of all cassette sizes, although additional care must be exercised when used with smaller cassettes to prevent damage to the grid.

Film-screen contact is tested by placing the cassette upon the flat surface of the radiographic table and placing a screen contact test grid upon the top of the cassette, carefully aligning the testing grid with the edges of the cassette. When used with smaller cassettes than the actual grid size, the grid should be centered on top of the cassette surface equally distant from all sides. The x-ray tube should be centered over the cassette to be tested and arranged so that a perpendicular beam of radiation is employed. The beam of radiation should be limited to the size of the cassette. A radiographic exposure employing a relatively high kilovoltage (60 kVp) and low mAs is selected in order to produce a radiographic image of the grid's test pattern possessing long scale contrast. A long scale contrast image will enable you to visualize greater density changes associated with poor film-screen contact. The exposure factors selected will have to be adjusted according to the speed of the film-screen recording system being tested. After processing, the radiographic image is examined throughout its entire image using a particular order of investigation. For example, you can start at the upper left-hand side of the image and inspect the image along the upper portion of the film going from left to right. This pattern should be continued until the entire image has been examined. If at any point an area of the image appears blurred or if the radiographic density of any portion of the image differs from the overall pattern or radiographic density, you should circle that area with a marking pencil or pen. If any areas of question were identified, the cassette should be taken out of service and examined more closely, comparing the specific areas of suspected poor contact with the areas within the cassette approximating those locations. Occasionally, it is possible to repair a cassette that demonstrates poor film-screen contact. However, if repairs are not possible, the cassette must be removed from service and not employed for any additional radiographic procedures. (**Perform Experiment 7.**)

The implementation of a **quality assurance program** that includes the materials employed to record the image will assist in the efforts to reduce unnecessary repeat examinations due to equipment errors. An equipment maintenance program that includes at a minimum the following routine procedures is a positive example of the application of the principles of ALARA:

- Checking of cassettes for frame damage or warping, loose hinges or locks, artifacts on the surface of the cassette front, and light leaks
- Checking the surface of the intensifying screens for artifacts such as dust or embedded foreign matter, scratches, discoloration, and evidence of age
- Cleaning of the screens with an approved cleaner
- Checking and testing of the cassettes for proper film-screen contact

Equipment that does not meet the minimum standards required for radiographic imaging systems must be removed from service.

Poor film-screen contact is a more common problem associated with the larger sized cassettes since the hinge and locking apparatus of the cassette has to maintain an equal pressure on all surfaces of the film sandwiched between the intensifying screens over a much larger surface area. However, it is not impossible to find poor film-screen contact occurring in 8" × 10" (20 × 25 cm) cassettes if the cassettes were dropped, abused, or stored flat with a number of cassettes resting on top of them, creating considerable pressure on the cassette located at the bottom of the pile.

Greater importance should be assigned to the influence of film-screen contact and its effect on the radiographic quality of the image. Unlike motion unsharpness, which can be easily detected by its unsharp or blurred appearance throughout the entire radiographic image, unsharpness caused by poor film-screen contact is only evident in the immediate area around the poor contact. Its effect, however, can be disastrous. Contained within an otherwise detailed image, the limited, unsharp, blurred effect resulting from poor film-screen contact can potentially be mistakenly identified as an abnormal finding or pathology when masked by the clear, sharp images of the adjacent structures within the body part being examined.

Poor Film-Screen Contact Formula

The unsharpness produced by poor film-screen contact only compounds the existing unsharpness due to screen unsharpness and other factors. There is an inherent unsharpness associated with the use of intensifying screens. For example, the image detail loss associated with the application of a 200 speed conventional film-screen ($CaWO_4$) recording system produces a material unsharpness of 0.45 mm. With poor contact, an increased distance between the screen and the film enables the light emitted by the screen to further spread out in all directions before being recorded on the film. Thus, the light spreads and the unsharpness increases. The spreading of the light emitted by the screen is similar to the spread of the intensity of the beam associated with the SID. This influence can be determined through use of the inverse square law (see Chapter 4). Recognizing this influence, you can see that the unsharpness of the radiographic image is increased by the square of the poor contact distance created between the screen and the film. Thus, a formula paralleling the inverse square law can be employed to solve for the unsharpness attributed to poor film-screen contact. The new unsharpness (i.e., that resulting from poor film-screen contact) can be determined by simply multiplying the old unsharpness (the sum of all existing screen unsharpness, geometric unsharpness, and motion unsharpness) by the distance created between the film and the screen within the cassette squared:

$$\text{new unsharpness} = \text{old unsharpness} \times \text{poor contact}^2$$

We can easily solve for the new level of image unsharpness when the material unsharpness of 0.45 mm attributed to a 200 speed conventional film-screen ($CaWO_4$) recording system and the poor film-screen contact distance of 1.5 mm are both known (Figure 2-8). The old unsharpness (0.45 mm) times the poor film-screen contact distance of 1.5 mm squared (2.25 mm) equals the new unsharpness. In this example, 0.45 mm x 2.25 mm = a new unsharpness of 1.01 mm. It should be noted that in this example of poor film-screen contact, we have only investigated the material unsharpness attributed to the film-screen recording system. In actuality, the unsharpness would be greater than this due to other factors, such as the geometric unsharpness associated with the arrangement of the radiographic equipment, the patient, and the recording material and any unsharpness attributed to motion. In this example, the screen unsharpness factor of 0.45 mm has increased to an unacceptable level of 1.01 mm of image unsharpness. (**Perform Experiment 8.**)

Poor film-screen contact produces a significant loss of image quality and must be eliminated from the recording of a radiographic image. It multiplies the image unsharpness attributed to all of the other factors contributing to the loss of image

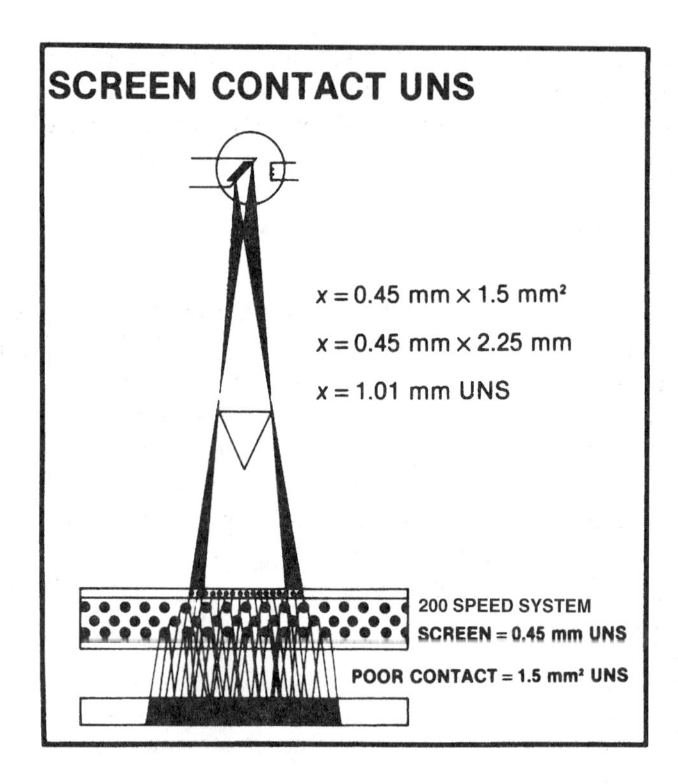

Figure 2-8
Example of Calculation of the Unsharpness (UNS) Attributable to Poor Screen Contact, and the (New) Total Level of Image Unsharpness That Results from It

detail by an amount equal to the square of the poor contact distance. Next to motion, poor film-screen contact contributes to the greatest amount of image detail loss.

Material Unsharpness Summary

1. X-ray films are capable of resolving far greater numbers of lines than visual acuity can perceive and are therefore not a significant contributing factor to the evaluation of overall material unsharpness.
2. Given the current state of the art and the concerns related to patient exposure, direct exposure radiography is contraindicated and discouraged as a method of producing radiographic images on human beings.
3. When film-screen recording systems are employed, an analysis of the needs of the examination, the condition and cooperation of the patient, and the available systems should be made. Often a trade-off in the materials employed will be required in order to reduce motion even though the higher speed system may not produce an image possessing the best possible image details.
4. It is important that the technologist be familiar with the manufacturer's recommendations and specifications related to the purchase and application of any film-screen recording system. Using the system to full advantage will enable the technologist to avoid repeat exposures and the errors inherent in the technical guessing game. To that end, read the literature provided, and, if necessary, meet with manufacturer's representatives in order to understand completely the system you are using.
5. Film-screen contact must be your primary consideration when using a film-screen recording system, since poor screen contact not only increases the

unsharpness attributed to the screens, but also increases the unsharpness related to the other factors affecting image detail. Poor film-screen contact adds to the unsharpness in direct relation to the square of the distance of the poor contact.

Geometric Unsharpness

Geometry is that branch of mathematics dealing with the properties, measurement, and mutual relationships of points, lines, surfaces, and solids. The radiographic image is actually made up of minute strands of silver that form the image's structural lines and can therefore be accurately measured by geometric methods. The formation of these lines and their recording on the film depend on certain geometric principles of image balance incorporated into the relationship between the x-ray beam, the structures being examined, and the position of the image receptor. An analysis of geometric unsharpness is an examination and measurement of these linear relationships. Geometric unsharpness is produced and influenced by the following factors.

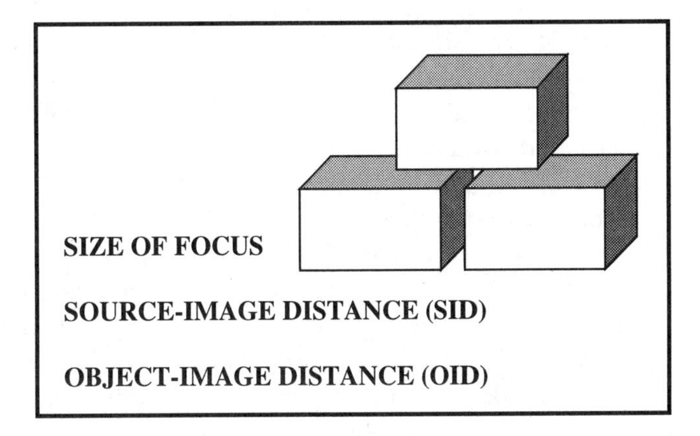

The factors contributing to geometric unsharpness are all measurable, and therefore the degree of unsharpness related to their arrangement and application can be determined, measured, and expressed. The following formula can be used to determine the amount of geometric unsharpness present within a recorded image:

$$\frac{\text{Size of focus} \times \text{object-image distance (OID)}}{\text{Source-object distance (SOD)}}$$

Geometric Unsharpness Factors

X-ray Production

X-rays are produced whenever a fast-moving stream of electrons undergo a sudden slowing down or stoppage. The x-ray tube is a device that is designed to produce x-rays. Within the cathode section of the x-ray tube, a filament in the form of a spiral wire is energized, causing the release of electrons. The filament is recessed in a conical-shaped cup in order to keep the released electrons contained within the area surrounding the filament. A high-energy electrical current is conducted between the cathode and anode of the x-ray tube, causing these released electrons to move at rapid speeds from the cathode to the anode. At the anode of the x-ray tube, an area referred to as the **target** is located in the path of the high-speed electrons. On the target area of the anode is a smaller area called the **actual tube focus,** where the high-speed electrons are actually going to strike. The size of the electron stream is dependent on the size of the filament and the amount of electricity applied to the filament, which affects the number of electrons released and ultimately the number of x-rays produced. The energy level of the x-rays produced depends on the speed of the electron stream, which is controlled by a separate high-energy current applied between the cathode and anode of the x-ray tube.

Effective Focus Size

The filament within the x-ray tube has a measurable length and width, and therefore for any given current applied to the filament, the cloud of electrons released also has a given size. Most x-ray tubes employed in radiography have at least two filaments of differing sizes. As a result of a defined size of the electron stream traveling from the cathode to the anode of the x-ray tube, x-rays are not produced from a single point of emission. They originate at various points along the x-ray tube's actual focus. This area upon the target where the electron beam is focused and strikes can be measured and has a specific length and width. The beam of radiation that ultimately leaves the tube originates and spreads out in all directions from these different points of origin on the x-ray tube's actual focus. The target area of the x-ray tube is designed so that the surface facing the electron stream is angled or beveled from the perpendicular. As a result of this design, a measurement of the beam of x-radiation produced at the actual focus of the tube's target and directed down toward the patient will demonstrate a size that is smaller than the actual focus area on the tube target. The smaller area associated with the beam of radiation emerging from the tube is referred to as the **effective focus**. The tube design allows for a much smaller effective focus within the beam of radiation emerging from the tube than the size of the actual area on the target of the tube where the electron stream strikes. This smaller effective tube focus helps to reduce the geometric unsharpness of the beam of radiation directed toward the image receptors. It is the effective focus that we measure when we examine the influence of geometric unsharpness upon the recorded details of the image. The lead-lined tube housing and beam limitation devices attached to the tube housing effectively absorb radiation produced and traveling in directions other than those emerging from the tube and directed toward the patient. Thus, although a beam of radiation cannot be focused, it can be controlled so that only the beam traveling in a useful direction will be allowed to exit from the tube.

X-rays produced in this manner are heterogeneous. That is, the energy of the beam of radiation produced by the x-ray tube is not homogeneous. Within any x-ray beam, there will be a wide range of wavelengths and energy levels. Therefore, the x-radiation produced by the x-ray tube is actually a complex beam of various energies that originate at multiple points on the target of the x-ray tube and spread out in different directions and at varying angles and positions with respect to one another. As the beam of radiation enters a structure and interacts with matter, there will be multiple photon interactions within the tissues that occur at slightly different positions and angles to one another. The complexity of the different energies and directions of the x-ray beam after it has traversed a structure produces a geometric unsharpness factor known as **penumbra.** Penumbra refers to the unsharpness in the recorded edges and lines within an image produced as a result of the actual size and shape of the tube focus. Several factors related to the production of x-rays affect the geometric unsharpness of the recorded image.

Size of Focus

The main factor affecting the geometric unsharpness of the recorded image related to the production of x-rays is the size of the tube focus. Obviously, the larger the focus area on the tube target, the larger and greater the number of points of origin for the formation of the radiation beam and, thus, the greater the loss of image detail. The ideal beam for any given tube focus, established as the midpoint of the central axis of the beam, is referred to as the **central ray.** Because of its position at the midpoint of the beam, the central ray suffers from the least amount of penumbra. The beam of radiation within the central ray traverses the structure in an almost perpendicular direction, allowing for little, if any, unsharpness. However, as you move further from the central axis (central ray) of the beam, the beam of radiation is directed at greater and greater angles from the perpendicular, and, therefore, there would be greater unsharpness recorded near the edges of the beam of radiation. As an example, there would be greater geometric unsharpness attributed to the size of the focus at the recorded edges of a circle that measures 6" in diameter than at the edges of a circle that measures 3" in diameter.

The size of the focus strongly influences the recorded detail of the image. X-ray tubes containing two filaments are designed so that each filament when energized focuses its electron stream on a different area of the tube target. Since the actual sizes of the filaments in such x-ray tubes are different, double-focus tubes enable the technologist to choose between a small or a large focus to produce a radiographic image. In most instances, the choice of which focus to use is made relative to the intensity of the exposure required for the examination and the limitations identified on the tube rating chart. In general, the use of the tube's small focus imposes a greater limitation on the intensity of exposure that you can employ for a given set of exposure factors. For example, the selection of a small focus may limit you to a milliamperage selection of 200 mA. If your procedure requires an exposure of 400 mAs, you would have to employ a 2-sec time of exposure. This could create motion unsharpness. Selecting the larger focus, you may be able to select a milliamperage of 500 mA. For an exposure of 400 mAs, you would only require a 0.8-sec exposure time. The faster time of exposure may help to eliminate motion unsharpness. However, by selecting the larger focus, you would significantly increase the geometric unsharpness of the image.

Let's compare the geometric unsharpness produced by a beam of radiation using a 0.6-mm focus with 4" of OID and an SID of 40" with the geometric unsharpness produced by a beam of radiation using a 2.0-mm focus, with the other factors remaining the same. Applying the geometric unsharpness formula, we can determine that the unsharpness factor using a 0.6-mm focus is 0.06 mm ($0.6 \times 4 \div 36$), whereas using a 2.0-mm focus the unsharpness level is 0.22 mm ($2.0 \times 4 \div 36$). (See Figure 2-9.) The geometric unsharpness attributed to focus size will differ for different focus sizes and in every examination depending on the source-image distance and the distance that the object is from the imaging receptor. Comparing two different focus sizes establishes a geometric comparison between two triangular figures as seen in Figure 2-9. The recorded detail of the image is always improved by a reduction in the size of the focus. Therefore, in all radiographic procedures, use the smallest available focus consistent with the energy levels required and governed by the tube rating chart. (**Perform Experiment 9**.) Choosing the small focus whenever possible enables you to produce an x-ray beam emerging from the tube that possesses the greatest geometric sharpness potential that the x-ray tube is capable of achieving and starts your radiographic procedure in a positive direction. All subsequent technical factors related to unsharpness will be influenced and exacerbated by the focus size employed.

Evaluation of Focus Size

The focus size of a double-focus tube is quoted by the manufacturer and should be used only as a guide when accurately assessing the modulation transfer function of a particular imaging system. These focus sizes provided by the manufacturer are generally computed using a very low milliamperage and, therefore, are only nominal values. A more practical evaluation of the focus size utilizing the exposure values employed in diagnostic radiography can be determined by using a pinhole camera as described in ICRU Report No. 10f or by using a star resolution test pattern. Either test provides reasonably accurate results. It is important to note that the actual focus size increases with increasing milliamperage applied to the tube. This increase in the size of focus attributed to milliamperage increases is known as **blooming** of the focus. It results from the greater number of released electrons and, thus, larger space charge (electron cloud) around the filament, produced by the higher milliamperage. In general radiography, the blooming effect is of little significance. However, in specialized angiographic procedures, which frequently employ high-intensity short exposures for macroradiography (magnification), the blooming effect can be a significant factor contributing to the loss of image resolution.

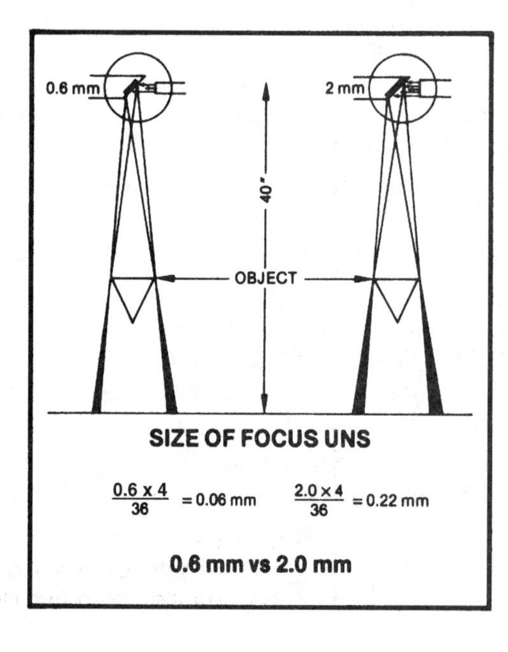

Figure 2-9
Comparison of Geometric Unsharpness (UNS) Resulting from Two Different Focus Sizes

Object-Image Distance (OID)

In an examination of the OID, we find that x-radiation can be compared with visible light since it follows many of the same principles governing the formation of a shadow produced by the emission of light. A light beam casts a shadow of an object on a wall similar to the way that an x-ray beam records an image on a film. In addition to the size of the source of the light, the clarity (sharpness) of the projected shadow depends on the distance of the object from the wall. The same principle is true for the x-ray beam and its recorded image.

Let's compare the geometric unsharpness produced by a beam of radiation directed at an object lying 4" from the film OID employing a 2.0-mm focus size and 40" of SID with the geometric unsharpness produced by a beam of radiation directed at an object lying 18" from the film OID, with the other factors remaining the same. Applying the geometric unsharpness formula, we can determine that the geometric unsharpness produced with an OID of 4" is 0.22 mm, whereas the geometric unsharpness produced with an OID of 18" is 1.6 mm (see Figure 2-10).

Recorded detail of the image is always improved by a reduction of the OID. Therefore, in all radiographic procedures, unless contraindicated by the circumstances of the examination (magnification radiography, for example), the part to be examined should be placed as close to the film as possible. With thick body parts containing a multiplicity of structures, identify the structures of interest to be recorded within the body part and place them as close to the film as possible. For certain examinations, this may require a posteroanterior projection of the body part rather than an anteroposterior projection. Obviously, placing the part to be examined directly on top of the cassette (tabletop radiography) would be preferable to Bucky radiography. Most upper and lower extremity examinations from the fingers up to and including the elbow and from the toes up to and occasionally including the leg can be performed as a tabletop procedure with a minimum OID. Larger, thicker body parts require increased exposure and higher kilovoltage to penetrate the part properly. Thick body parts produce significant amounts of secondary and scattered

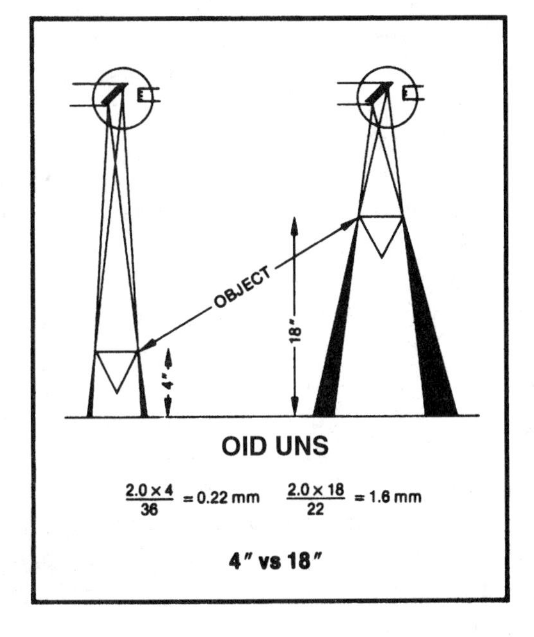

Figure 2-10
Comparison of Geometric Unsharpness (UNS) Resulting from Two Different OIDs

radiation. These secondary and scattered radiations will severely affect the quality of the recorded image, and, therefore, the examination of larger, thicker parts is normally performed using radiographic grids in a Bucky apparatus located beneath the tabletop. Procedures performed utilizing the Bucky apparatus will automatically increase the OID as a result of the distance created between the tabletop and the location of the cassette placed within the tray of the Bucky apparatus. The OID has a major impact on the geometric unsharpness levels produced in the recorded image. A careful review of the requirements of the examination can enable the technologist to adjust the equipment, position the patient, and employ the best possible technical factors to reduce or minimize the influence of the OID. (**Perform Experiment 10.**)

Source-Image Distance (SID)

Image sharpness in radiography is also a function of the SID selected. Sharpest image details will be produced when the SID selected is at the most reasonable, maximum distance. For most routine radiography, this distance has been established and standardized at 40" (102 cm). Some procedures, including chest radiography, are routinely performed at 72" (183 cm), and, except in rare instances, little is to be gained from an increase beyond 72" (183 cm). Although 72" (183 cm) would be more ideal for all radiographic procedures from the standpoint of image sharpness, the increased exposure required at this distance for most routine radiographic examinations makes the problems associated with potential motion unsharpness of much greater concern than the geometric unsharpness produced at a lesser SID.

Let's compare the geometric unsharpness produced by a beam of radiation using an SID of 40" (102 cm), a 2.0-mm focus size, and an OID of 4" with the geometric unsharpness produced by a beam of radiation using 72" (183 cm) of SID, with the other factors remaining the same. Applying the geometric unsharpness formula, we can determine that the geometric unsharpness at 40" (102 cm) SID is 0.22 mm compared with a geometric unsharpness of 0.11 mm at an increased distance of 72" (183 cm) SID (see Figure 2-11).

Recorded detail of the image is always improved by an increase in the SID. Therefore, always use the greatest practical distance as governed by the requirements of your examination. Consider the problems associated with motion unsharpness resulting from the increased exposure necessary at an increased SID. (**Perform Experiment 11.**) Overall, the SIDs employed in radiography have been standardized. This enables the technologist to develop accurate technique charts and to standardize a number of technical factors involved in the performance of the examination. The standardization of technical factors helps to avoid the problems associated with the technical guessing game.

A reduction in the SID, although increasing the geometric unsharpness of our image, can be an advantageous maneuver when you are confronted with the problems associated with motion unsharpness. A reduction of the SID enables the technologist to reduce the time of exposure according to the principles of the inverse square law. At an SID of 40" (102 cm), a reduction of 10" (25 cm) enables you to reduce your exposure factors by 44%. Therefore, a 100-mAs exposure at an SID of 40" (102 cm) would only require a 56-mAs exposure at a reduced SID of 30" (77 cm).

A reduction in the SID would also have an influence on the size and shape of the recorded image. A reduced SID would increase the size and shape distortion of the recorded image, thereby reducing the overall radiographic quality of the image. The effect of the SID on image distortion will be examined in Chapter 3.

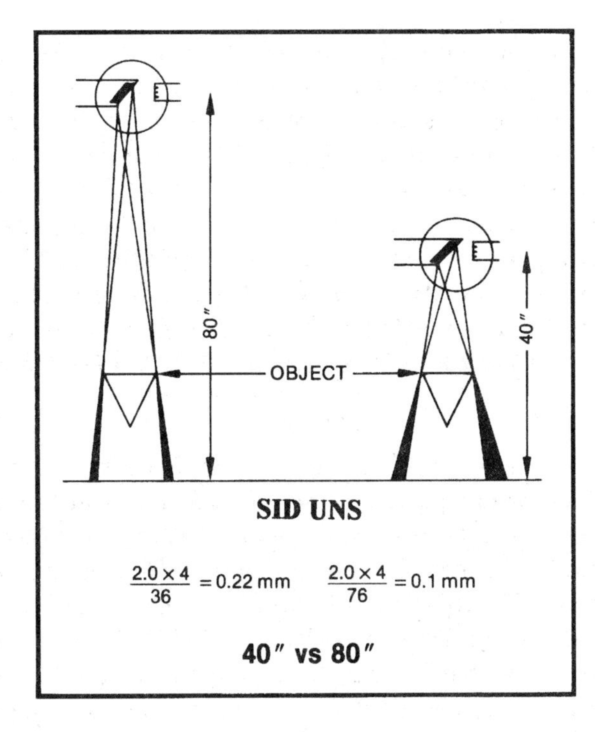

Figure 2-11
Comparison of Geometric Unsharpness (UNS) Resulting from Two Different SIDs

Geometric Unsharpness Summary

You can see by the relationships demonstrated in this section that the size of the x-ray tube's focus, the location of the object to be examined relative to the image recording medium (OID), and the SID play important roles in the formation of a detailed image. Of equal importance is the fact that these influences and the amount of geometric unsharpness produced can be accurately evaluated and measured. Choosing the most appropriate combination of these three factors for a given examination will enable the technologist to record an image that possesses the least amount of geometric unsharpness possible.

What you must further recognize is that these three factors influencing geometric unsharpness cannot be considered separately but must be considered together as combined, interrelated influences in their radiographic effect on the recorded detail of the image. Let's consider these interrelated influences from a practical point of view in an examination of the clavicle:

AP projection of the clavicle
Large focus (2.0 mm)
8" OID
40" (102 cm) SID

$$\frac{2.0 \text{ mm} \times 8''}{32''} = 0.50 \text{ mm UNS}$$

PA projection of the clavicle
Small focus (0.6 mm)
3" OID
40" (102 cm) SID

$$\frac{0.6 \text{ mm} \times 3''}{37''} = 0.04 \text{ mm UNS}$$

You may recall that the human eye can begin to detect unsharpness in a radiographic image when the distinct lines of structural details are recorded wider than 0.2 mm at a visual acuity of 5 LP/mm. In this comparison, the anteroposterior clavicle examination employing a large focus of 2.0 mm and an OID of 8" at an SID of 40" (102 cm) produces a geometric unsharpness level of 0.50 mm. Thus, prior to the addition of any material unsharpness associated with the film-screen recording system, this image will have already reached a level of unsharpness that could be visible and intolerable. This level of unsharpness primarily resulted from the increased OID associated with the anteroposterior projection.

The posteroanterior clavicle examination with the reduced size of focus and reduced OID resulting from the improved positioning of the patient, while still employing the same SID, produces a geometric unsharpness level of only 0.04 mm. The amount of geometric unsharpness produced in the posteroanterior clavicle is far below the level where it will begin to become intolerable and visible within the recorded image. The increased source-object distance (SOD) created as a result of reducing the OID was a major factor in this reduction of geometric unsharpness. This example demonstrates the interrelationship between the factors contributing to geometric unsharpness. The factors influencing geometric unsharpness are listed below in order, from the most critical or influential factor producing geometric unsharpness to the least critical or influential factor.

- **Object-image distance**. The OID is the most critical influence contributing to geometric unsharpness. Increases in OID automatically alter the relationship of the source-object distance in the geometric unsharpness formula. Therefore, an increase in the OID will have a twofold influence on the production of geometric unsharpness. Another reason the OID is considered the prime factor related to geometric unsharpness is that some degree of OID is present in all radiographic procedures and can never be totally eliminated and that, because of the patient's position, the specific structures being examined, or the requirements of the procedure, you may be unable to reduce the OID at all. There are many procedures in which an increased OID must be accepted together with the increased geometric unsharpness it produces.
- **Size of focus**. X-ray tubes are available in a wide range of focus sizes. Some geometric unsharpness is attributable to even the smallest focus available. Other than the selection of a small or large focus when you energize your x-ray tube, this factor does not permit any further adjustments. You are literally stuck with the level of geometric unsharpness associated with the size of focus employed. However, since its effect on geometric unsharpness is known, it can be easily evaluated and standardized utilizing the geometric unsharpness formula.
- **Source-image distance**. The SID is the least critical factor influencing geometric unsharpness since it is the factor that is most seldom changed. The SID employed for most radiographic procedures has been standardized. In addition, considerable adjustments (increases) in the SID would be necessary to produce a visible improvement in the recorded detail of the image. Significant changes in the SID in most instances would be impractical both as a result of the physical limitations of the x-ray examination room and the increased exposure necessary with increases in the SID.

Minimal Geometric Unsharpness Radiography

1. Plan your procedural setup in advance in order to use your equipment to advantage for the specific procedure and projection you are performing.
2. Place the part to be examined as close to the film as possible. Consider your positioning of the part related to the OID produced and adjust your position whenever it is possible to reduce the OID.
3. Perform examinations as tabletop procedures whenever possible, since this reduces the OID to a minimum. This method will be subject to the thickness/opacity of the part being examined and the possible need for a Bucky device. Using a Bucky device automatically increases the OID by 3" to 4" (8 to 10 cm).
4. Select the small focus whenever possible. The small focus can be utilized in the majority of examinations. Check your tube-rating chart to determine whether the exposure selected will be operating within the safe limitations of the tube-rating chart.
5. Standardize your SID. In procedures with an increased OID, consider increasing your SID to reduce the severity of its effect. Analyze the improvement carefully, since an increased SID will require an increase in the total exposure required. If an increase in the exposure requires an increase in the time of exposure, you must consider whether motion and the recording of motion unsharpness may become a greater concern than the improvement of geometric unsharpness associated with the increased SID.

As you can see, the arrangement of the radiographic equipment related to the geometric factors for a radiographic procedure is not a simple matter, but a complex technical procedure. You must consider the multiple effects that the differing geometric factors will have on the radiographic quality of the recorded image and choose the geometric arrangement that will produce the least amount of geometric unsharpness. When adjusting these factors, you must consider the influence those changes will produce and choose those influences that will be the least detrimental to the recorded image.

Structural Shape Unsharpness

The anatomical structures contained within a body part vary considerably in size, shape, and thickness. In addition, within any body part, there are a multiplicity of different structures that overlap and are superimposed upon one another. The multiple structures contained within a body part lie at different levels within the thickness of the part and may not lie parallel to one another or have any particular relationship to one another. These varying structural shapes of the body part being examined introduce additional unsharpness influences over which the technologist has little control. In fact, the unsharpness attributed to these multiple structural relationships may even exceed the overall influence of geometric unsharpness.

If the structure to be examined has an overall trapezoidal shape and is accurately aligned to the limits of the beam's edge, there will be less structural shape unsharpness produced (see Figure 2-12). The edges of the structure would exactly parallel the edge of the radiation beam. Thus, the edges would be recorded with

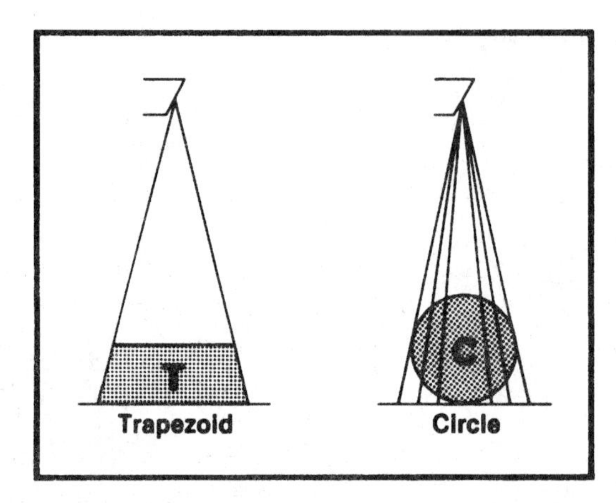

Figure 2-12
Example of Unsharpness Resulting from Different Structural Shapes

sharpness and minimal density gradient differences. However, if the structure to be examined is circular in shape and accurately aligned to the limits of the beam's edges, the recorded pattern will differ considerably from the structure within the body (see Figure 2-12). With a circular-shaped structure, the edge relationship of the structure would always be in the shape of an arc, and the beam's edge would be unable to record it accurately. Additionally, since the thickness of the structure would differ at every point from its edge to its center, there will be a wide pattern of different density gradients throughout the recorded image.

The complexity of structural shape unsharpness is difficult to comprehend and impossible to eliminate. Considering the multiplicity of different sizes and shapes of structures contained within a body part, the technologist can attempt to minimize structural shape unsharpness by identifying the specific **structures of interest** within the body part being examined. Attempts to reduce structural shape unsharpness include the proper centering and alignment of the structures of interest within the part with respect to the central ray and the direction of the beam of radiation and the alignment of the radiographic film with the structures of interest. Assessment of the success of these attempts is investigated in Chapter 3.

Recorded Detail Summary

We have now investigated the three major factors that contribute to the recorded detail of the image: (1) motion, (2) material unsharpness, and (3) geometric unsharpness. The following list positions these factors in decreasing order of influence on recorded detail. (Structural shape unsharpness, being impossible to quantify or calculate, is not considered in this summary.)

1. Motion
   - Voluntary
   - Involuntary
2. Material unsharpness
   - Film-screen contact
   - Intensifying screens

3. Geometric unsharpness
   - OID
   - Size of focus
   - SID

Motion is always the first consideration in any investigation of the loss of image details. Material unsharpness is identified as the next influence even though the actual unsharpness attributed to the materials themselves may be less than the geometric unsharpness produced by the arrangement of the equipment and the patient for the procedure. The reason material unsharpness is identified after motion as the next greatest influence is that any geometric unsharpness produced will be further exacerbated and intensified as a result of the unsharpness produced by the materials selected for the procedure.

Except for motion, all of these influences can be accurately measured and evaluated. Within this chapter, we have examined the factors separately in order to investigate the influence of each on the sharpness of the recorded details. In actuality, these factors are interrelated, and their effects are often compounded by this interrelationship. The geometric unsharpness produced as a result of the OID, size of focus, and SID is further exacerbated by the image detail loss produced by the materials selected to record the image. The total unsharpness will result from the integrated influence attributed to each factor. Obviously, if motion were added to this equation, the recorded image would simply be a blur. To fully understand the interrelationship of these influences, we have to combine the total effect of all of these factors together. In the diagrammatic representation of the formation of the recorded image demonstrated in Figure 2-13, you can see the interrelationship of these multiple factors. The specific relationships of the technical factors chosen for this example have been exaggerated in order to demonstrate the severity of the individual influences on the combination of factors employed.

In this example, a large focus of 2.0 mm has been selected, the midpoint of the part to be examined has been placed 18" (48 cm) from the image receptor, and the source-image distance selected is 36" (91 cm). The geometric unsharpness produced as a result of this arrangement is 2.0 mm. A 200 speed film-screen recording system has been chosen that has an intensifying screen unsharpness of 0.45 mm. Adding the intensifying screen and geometric unsharpness, the total unsharpness is now 2.45 mm. A poor film-screen contact distance measuring 1.5 mm is present in the cassette, which produces an additional unsharpness related to the square of the poor film-screen contact distance. The total unsharpness would be determined by multiplying the sum of the geometric and intensifying screen unsharpness by the square of the poor film-screen contact distance. In this example, 2.45 mm × 2.25 mm (1.5 mm^2) = 5.51 mm total unsharpness. When the width of a recorded line pair measures more than 0.2 mm wide, the image has already begun to demonstrate unsharpness. The unsharpness level produced and represented by this diagram (5.51 mm) is totally unacceptable and intolerable. Such an image would certainly lack radiographic quality and may not even be of diagnostic value.

In this exaggerated example, you can see that the film-screen contact is the major contributor to the loss of image details. However, in actual practice, the ratio of the specific influence of any of the multiple factors that contribute to unsharpness within

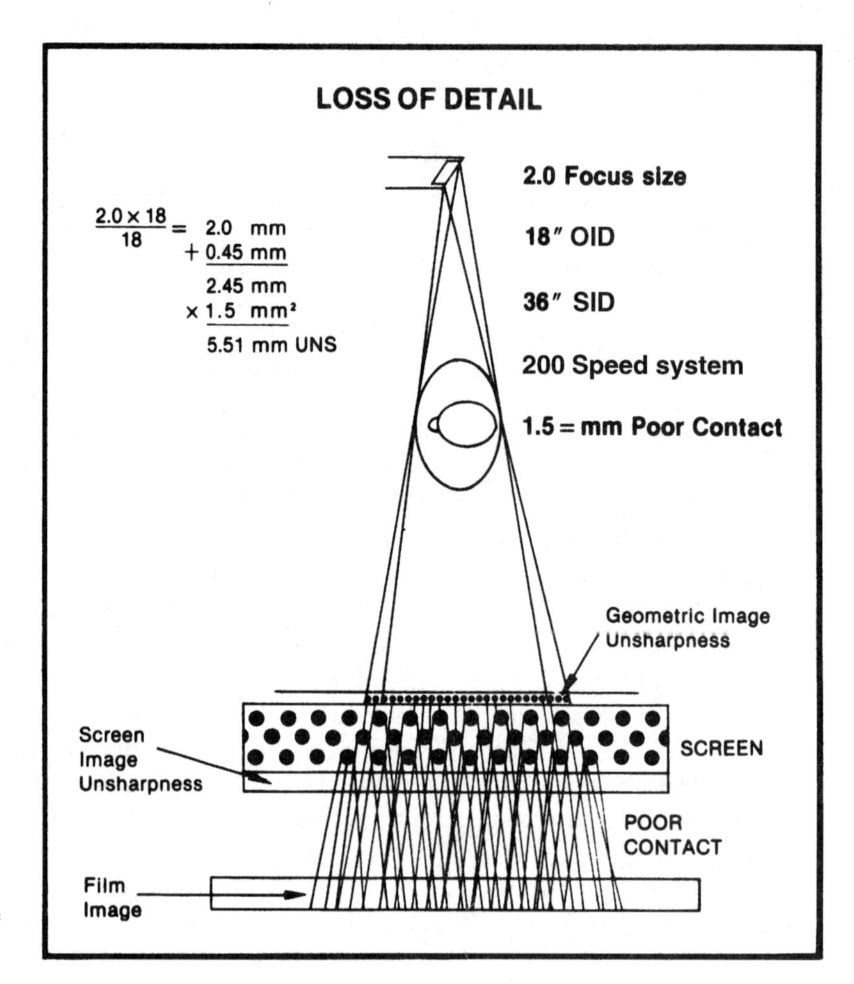

Figure 2-13 Diagrammatic Representation of the Loss of Detail in the Formation of an Image

the recorded image depends on the interrelationships between those influences and the actual arrangement of the equipment and selection of technical factors by the technologist. These arrangements are often imposed by the limitations of the equipment, the needs of the patient, and the requirements of the procedure. Experiment 12 will help you to recognize and understand the complex relationship between the multiple factors contributing to the recorded detail of the radiographic image. (**Perform Experiment 12.**)

Recorded Detail Review Questions

Name ______________________________ Date ______________

1. Discuss the implications of the statement: It is more practical to identify the degree of unsharpness that can be tolerated within the recorded image than to attempt to identify the amount of recorded detail that has been achieved.

2. Optimum visual acuity at 18" to 24" (48–61 cm) viewing distance is represented by the ability to distinguish a minimum of _________ line pairs within a millimeter (LP/mm).

3. Actual visual acuity is limited. The level of unsharpness that is visibly perceptible begins at _________ and gets progressively worse as it increases.

   a. 0.05 LP/mm
   b. 0.10 LP/mm
   c. 0.20 LP/mm
   d. 0.50 LP/mm

4. The most detrimental factor contributing to image unsharpness is _________.

   a. use of a 1000 speed film-screen recording system
   b. an increased object-image distance
   c. motion
   d. poor film-screen contact

5. Define the terms **involuntary motion** and **voluntary motion** and differentiate between their influence on the recording of image detail.

6. After explaining the procedure and gaining the understanding and cooperation of the patient, how would you deal with the problem of physiological motion occurring during a chest examination for lung pathology? Indicate at least two technical factor adjustments, and describe how you would apply them to eliminate motion unsharpness.

7. Indicate the information you require in order to determine the exposure time required to eliminate unsharpness due to the physiological motion of the structure.

8. Describe how you would explain the requirements for a chest examination (1) to a deaf patient and (2) to a 3-year-old child accompanied by his mother.

9. Describe how you would handle problems associated with voluntary motion.

10. Describe how you would control the motion of a frightened adult patient suffering from a recent injury (suspected fracture) who appears unable to hold still for an examination of the wrist.

11. Describe a method for whole-body immobilization for an abdominal examination on an uncooperative 18-month-old child.

12. The purpose of manipulating the technical factors because of the problem of motion is to produce an image without any appreciable loss of __________, while maintaining the overall radiographic __________ of the recorded image.

    a. quality; mAs
    b. mAs; image detail
    c. image detail; exposure
    d. image detail; density

13. An abdominal radiograph was performed using the following technical factors. In an effort to eliminate the problem of motion, adjust your factors as needed to maintain the radiographic density of the image.

    100 mA
    1.2 sec exposure time
    48" (122 cm) SID
    80 kVp
    12:1 ratio, 100-line grid Bucky
    100 speed film-screen recording system
    Suspended respiration
    Collimated to the structure of interest

A. *mAs consideration*
(mA × exposure time = mAs) The mAs represented by the above exposure is __________. If you increase your mA to 300, your new time of exposure will be __________.

B. *Materials consideration*
Using the technique determined in Part A above, change to a 400 speed film-screen recording system. The new mAs required would be __________. Maintaining 300 mA, the time of exposure needed to produce the new mAs using the higher speed recording system would be __________.

C. *SID consideration*
new mAs = old mAs × new distance^2 divided by the old distance^2
Using the technique determined in Part B above, reducing your SID to 40" (102 cm) enables you to adjust your mAs further. The new mAs required to maintain radiographic density at the new distance is __________. Choosing a new mA of 500 mA, the time of exposure required to produce the new mAs would be __________.

a. 0.02 sec
b. 0.04 sec
c. 0.067 sec
d. 0.134 sec

D. *kVp/mAs relationship consideration*
Using the mAs determined in Part C above, if you were to increase your kVp by 15% to __________kVp, you could reduce your mAs to __________. Using 500 mA, the time of exposure required for this new mAs would be __________ sec.

E. *Summary of technical factor changes*
Your final exposure factors would be __________mAs, using __________ mA at a __________sec time of exposure; __________kVp; __________SID; with a __________speed film-screen recording system.

14. Changing from a 100 speed to a 400 speed film-screen recording system permits you to reduce the time of exposure while maintaining radiographic density. However, this adjustment affects another of the four factors that contribute to radiographic quality. Identify the factor and describe the influence that this change has on this factor.

15. Decreasing your SID will allow for a reduction in the time of exposure, but this change also affects the radiographic quality factors of recorded detail and distortion. Describe the influence this change has on each of these factors.

16. Increasing your kVp by 15% will permit you to reduce your mAs by 50%, while maintaining the radiographic density of the image. Of the technical factor adjustments utilized to reduce your time of exposure, the kVp/mAs relationship would be the least recommended method. Why?

17. Describe the MTF as it relates to image details and radiographic quality.

18. What is meant by image resolution? Indicate how it is identified and measured/evaluated.

19. What is meant by the term **line spread function (LSF)**? Would this function be more applicable to image resolution or image sharpness?

20. Radiographic film unsharpness is considered a (*circle one*) significant/insignificant influence affecting the recorded detail of the image. Explain the reasons for your choice.

21. Define the term **crossover** related to material unsharpness attributed to radiographic film and intensifying screens and indicate how this problem is solved by the film and screen manufacturers.

22. Describe the positive application of the principles of ALARA related to the selection of film-screen recording system radiography versus direct exposure radiography.

23. Describe the basic structure of a radiographic film.

24. Describe the basic structure of an intensifying screen.

25. Differentiate between the qualities of fluorescence and phosphorescence related to the phosphor materials utilized in intensifying screens.

26. Intensifying screen speeds are determined by the intensification factor. Describe how you would determine the intensification factor of an intensifying screen.

27. Identify the inherent factors of intensifying screens that affect screen speed, and describe the influence of each.

28. Identify the technical factors that affect screen speed, and describe the influence of each.

29. Identify the inherent factors of intensifying screens that affect the material unsharpness of the recorded image, and describe the influence of each.

30. Why are rare earth phosphor screens recommended over conventional phosphor intensifying screens?

31. Compare the light conversion efficiency of conventional phosphors with rare earth phosphors used for intensifying screens.

32. Compare x-ray energy absorption levels of conventional phosphors with rare earth phosphors used for intensifying screens.

33. Define quantum mottle. How does this factor affect the quality of the radiographic image? Would you anticipate encountering this effect utilizing a slow or a fast imaging system? Explain.

34. Changing from an 800 speed film-screen recording system to a 400 speed system will enable you to reduce _____ unsharpness.

    a. geometric
    b. motion
    c. material
    d. all of the above

35. Describe the influence of poor film-screen contact on the sharpness of the recorded detail of the image.

36. Describe the testing of a cassette for film-screen contact.

37. Discuss the importance of a quality assurance program to the reduction of the material unsharpness associated with poor film-screen contact.

38. Using the formula provided below, solve the following problems.

    new unsharpness = old material unsharpness × poor contact^2

    A. Comparing a radiograph taken with absolute contact with one having a poor film-screen contact distance of 1.5 mm, indicate the material unsharpness produced when utilizing a 100 speed film-screen recording system having a screen unsharpness factor of 0.35 mm.

    a. 0.58 mm
    b. 0.66 mm
    c. 0.78 mm
    d. 0.87 mm

    B. Considering the material unsharpness attributed to the poor film-screen contact in A above, how many lines would you theoretically be able to visualize in the space of 5 mm? Remember, visual acuity is represented by the ability to visualize up to 5 LP/mm.

    a. 5 lines
    b. 10 lines
    c. 15 lines
    d. 20 lines

39. List the three factors contributing to geometric unsharpness.

40. Describe how x-rays are produced in an x-ray tube.

41. Most x-ray tubes are double-focus tubes. Describe the purpose and limitations of the smaller focus. Describe the purpose of the larger focus.

42. Describe the influence of the size of focus related to the production of geometric unsharpness.

43. Describe the influence of the OID related to the production of geometric unsharpness.

44. Describe the influence of the SID related to the production of geometric unsharpness.

45. A lateral examination of the lumbar spine was performed using the following technical factors:

2.0-mm focus size	200 mA
12" (30 cm) OID	1.5-sec exposure time
36" (91 cm) SID	80 kVp

Applying the geometric unsharpness formula provided below, solve the following problems.

$$\text{geometric unsharpness} = \frac{\text{size of focus} \times \text{OID}}{\text{SOD}}$$

A. Determine the geometric unsharpness attributed to the above examination. Geometric unsharpness __________.

B. Reducing the size of focus to 0.6 mm, determine the new geometric unsharpness. The new geometric unsharpness is __________.

C. Rather than changing to a small focus as indicated in Part B above, change your SID to 72" (183 cm) instead. The geometric unsharpness at this new distance is __________.

D. Considering the distance change in Part C above, what does the mAs have to be changed to in order to maintain radiographic density? The new mAs would be __________.

E. Using the same mA indicated by the original examination, what would the new time of exposure be in order to produce the mAs required by this change? The new time of exposure would be __________. What additional problems related to radiographic quality could this change produce?

F. Using the original technical factors, determine the geometric unsharpness if you could reduce your OID to 8" (20 cm). The new geometric unsharpness would be __________.

G. Determine the geometric unsharpness if you were able to perform the examination after adjusting the size of focus to 0.6 mm, the SID to 72" (183 cm), and the OID to 8" (20 cm). The new geometric unsharpness would be __________.

46. Indicate the total image unsharpness (UNS) in the following procedures.

    total unsharpness = geometric UNS + screen UNS × poor contact UNS²

    A. 2.0-mm focus
    10" (25 cm) OID
    40" (102 cm) SID
    200 screen speed
    ($CaWO_4$) (0.45-mm UNS)
    2-mm poor film-screen contact
    Total UNS ________ mm

    B. 0.6-mm focus
    10" (25 cm) OID
    40" (102 cm) SID
    200 screen speed
    ($CaWO_4$) (0.45-mm UNS)
    2-mm poor film-screen contact
    Total UNS ________ mm

    C. 0.6-mm focus
    2" (25 cm) OID
    40" (102 cm) SID
    200 screen speed
    ($CaWO_4$) (0.45-mm UNS)
    2-mm poor film-screen contact
    Total UNS ________ mm

    D. 0.6 mm focus
    2" (25 cm) OID
    60" (152 cm) SID
    200 screen speed
    ($CaWO_4$) (0.45-mm UNS)
    2-mm poor film-screen contact
    Total UNS ________ mm

    E. 0.6-mm focus
    2" (5 cm) OID
    60" (152 cm) SID
    100 screen speed
    ($CaWO_4$) (0.35-mm UNS)
    2-mm poor film-screen contact
    Total UNS ________ mm

    F. 0.6-mm focus
    2" (5 cm) OID
    60" (152 cm) SID
    100 screen speed
    ($CaWO_4$) (0.35-mm UNS)
    Absolute film-screen contact
    Total UNS ________ mm

    G. 0.6-mm focus
    2" (5 cm) OID
    60" (152 cm) SID
    nonscreen/cardboard holder
    (direct exposure)
    Total UNS ________ mm

47. To select technical factors effectively, you must be able to analyze the influence of different changes in order to choose the best adjustment for the needs of the examination. Solve the following:

    A. 1.5-mm focus
    40" (102 cm) SID
    8" (20 cm) OID

    These factors would produce a total geometric UNS of ________ mm.

    You can change to a 0.6-mm focus or increase your SID to 60" (152 cm). Identify the more effective change related to the reduction of geometric unsharpness.

    New focus UNS ________ mm   New SID UNS ________ mm

B. 1.5-mm focus
40" (102 cm) SID
10" (25 cm) OID

These factors would produce a total geometric UNS of _______ mm.

You can change to 0.6-mm focus or reduce your OID to 4" (10 cm). Identify the more effective change related to the reduction of geometric unsharpness.

New focus UNS _______ mm New OID UNS _______ mm.

C. 1.5-mm focus
40" (102 cm) SID
6" (15 cm) OID

These factors would produce a total geometric UNS of _______ mm.

You can change your SID to 80" (203 cm) or reduce your OID to 2" (5 cm). Identify the more effective change related to the reduction of geometric unsharpness.

New focus UNS _______ mm New OID UNS _______ mm

48. The greatest amount of geographic unsharpness is usually attributed to _____.

a. size of focus
b. SID
c. OID
d. SOD

49. Manipulation of the kVp/mAs relationship in order to reduce the time of exposure due to motion unsharpness (*circle one*) will/will not directly affect the recorded details of the image. Describe the effect, if any, that a manipulation of the kVp/mAs relationship has on the radiographic image.

50. Number the following factors that influence the recorded detail of the radiographic image in proper sequence from the greatest influence (1) to the least influence (8).

_____ SID
_____ Intensifying screen speed
_____ Involuntary motion
_____ Radiographic film
_____ Film-screen contact
_____ Size of focus
_____ Voluntary motion
_____ OID

Recorded Detail Analysis Worksheet

Let's review our analysis of recorded detail with the following exercise. A satisfactory diagnostic radiograph of the pelvis was produced using the following technical factors:

100 mA
1 sec (time of exposure)
80 kVp
40" (102 cm) SID
1-mm focus size
Minimum OID
Regular screen-type film
100 speed film-screen system
8:1 ratio, 100-line moving grid
Proper collimation
2.5-mm aluminum filtration
92°F development at 90 sec
Normal thickness and opacity of the part

Without compensation, the changes listed in the table below are made one by one. Indicate the effect, if any, each change has on the sharpness of the recorded details of the radiographic image.

1. If the sharpness of the recorded details is improved/increased, mark a plus (+) in the space provided.
2. If the sharpness of the recorded details is reduced/decreased, mark a minus (–) in the space provided.
3. If the sharpness of the recorded details is not affected by the change, mark a zero (0) in the space provided.

Recorded Detail

Proposed change	Effect	Proposed change	Effect
Use a 2.0-mm focus	_____	Change to a 5:1 ratio, 100-line moving grid	_____
Use 10" (25 cm) OID	_____	Change to a 12:1 ratio, 100-line moving grid	_____
Reduce the SID to 25" (63 cm)	_____	Remove all collimation	_____
Increase the SID to 48" (122 cm)	_____	Remove all filtration	_____
Use screen-type film in a cardboard holder	_____	Develop at 100°F for 90 sec	_____
Use a 200 speed film-screen recording system	_____	Body thickness reduced by an atrophic condition	_____
Use 200 mAs	_____	Pathological condition reduces opacity of body tissues	_____
Increase the kVp to 90	_____		
Omit the use of a grid	_____		

3 Distortion

In our analysis of recorded detail, we discovered that some level of unsharpness is present in every recorded image. The same relationship is true for the radiographic quality property distortion. The image produced on the radiographic film is not an accurate or exact recording of the structures of interest but differs from them in varying degrees of **size** (magnification) and **shape** (elongation or foreshortening) distortion. Similar to the challenge presented by image unsharpness, the misrepresentation of the size or shape of the structures of interest (distortion) cannot be totally eliminated from the recorded image. Ordinarily, distortion has a detrimental effect on the radiographic quality of the image, and a variety of methods are utilized to minimize this influence. There are circumstances, however, when distortion is intentionally produced and used to advantage in radiographic procedures. In this chapter, we will introduce and examine the factors that produce and influence distortion. We will learn to control and minimize the negative influences of distortion and discover how to use it to advantage when applicable to the radiographic procedure.

It is important to recognize that a relationship between the recorded details of the image and distortion exists. Basically, as the geometric unsharpness of the image detail increases, the distortion of the recorded image also increases. Conversely, as the technologist makes changes in the technical factors in order to decrease the distortion of the image, the geometric unsharpness of the recorded image will decrease. The reason for this interrelationship and interdependency between these two major properties of the recorded image that constitute the geometric properties of the image will become evident as we investigate the influences of distortion. The factors that influence the geometric unsharpness of the recorded detail, object-image distance (OID), and source-image distance (SID), are some of the same factors that influence image distortion.

Size Distortion

Size distortion refers to the misrepresentation of the actual size of the structures of interest recorded as an image upon the radiographic film. For the purposes of our analysis, we shall refer to size distortion as the magnification of the image. Every structure visualized within the recorded image is actually a magnified representation of the structures contained within the body part being examined. There is a percentage of magnification in every radiographic image. As the percentage of magnification increases, the level of geometric unsharpness produced and visualized within the image also increases. The interrelationship between these properties of the recorded image assist us in determining a level of magnification that would be identified as unacceptable. Basically, when the percentage of magnification of the image reaches the point where the level of geometric unsharpness of the recorded detail is visible and detracts from the radiographic quality of the image, size distortion is considered intolerable and unacceptable.

Size distortion is influenced by the geometric relationship between the OID and SID. These two technical factors determine the amount of magnification in the recorded image. To evaluate the amount of magnification in the recorded image, it is necessary to compare the size of the recorded image with the actual size of the structures of interest being examined. As with our investigation of recorded detail, you will be utilizing geometric principles to determine the amount of size distortion present within the image. The following formula enables you to determine the amount of magnification within a recorded image:

$$\frac{\text{Image Width}}{\text{Object Width}} = \frac{\text{SID}}{\text{SOD}}$$

The source-object distance (SOD) represented in this formula is the distance from the focus of the tube to the structures of interest being examined. Although the terms **image width** and **object width** are specified in this formula, a measurement of the length of an image or structure could be substituted within the formula and used to determine both the length and width (total area) of the recorded image or the true size of the structures of interest.

The formula will enable you to determine the recorded image size, when the actual size of the structures of interest is known. However, more frequently, the formula is utilized to determine the unknown size of the structures of interest within the body part compared with the measurement of the actual length and/or width of the recorded image by simply substituting the factors within the equation.

A practical application of this formula would be the measurements of the radiographic image of the pelvis for a pelvimetry examination. In this procedure, the formula is used to determine the size of the pelvic inlet and outlet compared to the cranial size of the fetus to determine whether the size of the birth canal is adequate for the passage of the fetus during childbirth. Another example of the application of the magnification formula would be for the orthopedic evaluation of the true lengths of long bones related to bone growth, injury, or the pre- and postoperative comparisons of the lengths of the injured versus the noninjured extremity.

Once the magnification of the structures has been identified, the percentage of magnification can also be determined using the following formula:

$$\text{Percentage of Magnification} = \frac{\text{Image Width} - \text{Object Width}}{\text{Object Width}} \times 100$$

In the example provided in Figure 3-1, the object size (width) is known to be 4" (10 cm) long. Using the magnification formula, it was determined that the recorded image size (width) would be 8.88" (23 cm). Applying the percentage of magnification formula, we discover that the percentage of magnification related to the recording of this image is 122%. The percentage of magnification associated with this image would be totally unacceptable, since the geometric unsharpness in the recorded detail would be intolerable. For this example, we have exaggerated the representation of the size distortion relationship in order to demonstrate this influence more graphically. However, it must be kept in mind that as the OID increases, the percentage of magnification in the recorded image will increase, and since there is an interrelationship between distortion and the recorded detail of the image, the geometric unsharpness of the image will also increase.

The actual size distortion produced in every radiographic image is much more subtle and complex. The radiographic image has the same appearance as the structures of interest, only larger. Depending on the percentage of magnification, even the appearance of size distortion may not be obvious. The radiographic examination frequently involves the examination of thick, three-dimensional structures (i.e., the thorax, abdomen, or pelvis) and records these structures upon a flat, two-dimensional sheet of radiographic film. Contained within these thick body parts is a multiplicity of different structures having different thicknesses, located at many different depths within the body part. These structures are therefore located at many different OIDs. Thus, when examining a radiographic image, we must remember

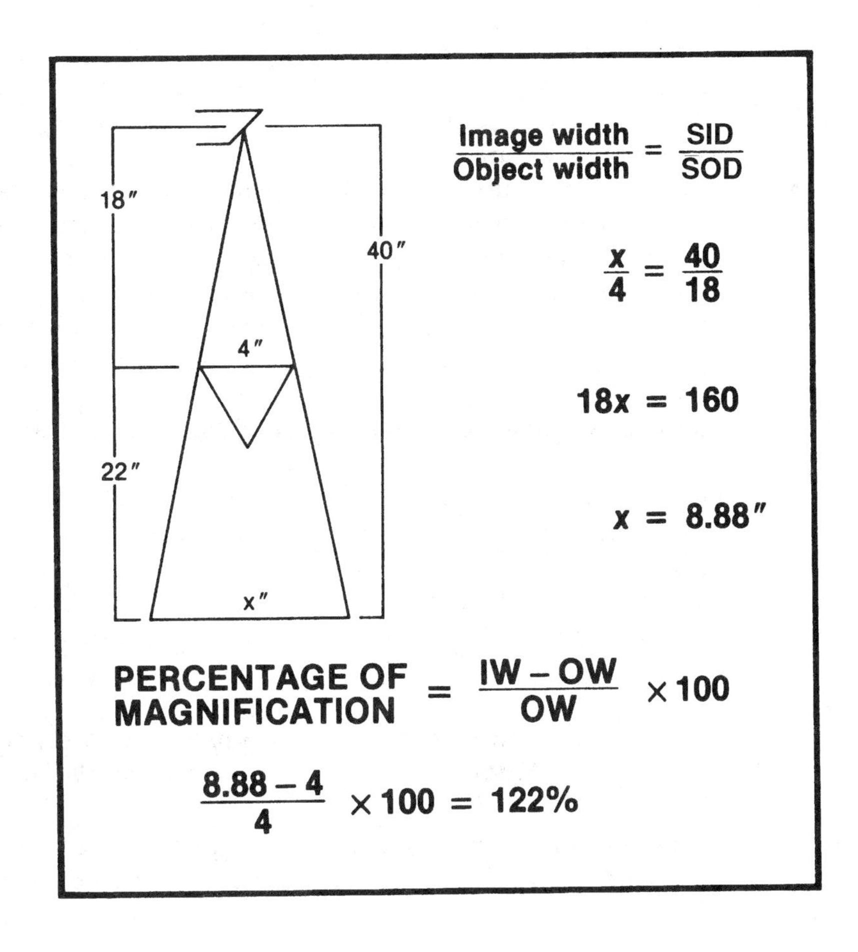

Figure 3-1 Example of Calculation of the Amount of Magnification of an Image (IW, Image Width; OW, Object Width)

that it is composed of a complex relationship between many different recorded structures, each of which will have a different percentage of magnification depending on its distance from the radiographic film. This complex relationship becomes even more apparent when we consider the number of structures that are superimposed and overlapping one another within any body part.

Object-Image Distance

Since the multiple structures contained within any body part lie at many different distances from the radiographic film, the amount of magnification for every structure recorded upon the film will be different. The further the structure is located from the surface of the image receptor (film), the greater the magnification of the recorded image of that structure. For example, when examining the thorax using an anteroposterior projection, the anterior portions of the upper pairs of ribs lie at a considerable distance from the film. When recorded, they overlap and are superimposed over the recorded images of the posterior portions of the ribs located much closer to the film. As we begin to consider these complex structural relationships, the more subtle influence of size distortion and the difficulty of its evaluation become more apparent.

The examination of thin body structures (i.e., an extremity such as the forearm) presents much less of a problem associated with size distortion than does the examination of a thick body part. In the forearm, the thickness of the overall part is not significantly different throughout its length, measuring only 1.5" (4 cm) at its distal end and 2.5" (6 cm) at its proximal end. The structures lie fairly parallel and are therefore almost equidistant from the film throughout its entire length. In most instances, the examination is performed as a tabletop procedure and the forearm can be placed very close to the film surface; therefore, the OID would be at a minimum. An examination of the forearm would present a minimum amount of size distortion (see Figure 3-2).

Figure 3-2 represents the diagram of a bone 4" (10 cm) in length lying parallel with the film surface. In the left diagram, the bone lies 2" (5 cm) from the film. In the diagram to the right, the OID is 12" (30 cm). The SID is 40" (102 cm) in both diagrams. If an injured forearm is wrapped in a pillow or an inflatable plastic bag to immobilize the part and stabilize the possible fracture, the OID represented by Figure 3-2 would not be unreasonable. Applying the magnification formula, we have determined that the bone lying closer to the film, with an SOD of 38" (96 cm), would increase from its actual size of 4" (10 cm) to a recorded image length of 4.2" (11 cm). The bone lying at the increased OID has an SOD of 28" (71 cm) and would have a recorded image length of 5.7" (14 cm). The OID of 2" (5 cm) produces a 5% magnification of the bone, whereas the OID of 12" (30 cm) produces a 42.5% magnification of the bone. Depending on the size of focus utilized and the material unsharpness attributed to the speed of the film-screen recording system selected, this could easily represent a percentage of magnification that would produce an unacceptable level of image detail loss. (**Perform Experiment 13**.)

It is obvious from this example that the least amount of OID should always be employed in an attempt to minimize size distortion. By arranging the structures of interest as close to the film as possible, not only do you reduce the magnification of the image, you also increase the sharpness of the recorded detail of the image.

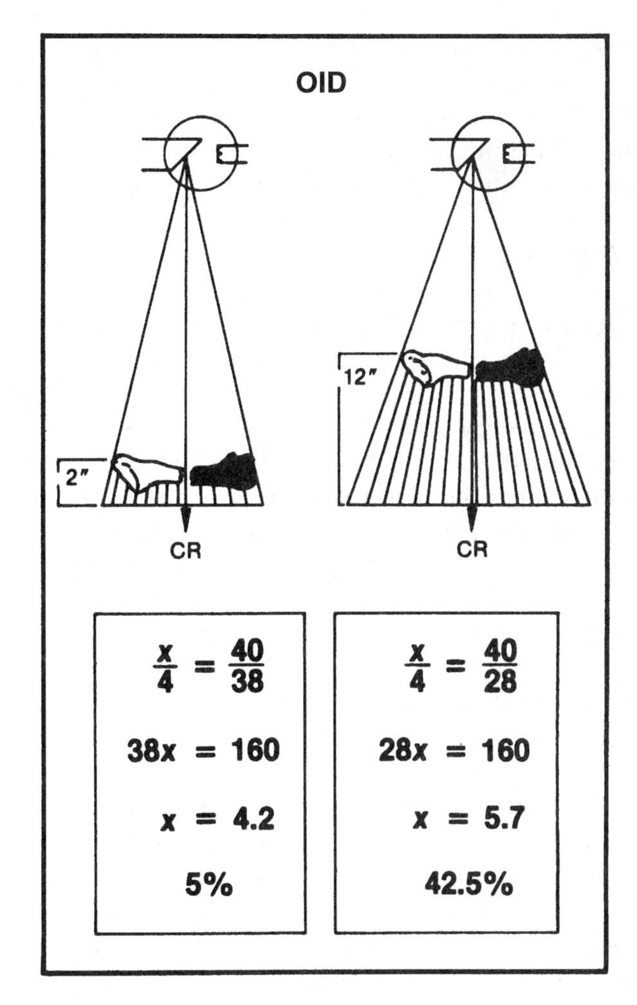

Figure 3-2
Magnification Resulting from Different OIDs (CR, Central Ray)

Reduction of the OID is frequently accomplished by adjusting the arrangement of the radiographic equipment or by changing the patient's position (e.g., a posteroanterior projection of the clavicle vs. an anteroposterior projection of the clavicle). Of the two factors influencing size distortion, the OID is the most critical. You may recall that, of the geometric factors influencing the recorded detail of the image, the OID was also identified as the most critical.

Source-Image Distance

The distance from the tube focus to the film (SID) also influences the size distortion of the recorded image. As suggested in the analysis of recorded detail, the SID should be standardized to avoid potential exposure factor errors associated with the maintenance of radiographic density utilizing the inverse square law when the SID is changed. Most radiographic procedures are performed at 40" (102 cm) SID. The one major exception is chest radiography, which is normally performed using a 72" (183 cm) SID. Therefore, since the SID is nearly always constant, it is less of an influence in the consideration of size distortion.

There are, however, a limited number of examinations that advocate the use of an increased or decreased SID, and, therefore, its influence on the magnification of the image must be examined (see Figure 3-3). In comparing the diagrams in Figure 3-3, we can see that the OID of both diagrams is 2" (5 cm), and the length of bone measures 4" (10 cm). In the left diagram, the SID is 80" (203 cm), producing an SOD of 78" (198 cm). In the right diagram, the SID is 20" (51 cm), producing an SOD of 18" (46 cm). Using the magnification formula, we have determined that the image of the bone utilizing the increased SID would measure 4.1" (10 cm), for a total magnification of 2.5%. The image of the bone utilizing the decreased SID would measure 4.4" (11 cm), for a total magnification of 10%. (**Perform Experiment 14**.)

The example provided in Figure 3-3 is an exaggerated representation of the application of SID. There are no actual applications in radiography where you would routinely select an 80" (203 cm) SID. Similarly, there are no situations where you would reduce your original SID by a factor of 4 to a 20" (51 cm) SID. However, this example can help you to recognize that the SID is not a major influence related to size

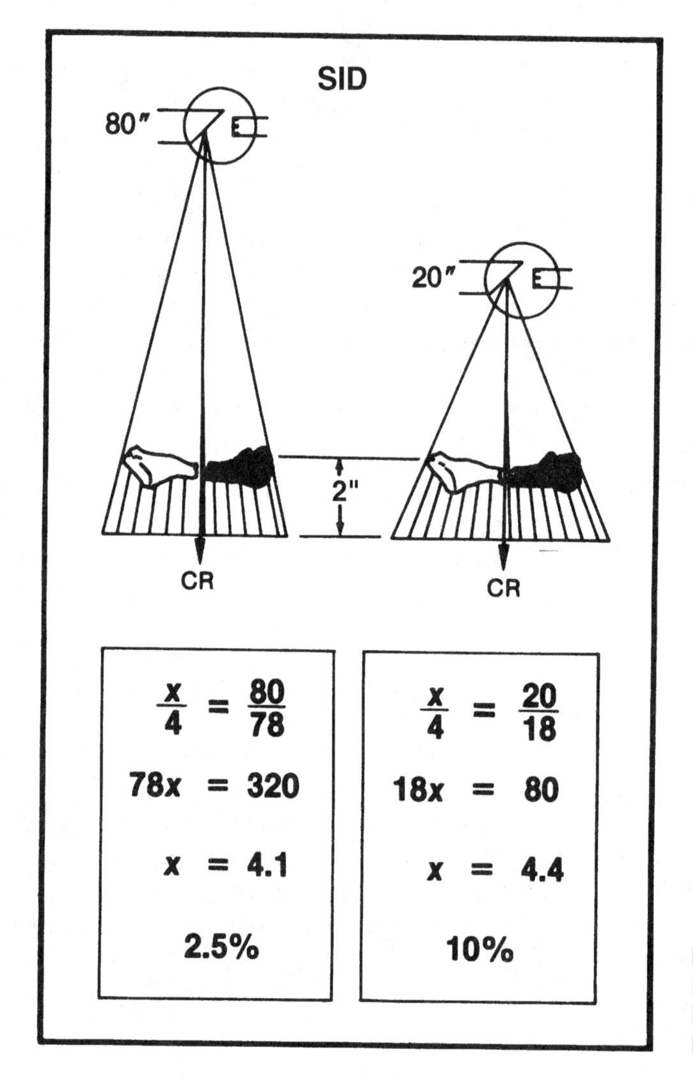

Figure 3-3
Magnification Resulting from Different SIDs (CR, Central Ray)

distortion compared with the influence of OID. Review the diagrams in Figures 3-2 and 3-3, and consider the amount of magnification produced by the influence of each factor in the production of size distortion.

Our examination of the SID factor does demonstrate that the SID influences the magnification of the recorded image and that the ideal situation would result from using a maximum SID in all procedures. A quick review of the influence of the SID related to the maintenance of radiographic density will provide ample reason why the maximum SID is not recommended in routine radiography. Review the influence of the SID on radiographic density in Chapter 4. If in the example provided by our diagram in Figure 3-3 a radiographic exposure of 50 mAs was required to produce a desired radiographic density at 20" (51 cm) SID, the required mAs at 80" (203 cm) SID would have to be 800 mAs in order to maintain the same radiographic density of the recorded image. Considering potential tube load limitations, the need to select the larger tube focus, and the possibility of having to increase the time of exposure in our attempt to maintain the radiographic density of the image, the problems associated with motion and geometric unsharpness come quickly to mind.

Minor changes in the SID may be considered in specific examinations, such as a lateral lumbar spine examination where the OID is considerably increased. In this instance, an increase in the SID from 40" (102 cm) to 48" to 50" (122–127 cm) could help to reduce the size distortion and the geometric unsharpness produced by the increased OID of the lumbar vertebrae. The increase in exposure required to maintain radiographic density may not represent an unreasonable choice. In another example, the open-mouth anteroposterior projection of the upper cervical spine (atlas and axis) can employ a reduction in the SID to produce a controlled magnified image. Reducing the SID from 40" (102 cm) to 25" to 30" (63–76 cm) will create a controlled magnification of structures further from the film. In this example, the recorded image of the open mouth will be considerably more magnified than the structures of interest, the cervical spine, which lie at a much closer distance to the film. Since the purpose of this examination is to project the atlas and axis within the space created by the opening of the patient's mouth, the reduced SID will frequently permit the technologist to perform this procedure more successfully. The open-mouth anteroposterior projection of the upper cervical spine is an example of producing a controlled level of size distortion and using it to improve the recording of the structures of interest.

Total Size Distortion

As with our investigation of geometric unsharpness related to the radiographic quality property recorded detail, the size distortion of the recorded image results from a combination of several factors rather than the association of a single factor. Size distortion results from the combined influences of both the OID and the SID. The ideal situation exists when the greatest practical SID and the least OID are employed.

We have already introduced the problems associated with the recording of a thick body part due to the multiplicity of structures of different thickness lying at many different levels of depth within the body part. A further investigation into the complex nature of size distortion will help us to understand this factor better (see Figure 3-4). In Figure 3-4, objects of different sizes are placed at different OIDs from the film. In this example, the objects include a dime, a nickel, and a quarter. The

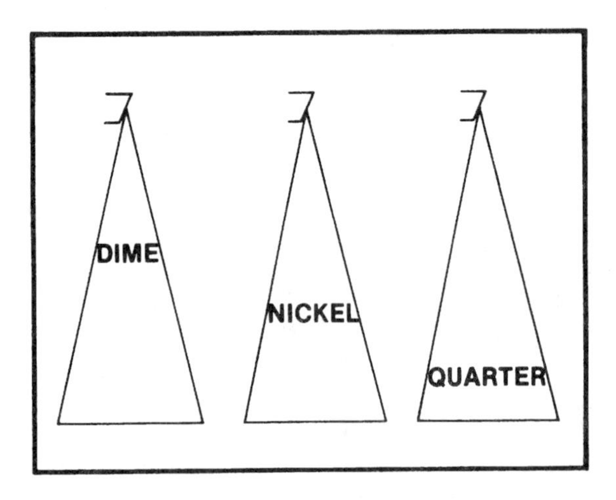

Figure 3-4
Examples of Different OIDs Producing the Same Image Size for Different-sized Objects

radiographic evaluation of these three coins would be difficult because their radiographic appearance would demonstrate that they are all of the same size. Only the appearance of the profiles of Lincoln, Roosevelt, and Washington in the different coins would enable us to determine the true identity of the images. Consider a more complex situation within the body related to the radiographic image of the symphysis pubis, the ischial spines, and the ischial tuberosities of a pelvis examination. All three structures lie at different OIDs, and, therefore, the recorded image of each would have a different percentage of magnification. In the pelvimetry examination, the need to determine the actual sizes of structures and the distances between structures is vital to the diagnostic interpretation of the recorded image. In these examples, the structures or objects being examined lie adjacent to one another. The complex nature of size distortion becomes even more apparent when we consider that, within the body part, numerous structures overlap or are superimposed upon one another and therefore maintain that relationship when recorded on the film (see Figure 3-5). In Figure 3-5, the same three coins are maintained at the same different OID, but in this instance, they are also placed directly above one another. Not only are the radiographic images of the three coin sizes similar, but they are superimposed upon one another when recorded on the film. The difficulty of interpreting the radiographic image and evaluating each object separately becomes obvious. Within the body, the diversity of structures, the multiplicity of their locations, and the many different OIDs represented by them all add to the complex nature of the problem of evaluating size distortion.

One method utilized to assist in the evaluation of the multiplicity of structures contained within a body part and the problems associated with the overlapping and superimposition of these structures is to perform the radiographic examination using multiple projections. In nearly every examination, a minimum of two projections taken at 90° from one another (e.g., an anteroposterior and lateral projection) is required to investigate and evaluate fully the structures recorded in the radiographic images. The second projection taken at 90° from the first enables the physician to evaluate the appearance of both images and relate it to the actual three-dimensional nature of the body part being examined. Due to the complex nature of the structures, it is not unusual for additional projections utilizing various degrees of obliquity to be included in a full examination of a body part. The additional

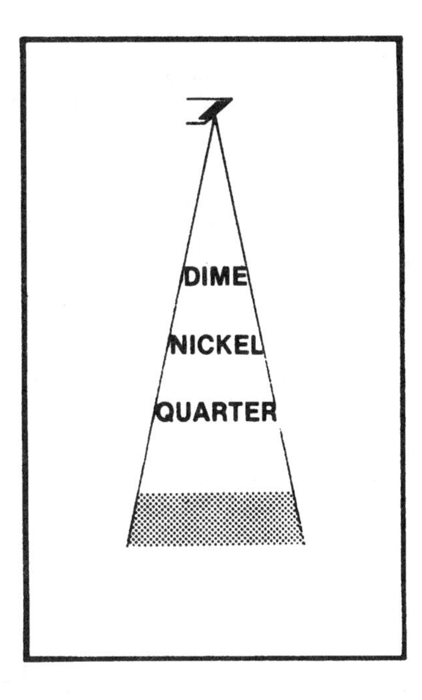

Figure 3-5
Example of Additional Complexity in an Image Resulting from Superimposition of Different-sized Objects

projections will permit a more complete evaluation of the structures of interest free from superimposed structures and the overlapping of close, adjacent structures.

Size Distortion Used to Advantage

Size distortion can be used to advantage in a number of radiographic examinations. We have already indicated several instances where the planned application of the OID or SID can be used to produce a desired effect. In other instances, the use of direct magnification techniques are employed to demonstrate structures that are not clearly visualized using routine technical factors. This technique is referred to as **macroradiography.**

In angiography, circulatory vessels important to the evaluation of the patient's condition are often too small to be resolved adequately by the imaging system utilized. By applying macroradiographic techniques, we can enlarge these vessels sufficiently so they can be recorded and visualized within the radiographic image. Such an image provides enhanced interpretive information and can assist in evaluating the patient's condition.

In skeletal radiography for the investigation of early osteolytic or osteoblastic changes associated with pathological changes in the bones or for the investigation of early metabolic changes associated with genetic disorders, macroradiographic techniques can enlarge these early osseous changes to the level where they can be recorded and visualized. In these instances, the diagnosis of skeletal pathology, metabolic disorders, or genetic conditions can frequently be made in the earliest stages of the disease before considerable change or destruction has occurred. Earlier diagnosis can lead to earlier treatment and intervention that may make a considerable difference in the patient's overall condition and prognosis.

There are occasions when macroradiography is employed to verify the presence of a suspected fracture. The scaphoid bone of the wrist is one example of the use of macroradiography to investigate the presence of a fracture. Because of its shape (like an upside-down canoe), and the overlapping and superimposed relationships of the other bones contained in the wrist, it is a difficult bone to examine. Fractures of the scaphoid bone are common, but the appearance of the fracture site is difficult to record both due to the small size of the bone and the fact that the fractured bone segments are often nondisplaced. However, if not discovered and left unattended, it is not uncommon for one of the segments of the fractured bone to die as a result of a lack of an adequate blood supply. Macroradiographic techniques can enlarge the edges of the fractured bone segments of the scaphoid bone to the point where they can be visualized, and the proper diagnosis can be made. The management of the fracture can now be undertaken.

Macroradiography

Macroradiographic techniques are accomplished by creating a controlled OID and maintaining it during the examination. Special equipment has been designed to accomplish this purpose. The usual technique is to place the structures of interest to be examined equally distant between the focal spot of the x-ray tube and the radiographic film. When employing a 40" (102 cm) SID, for example, the structure to be examined would be placed at an OID of 20" (51 cm). This arrangement would result in a 2× magnification factor creating a 100% increase in the size of the recorded image. This is the most common magnification factor employed with macroradiographic techniques.

Because of the significant increase in the OID, the influence on the production of geometric unsharpness in the recorded detail of the image is a major concern. Macroradiographic procedures cannot be performed successfully using general radiographic equipment. If general radiographic equipment with a small focal spot size of 0.6 mm or 1.0 mm is used, the geometric unsharpness attributed to the recorded image will be intolerable and the image detail unacceptable. Macroradiography requires the use of "fractional" focal spot sizes. Special x-ray tubes that produce an effective focus of 0.3 mm or less and that are capable of improved heat dissipation rates have been designed for macroradiography. By using a fractional focal spot size of 0.3 mm or less, the increased OID necessary to produce the required magnified image will be compensated for, and an image of acceptable recorded detail can be produced.

Fractional focal spot tubes utilize a smaller filament size than conventional x-ray tubes. As a result, they create a smaller, focused stream of electrons traveling from the cathode to anode side of the x-ray tube. As the stream of electrons is more concentrated and focused on a smaller point upon the tube target, the actual focus and effective focus of the tube produced are smaller. However, greater amounts of heat are produced in the smaller focus area on the target, and, therefore, one must carefully adhere to the limitations of heat tolerance of the tube. Fractional focus x-ray tubes are designed with larger, thicker tube targets to conduct heat away from the focus better and have added circulation fans within the tube housing to increase the rate of heat dissipation. When using fractional focal spot tubes, the tube rating chart must be reviewed carefully prior to performing macroradiography to avoid excessive tube loading and possible damage to the tube.

Shape Distortion

Often referred to as **true distortion,** shape distortion is another form of image misrepresentation. Unlike size distortion, which only enlarges the appearance of the structures of interest but magnifies them in the overall same relationship as they appear within the body part, shape distortion produces a true perversion of the structures of interest that can make the recorded image totally unrecognizable from the original structures of interest.

To recognize and understand the problems associated with shape distortion, we must review the complex nature of the human body. Each body part as well as the individual structures of interest differ in size, shape, and thickness both from each other and within the same structure. The individual structures of interest contained within a body part also lie at different depths throughout the body part. We examined the complex nature of the human body in our investigation of size distortion and discovered that it was a major factor in the subtle nature of the size distortion produced in the radiographic image. The factors and influences that produce shape distortion are also affected by the relationships of the structures of interest contained within the body part. As a result of these influences, shape distortion can produce a much more pronounced effect on the appearance of the structures of interest. The structures of interest can appear misshaped, elongated, or foreshortened in the radiographic image. Shape distortion can also cause the displacement of structures from their normal position and relationship with other structures contained in the body part to a new position creating overlapping and superimposition of structures where no such overlapping or superimposition exists. In its severest representation, the radiographic image loses all sense of anatomical recognition. Shape distortion is a far more complex principle to examine and evaluate than size distortion. A recognition and understanding of the nature of shape distortion will enable the technologist to minimize its influences and, like size distortion, even use the principles related to the production of shape distortion to advantage.

Since you are examining a complex organism with a multiplicity of structures of differing sizes and shapes as well as considerable overlapping and superimposition of these structures, you must recognize that it is impossible to eliminate shape distortion from the radiographic image. There is considerable shape distortion present in every radiographic image, and you will find that as you adjust the positioning of your patient or the alignment of the equipment to reduce the shape distortion of one structure, you will increase the distortional effect of other structures within the same body part.

Herein lies the key to an understanding of shape distortion and ways to begin to manage and minimize its effect. In every radiographic procedure the technologist performs, there is a purpose to the examination. There are specific structures or information that must be recorded as accurately as possible. The technologist must analyze the procedure and the body part to be examined and identify which specific structures or plane within the body part is of greatest importance. Identifying the structures of interest or the plane within the body part where the structures of interest are located, the technologist can arrange the equipment, the positioning of the patient, and the materials used for the examination to minimize the shape distortion of the structures of interest even though adjacent structures may be recorded with considerable shape distortion.

For example, when performing an examination of the abdomen during an intravenous pyelography (IVP) procedure, we would accept the shape distortion associated with the appearance of the ribs and the pelvis in order to demonstrate the structures of interest, the kidneys, ureters, and bladder, with minimal shape distortion. In fact, the routine projections established for the IVP procedure are designed to demonstrate the urinary excretory system to advantage. Many of the projections used for the IVP procedure will create significant shape distortion in adjacent structures. However, this is acceptable since the purpose of the examination is to demonstrate the urinary excretory system.

When examining a knee for a possible fracture, it is important to demonstrate the actual anatomical relationships of the bones. In arthrography of the knee, however, it would be more important to demonstrate the relationships of the structures contained within the knee joint space. This specific plane of interest should be demonstrated with the least amount of shape distortion and may require many different projections and angles of the central ray not normally performed on a routine knee examination in order to demonstrate the structures of interest to advantage.

Shape distortion is controlled by the proper alignment of the structures of interest with the plane/surface of the film and the proper geometric relationship of the beam of radiation (central ray) passing through the part to be recorded on the film.

Structure–Film Relationship

The ideal relationship exists when the structures or plane of interest within the body part is placed parallel to the plane of the film (see Figure 3-6). In Figure 3-6, the right diagram demonstrates the correct relationship of a broken bone placed parallel with the surface of the film. When the central ray is centered to this relationship using a perpendicular beam, shape distortion of the structure is minimized. The true alignment of the bone and the fracture fragments will be recorded. When this relationship is not achieved, overlapping and displacement of the various structures within the part being examined will be recorded in the radiographic image. In the left diagram, the broken bone lies on an angle of approximately 45° with the plane of the film. As a result, although the central ray is still centered to the part, and the beam of radiation is perpendicular, considerable shape distortion of the bone will occur. The actual site of the fracture and the true location of the fracture fragments will be impossible to ascertain. Parts of the bone will be displaced and superimposed over other portions of the bone. The required relationship is no longer true, and an accurate evaluation of the image will be difficult to achieve.

This example described the difficulties of arranging the proper relationship between the structure of interest and the film. Consider the problems associated with a more complicated structure, such as the thorax or abdomen. Unfortunately, the complexity of the different sizes and shapes of structures within a given body part does not always enable you to achieve this ideal relationship. (**Perform Experiment 15.**) Shape distortion will be at a minimum when the structures of interest or body plane has been arranged in a parallel relationship with the surface of the film.

Central Ray–Part–Film Centering

Whenever possible, the part to be examined or the structures of interest contained within the body part are centered to the midpoint of the film, and the central ray of

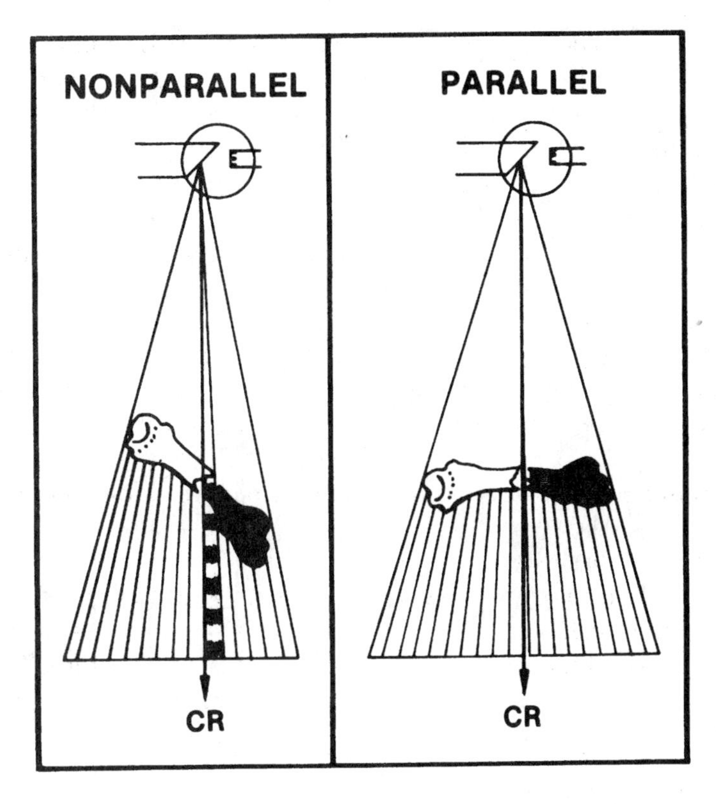

Figure 3-6
Examples of Incorrect (Nonparallel) and Correct (Parallel) Relationship of Structure/Plane of Interest to the Plane of the Film (CR, Central Ray)

the beam is aligned to the middle of the part–film relationship. The radiation beam consists of multiple photons of energy traveling in all directions that diverge in an ever-widening beam of coverage as it travels further from the focus of the tube. As a result, the central portion of the beam directed toward the film is more perpendicular to the part–film relationship than the ever-diverging periphery of the beam. In a 14" by 17" (36 × 43 cm) cassette, the x-ray beam reaching the film at the peripheral ends of the film is passing through the body part at an angle from the perpendicular. Recognizing this phenomenon, the ideal centering of the x-ray beam would be to align the midpoint (central ray) of the beam to the middle of the part–film relationship. In most procedures, this is also the center of the film. When the central ray of the beam is off-centered to the midpoint of the structures of interest, shape distortion will result. The distortion produced by an improper central ray–part–film centering will frequently be so subtle as to be overlooked. It can mean the difference between demonstrating the joint spaces between the intercarpal or carpometacarpal articulations or recording them as overlapped, foreshortened, or elongated structures. It can misrepresent the position of fracture fragments within a structure or their relationship with closely aligned adjacent structures. When the central ray–part–film centering is aligned correctly, shape distortion of the structures of interest is kept to a minimum (see Figure 3-7). In Figure 3-7, the left diagram represents the ideal arrangement with the structure of interest parallel with the film and centered to the middle of the film and the central ray directed with a perpendicular beam and centered to the part–film relationship. The right diagram demonstrates the subtle shape distortion that would occur when the central ray is off-centered from the part–

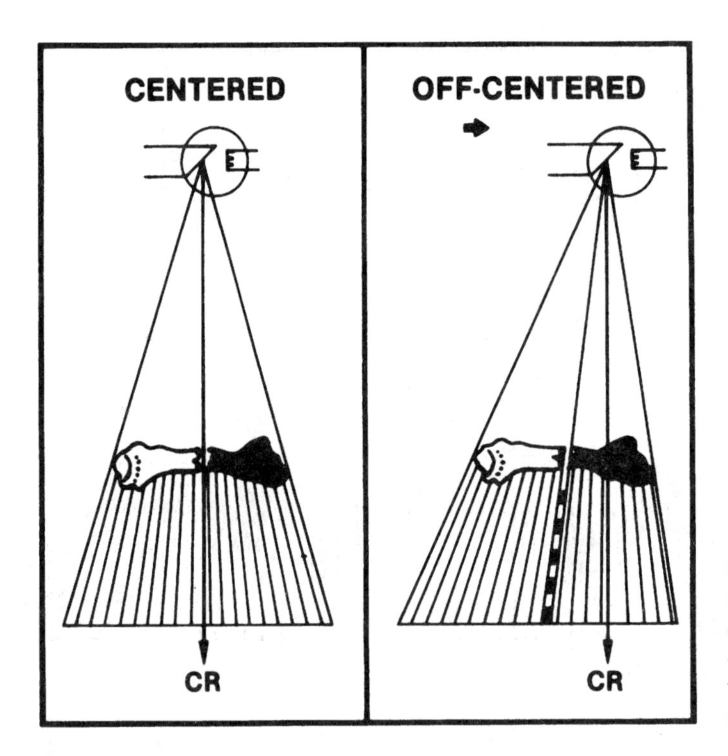

Figure 3-7
Examples of Good (Centered) and Poor (Off-centered) Alignment of the Central Ray (CR), the Body Part, and the Film

film relationship only a few inches. The fracture fragments would change from their true position within the body part and may appear as well aligned or poorly aligned related to the actual fracture of the bone. (**Perform Experiment 16.**)

Central Ray Direction

The central ray of the x-ray beam must always be directed at right angles to the structures of interest or plane of the body part being examined. If the part–film relationship is parallel, then the central ray will be directed perpendicular to the film. If the central ray is directed in anything other than a perpendicular relationship, severe shape distortion will result. The influence becomes more acute as the angle and direction of the central ray increases (see Figure 3-8). In Figure 3-8, the left diagram demonstrates the proper direction of the central ray, perpendicular and centered to the parallel part–film relationship. The right diagram demonstrates the significant shape unsharpness produced when the central ray is directed at a 25° angle to the parallel part–film relationship.

There are many instances in radiography where the technologist cannot achieve a proper relationship between the part and the film. The position of the patient or the structures of interest within the body part may not allow you to achieve the proper part–film relationship. However, if a proper part–film relationship cannot be achieved (review Figure 3-6), then an adjustment in the direction of the central ray will be necessary to minimize the shape distortion produced in the recorded image. The direction and angulation of the central ray depends on the position of the structures of interest or the plane within the part to be examined. In most instances, the central ray must be adjusted to produce a perpendicular relationship to the

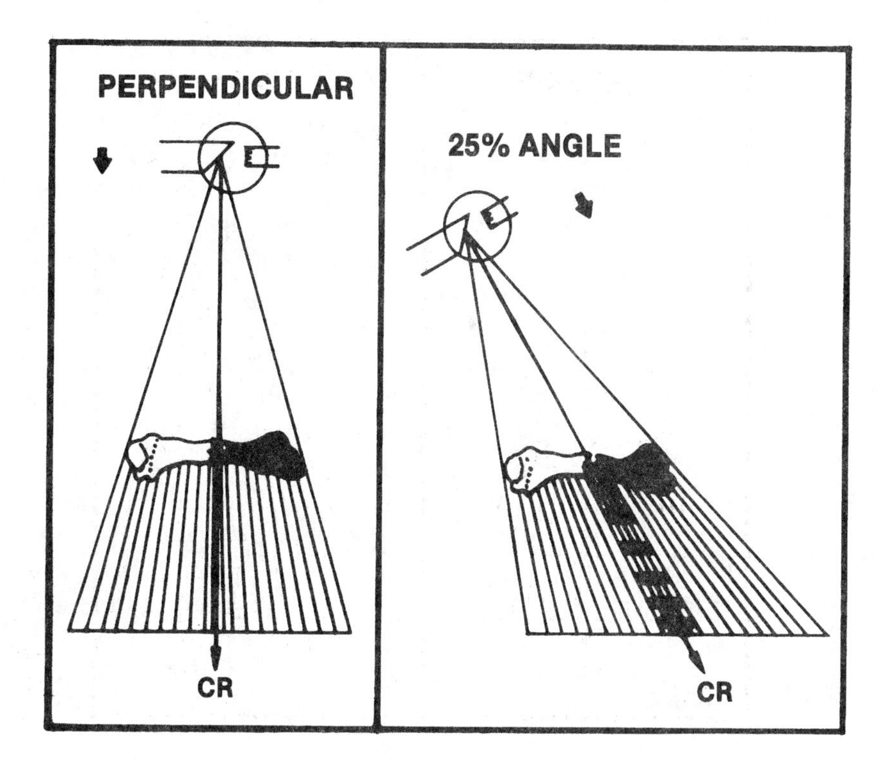

Figure 3-8
Examples of Proper (Left) and Improper (Right) Direction of the Central Ray Alignment to the Structure/Plane of Interest

structures of interest. Considerable overlapping of the multiple structures within the body part and severe displacement of structures from their true anatomical position and alignment with adjacent body structures will occur if the central ray direction cannot be adjusted properly to achieve a perpendicular direction to the plane of the structures of interest. At times, shape distortion can be so severe that the recorded image no longer has the anatomical appearance of the structure being examined. When the central ray is directed at right angles to the structures of interest, the least amount of shape distortion will occur. (**Perform Experiment 17.**)

Shape Distortion Used to Advantage

On occasion, the principles involved in the recording of the image on the film can be employed to produce a controlled form of shape distortion. This arrangement of the central ray–part–film relationship in order to produce a controlled form of shape distortion and to use shape distortion to advantage is utilized for two major purposes: (1) to eliminate or prevent the superimposition of structures and (2) to demonstrate specific anatomical structures or relationships.

To Avoid Superimposition

In the anteroposterior projection of the coccyx (Figure 3-9A), the central ray is angled in a 10° caudal direction to avoid the superimposition of the symphysis pubis, which lies anterior to and at the same level as the coccyx. By angling the central ray in this manner, the symphysis pubis will be projected far more inferior within the recorded image due to its increased OID compared to the recording of the image of the coccyx, which lies much closer to the film. The use of a 10° central ray angle has effectively prevented the superimposition of the symphysis pubis over the coccyx and allows for the clear visualization of the coccyx for evaluation.

In another example (Figure 3-9B), the central ray is directed in a 30° to 45° caudal direction in the posteroanterior projection of the clavicle to avoid the superimposition of the thoracic structures of the ribs and scapula, which lie posterior to the clavicle. The radiographic image of the scapula will have some shape distortion as a result of this procedure. However, since it is impossible to produce two projections at right angles to one another for a clavicle examination, the routine projections are normally the posteroanterior and the 30°- to 45°-angled posteroanterior projections. With possible clavicular fractures, it is important to project the clavicle free from superimposed structures in order to visualize the clavicle as a separate structure, and this procedure will enable you to do that. (**Perform Experiment 18.**)

To Demonstrate Anatomy

In many instances, it is impossible to place the structure of interest parallel with the surface of the film, and, as a result, if the central ray was directed perpendicular with the film, considerable shape distortion would result. An evaluation and identification of the structures of interest or plane of interest within the body part will enable the technologist to arrange the procedure to take advantage of and to use the direction of the central ray to advantage to demonstrate specific structures or anatomical relationships.

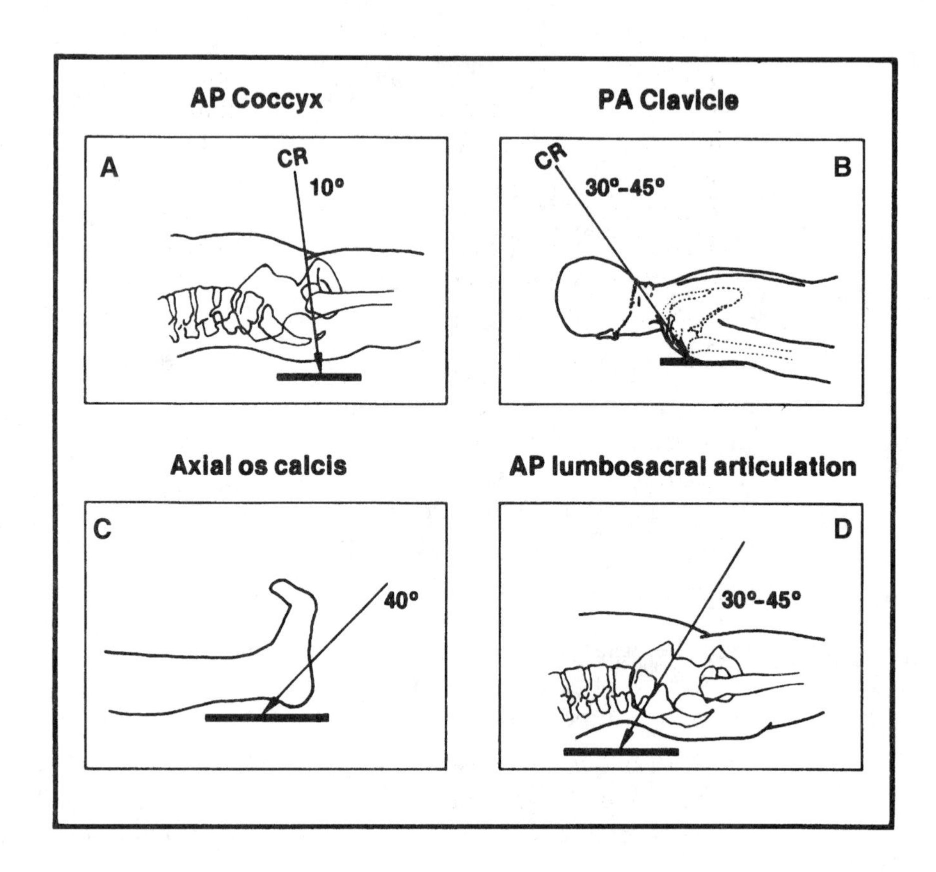

Figure 3-9 Examples of Arrangement of Central Ray–Part–Film Relationship To Produce Distortion That Avoids Superimposition and Demonstrates Anatomy (AP, Anteroposterior; PA, Posteroanterior)

In the axial projection of the os calcis (Figure 3-9C), the central ray is directed in a 40° cephalic direction in order to record the os calcis on the film with the least amount of anatomical distortion. The os calcis cannot be placed in a parallel relationship with the film, so by angling the central ray in this manner, the recorded image of the os calcis will be projected upon the film in an elongated shape that can be evaluated for possible fractures.

In another example (Figure 3-9D), the important plane or structure of interest is the lumbosacral articulation. To demonstrate the joint space, which in the body lies at a severe angle with the horizontal plane, the central ray must be adjusted to a 30° to 45° cephalic angle in order to approximate the angle that the joint space creates with the plane of the film. In this examination, it is also important that the central ray be centered to traverse through the body so that its axis lies even with the upper surface of the first sacral body and passes through the actual joint space created between the bottom of the fifth lumbar vertebra and the upper portion of the first sacral segment. The purpose is to enable the physician to demonstrate and evaluate the joint space. Only by passing the central ray through the body in a direction approximating that of the joint space and centered to the joint space itself will the technologist be able to accomplish this goal. In this anteroposterior projection, the other structures of the spine and pelvic girdle adjacent to and surrounding the lumbosacral articulation will be recorded with considerable shape distortion, but the structure of interest will be recorded to advantage. (**Perform Experiment 19.**)

Distortion Summary

We have now investigated the two types of image distortion: (1) size distortion (magnification) and (2) shape (true) distortion. The magnification of the image can have a detrimental effect on radiographic quality because of its interrelationship with image unsharpness and because the subtleties of its influence may not be obvious or apparent within the recorded image. Assessment of the specific size distortion (magnification) is very important in some examinations. Shape distortion can have devastating effects on the recorded image. Unlike the magnified image of the structure produced by size distortion, shape distortion causes the structure to be recorded with an altered appearance. In its simplest form, a round structure within the body may appear as an oval structure within the recorded image. In its severest form, the entire anatomical appearance will be changed and unrecognizable. We have learned that both size and shape distortion can be minimized by the proper application of the appropriate technical factors. We have also learned that in certain circumstances we can control and use size and shape distortion to advantage. The following portion of this summary reviews the technical factors that influence distortion of the radiographic image and how those factors are applied in order to minimize their effects.

Minimal Distortion Radiography

Size Distortion

- **Object-film distance.** Always place the structure to be examined as close to the film as possible. When applicable, tabletop radiography is better than Bucky radiography to accomplish this goal and to reduce the influence of OID.

- **Focal spot–film distance.** Use a standardized SID for all radiographic procedures. This should be the greatest distance consistent with the x-ray tube output and the requirements of the procedure with regard to motion control.

Shape Distortion

- **Structure–film relationship.** Consider the examination being performed. Determine the structures of interest or plane of interest within the body part in advance. Whenever possible, arrange the structures of interest or plane of interest within the body part in a parallel relationship with the film surface. When this is not possible, an angle of the central ray may be necessary to align and record the structures of interest properly and with the least amount of shape distortion.
- **Central ray–part–film centering.** The central ray should be arranged so that it is projected through the midpoint of the structure to be examined and the film centered to this arrangement.
- **Central ray direction.** The central ray must be directed at right angles (perpendicular) to the long axis of the structure or plane of interest. The arrangement is most ideal if the part to be examined is also arranged parallel to the surface of the film.

As you can see, the arrangement of the part to be examined with the central ray and the recording media is not a simple matter, but a carefully thought-out procedure requiring attention to all of the details that will affect the final recorded image. These factors must be correctly adjusted in order to minimize the loss of radiographic quality attributed to image distortion.

Distortion Review Questions

Name ______________________________________ Date________________

1. The misrepresentation of the ________ or ________ of the structures of interest observed within a radiographic image is known as distortion.

2. Of the two types of distortion, ________ distortion is referred to as **true distortion** and is considered the more critical influence affecting the radiographic image. Provide reasons for your choice.

3. In general, the distortion of the structures of interest observed within a radiographic image is considered to have a (*circle one*) degrading/beneficial effect on the radiographic quality of the image. Explain your answer.

4. Image magnification is another way to describe ________ distortion.

5. Describe the interrelationship between image magnification and the recorded detail of the image.

6. Identify the two influences that contribute to image magnification.

   1. ____

   2. ____

7. Of the influences listed in Question 6, which would have the greater influence on image magnification? Explain your answer.

8. At what level does image magnification become unacceptable?

9. Can image magnification be totally eliminated? Explain your answer.

10. Referring to the influences identified in Question 6, describe how you would minimize image magnification of the structures of interest.

11. Indicate the formula for determining the amount of magnification in a recorded image.

12. Indicate the formula for determining the percentage of magnification in a recorded image.

13. Provide an example of a radiographic examination where the true size of the structures of interest would be of critical importance. Describe the purpose of this examination.

14. Compare the influences of image magnification produced in an anteroposterior projection of a forearm with those produced in an anteroposterior projection of an abdomen. Which examination would demonstrate greater image magnification? Why?

15. Discuss the complex influences associated with image magnification related to the interpretation of the recorded image considering the overlapping and superimposition of structures in the posteroanterior projection of the chest.

16. Solve the following image magnification problems:

    A. A structure measures 12" × 4" (30 × 10 cm). It is located 8" (20 cm) OID from the film, and an SID of 40" (102 cm) is used. Draw a diagram to represent this arrangement, and solve for the following (area = length × width).

    Image length__________in/cm Object area__________square in/cm

    Image width__________in/cm Image area__________square in/cm

    Percentage of magnification__________%

    B. The structure measures 12" × 4" (30 × 10 cm). It is located 8" (20 cm) OID from the film; however, the SID is now 72" (183 cm). Draw a diagram to represent this arrangement, and solve for the following (area = length × width).

    Image length__________in/cm Image area__________square in/cm

    Image width__________in/cm

    Percentage of image magnification__________%

    C. The structure measures 12" × 4" (30 × 10 cm). The SID is 40" (102 cm), but you are able to reduce your OID to 3" (8 cm). Draw a diagram to represent this arrangement, and solve for the following (area = length × width).

    Image length__________in/cm Image area__________square in/cm

    Image width__________in/cm

    Percentage of magnification__________%

17. Using a ruler, the radiologist has measured the distance between the spinous processes of the ischial bones on a radiographic image of the pelvis and found it to be 9" (23 cm). You have performed this examination using a 40" (102 cm) SID. Knowing that the ischial spines lie at a level within the body that is 3" (8 cm) above the surface of the film, determine the actual (true) intraspinous distance.

18. Using general radiographic equipment, indicate an examination in which image magnification is used to advantage. Describe the purpose of the image magnification and how it is used to advantage for this examination.

19. How do multiple projections assist in the evaluation of multiple structures contained within a body part related to the overlapping and superimposition of individual structures?

20. What is the minimum number of projections required for a radiographic examination? Describe the relationship of those projections.

21. How would you minimize image magnification in a recorded image?

22. What is macroradiography?

23. How do macroradiographic techniques add to the diagnostic value of the recorded image in coronary angiography procedures?

24. The most common magnification factor utilized in macroradiographic procedures is __________.

25. Describe how the magnification factor in Question 24 is achieved by indicating the SID and OID utilized for a typical macroradiographic procedure.

26. Discuss the significance of a "fractional" focal spot x-ray tube to the success of macroradiography.

27. Would general radiographic equipment be employed for macroradiographic procedures? Explain the reason(s) for your answer.

28. "Fractional" focal spot x-ray tubes typically have a focus of __________ mm or less.

29. Identify the problems related to heat production and heat dissipation when utilizing macroradiographic equipment with a "fractional" focal spot x-ray tube.

30. What is meant by the **central ray?**

31. Describe how the structure–film relationship influences shape distortion, and identify the proper relationship to minimize this influence.

32. Describe how the central ray–part–film centering influences shape distortion, and identify the proper relationship to minimize this influence.

33. Describe how the direction of the central ray influences shape distortion, and identify the proper application of the central ray in order to minimize this influence.

34. It can be said that shape distortion can produce a perversion of the structures of interest within the recorded image. Explain this statement.

35. Describe the significance of identifying the structures of interest or the plane of interest within the body part related to the control of shape distortion produced within the recorded image.

36. In an anteroposterior projection of the abdomen, shape distortion would be more pronounced in the (*circle one*) central/peripheral portion of the recorded image. Explain the reason(s) for your choice.

37. Of the factors influencing shape distortion listed in Questions 31, 32, and 33, which factor contributes the least amount of shape distortion to the recorded image?

38. Identify the two major purposes to which shape distortion can be utilized to advantage.

    1. ______________________________

    2. ______________________________

39. In the anteroposterior projection of the coccyx, describe how shape distortion is used to advantage.

40. In the anteroposterior projection of the lumbosacral articulation, describe how shape distortion is used to advantage.

Distortion Analysis Worksheet

Let's review our analysis of distortion with the following exercise. A satisfactory diagnostic radiograph of the pelvis was produced using the following technical factors:

100 mA
1 sec (time of exposure)
80 kVp
40" (102 cm) SID
1-mm focus size
Minimum OID
Regular screen-type film

100 speed film-screen system
8:1 ratio, 100-line moving grid
Proper collimation
2.5-mm aluminum filtration
92°F development at 90 sec
Normal thickness and opacity of the part

Without compensation, the changes listed in the table below are made one by one. Indicate the effect, if any, each change has on the size or shape distortion of the radiographic image.

1. If the size or shape distortion of the recorded image is increased, mark a plus (+) in the space provided.
2. If the size or shape distortion of the recorded image is decreased, mark a minus (–) in the space provided.
3. If the size or shape distortion of the recorded image is unchanged, mark a zero (0) in the space provided.

Distortion

Proposed change	Effect	Proposed change	Effect
Use a 2.0-mm focus	_____	Change to a 5:1 ratio, 100-line moving grid	_____
Use 10" (25 cm) OID	_____	Change to a 12:1 ratio, 100-line moving grid	_____
Reduce the SID to 25" (63 cm)	_____	Remove all collimation	_____
Increase the SID to 48" (122 cm)	_____	Remove all filtration	_____
Use screen-type film in a cardboard holder	_____	Develop at 100°F for 90 sec	_____
Use a 200 speed film-screen recording system	_____	Body thickness reduced by an atrophic condition	_____
Use 200 mAs	_____	Pathological condition reduces opacity of body tissues	_____
Increase the kVp to 90	_____		
Omit the use of a grid	_____		

4 Radiographic Density

Image Visibility

The preceding two chapters analyzed the principles and factors related to the geometric properties of the recorded image, recorded detail, and distortion. Recorded detail and distortion contribute to the sharpness and accuracy of the recorded image. However, these geometric properties of the image identify only two of the four major factors contributing to radiographic quality of the image. An image of radiographic quality must possess a proper balance between the sharpness and accuracy of the recorded detail and the visibility of those details. Beginning with this chapter, we will now direct our attention to those factors that affect the photographic properties of the radiographic image (Figure 4-1). A review of Experiments 1 and 2 indicates that it is possible to produce a radiographic image that possesses excellent recorded detail and minimal distortion but that is still an unsatisfactory image. Unless the recorded detail is visible and demonstrated completely and clearly, the overall radiographic quality of the image will be considered unsatisfactory.

When considering the balance to be achieved between the geometric properties of the image and the photographic properties of the image, it would be appropriate to weight the scale in favor of the photographic properties. It is the photographic properties of the image, radiographic density and radiographic contrast, that determine the completeness and visibility of the recorded image. A radiographic image that possesses the proper level of radiographic density and an appropriate scale of radiographic contrast enables you to demonstrate optimal visualization of the recorded details regardless of the specific sharpness and resolution recorded or the amount of size or shape distortion produced within the recorded image. Remember, some degree of image unsharpness and distortion will be recorded in every radiographic image. Therefore, it is essential that optimal visualization of whatever recorded detail there is within the recorded image be demonstrated. Depending on the requirements of the procedure, the materials utilized to record the image, and the patient's condition, it may not be possible to record an image possessing a high degree of recorded detail. However, it is possible to demonstrate those details that you have been able to record with photographic properties that ensure maximum visualization of all of the structures of interest.

Figure 4-1
Examples of the Photographic Properties of the Radiographic Image

Radiographic Density

Of the photographic properties of the radiograph, radiographic density can be identified as the controlling factor. Proper radiographic density ultimately determines the completeness of the visibility of the recorded image. In fact, the analysis of all of the other film properties is dependent on the presence of proper radiographic density. Basically, a radiographic image must demonstrate (visualize) all of the structures of interest, or it is considered an unsatisfactory image. This does not mean that every structure recorded within the radiographic image will be visualized successfully. Again, the reference is to provide visualization of the structures of interest.

In an examination of the chest for the investigation of a possible rib fracture, the appropriate radiographic density and contrast scale will be quite different from the radiographic density and contrast scale required to investigate the presence of lung pathology. In the radiographic image for the examination of a possible rib fracture, the radiographic density of the lungs will appear overexposed (too dark), and the visualization of the lungs and other soft-tissue structures of the chest will be minimal. However, the purpose of the examination is to visualize the structures of interest, the ribs. Similarly, in a chest examination for lung pathology, the lungs would be recorded at the appropriate radiographic density to maximize their visualization and to determine the presence of disease, whereas the recording of the ribs would be inadequate to diagnose the presence of possible rib fractures. A proper evaluation of the purpose of the examination and the identification of the structures of interest is essential to the performance of every radiographic examination. The technologist can increase or decrease the radiographic density of the image and adjust the radiographic contrast to respond to the needs of the examination and the part that is being examined.

When changes in the exposure factors are being considered to meet the requirements of a particular procedure or patient condition, those changes must be analyzed for the effect they have on the radiographic density of the recorded image. The overall radiographic density of the recorded image is the one property of the radiograph that must remain constant. This principle is essential to the selection and adjustment of proper imaging materials and exposure factors in order to produce a film that is of radiographic quality regardless of the circumstances of the procedure you are performing. Changes in the state of health of the patient, differences in the thickness or opacity of the structure, and individual requirements of the procedure or of the physician may require adjustments in the equipment utilized, the recording materials selected, and the determination of the exposure factors to be applied. Considering the geometric properties of the recorded image, we know that changes in the set-up of the equipment, the recording materials utilized, and the exposure factors selected will frequently affect the sharpness and resolution of the recorded detail and the distortion of the structures of interest. Unfortunately, these influences cannot always be avoided, but, regardless of the adjustments required, the radiographic density, and therefore the completeness of the visibility of the recorded image, must remain constant.

Let's examine the radiographic density of the recorded image. The radiographic film is especially sensitive to the specific spectrum of light emitted from intensifying screens and to the direct effects of x-radiation. The quantity of radiation directed toward the radiographic film is referred to as the **radiation intensity.** As the intensity of the radiation produced increases, a greater quantity of radiation will be directed toward the film-screen recording system. The greater the intensity of the radiation produced, the greater the radiographic density of the recorded image. Basically, the recorded image gets darker as the intensity of the radiation (exposure) increases. In fact, you have probably heard radiographic density referred to as the overall blackening of the recorded image. This definition, while essentially correct, is not really appropriate for an analysis of the radiographic quality of the image. For our purposes, a more definitive approach to the subject of radiographic density and a more practical assessment of its influence on the radiographic quality of the image is required. After all, the technologist's purpose is not to produce an overall black radiographic image.

The technologist is more appropriately concerned with producing the proper radiographic density required to record the structures of interest. A radiographic image that possesses proper radiographic density provides appropriate visualization of all of the structures of interest. Any variance from this concept as applied to radiographic density is detrimental to radiographic quality. Insufficient density is just as degrading to radiographic quality as is excessive density—in fact, perhaps even more so. Although it is not recommended nor an example of the principles of radiographic quality, depending on the level of overexposure, a radiographic image that possesses excessive radiographic density may be able to be "bright-lighted" in an effort to provide a measure of diagnostic value to the image. Bright-lighting such an image may prevent the need for a repeat examination and additional radiation exposure to the patient. A radiographic image that possesses insufficient radiographic density cannot be investigated further by any means other than a repeat examination. The recorded information is simply not complete and able to be visualized.

To determine what the proper radiographic density of a recorded image is, and to evaluate whether you have achieved this goal, we must examine the multiple factors that influence this photographic property of the radiographic image. By investigating these factors individually, we can analyze their influence and formulate successful methods to produce, control, and adjust for the proper radiographic density of the radiographic image. As our investigation progresses, we will discover the interrelationships that exist between the multiple factors and influences contributing to the radiographic density of the recorded image. As with the factors contributing to the geometric properties of the recorded image, we will discover which factors and influences are the most significant and ways to utilize them to advantage in producing an image of radiographic quality.

Radiographic density determines the completeness of the image recorded upon the radiographic film, the overall blackening of the film caused by the intensity of radiation, and the visible light emission produced by the intensifying screens. However, it is important to recognize and understand that the photographic properties of the recorded image are made up of the combined interrelationship between the radiographic density and the radiographic contrast. The ability to discriminate between the different radiographic densities, so essential to the visibility of the structures of interest reflected by the contrast scale produced, will be examined in Chapter 5.

Patient Factors

The greatest influence on the production and control of proper radiographic density is the patient. The thickness of the part to be examined and the opacity of tissue comprising that part are the basic patient factors that must be considered in the production and control of proper radiographic density, regardless of the procedure to be performed or the equipment to be employed.

As with the other properties related to radiographic quality, it is essential to identify the structures of interest within the body part. In addition, it is important to evaluate the examination request to determine the reason or purpose for the examination. The examination request should also include information related to the patient's diagnosis. This information about the patient and the purpose of the examination will enable the technologist to determine the proper photographic properties required for the procedure. As we examine the patient factors, we will learn the significance of this information to the selection of radiographic materials and exposure factors and to the proper performance of the radiographic procedure.

Unfortunately, we have little influence or control over these patient factors. No two patients are identical. Not only do the patients' physiques differ in general characteristics, but individual body parts within the same patient vary considerably in size, shape, tissue thickness, and tissue opacity. The radiograph that possesses proper radiographic density must demonstrate all of the body tissue differences within the structures of interest with sufficient visibility so that consistent quality standards can be achieved and maintained. That is, the proper radiographic density having been established for a body part or for a specific examination, the technologist must be able to achieve that visibility level whether examining an infant, a child, or an adult; a young, muscular patient or an elderly, disabled patient; a small, thin patient or a large, obese patient. In addition, the technologist must be able to achieve the proper levels of radiographic density when applying those techniques to the performance of different examinations using a variety of radiographic equipment and materials.

Tissue Thickness

In any discussion of the thickness of the part to be examined, we are actually determining the quantity (amount) of the various tissues and structures contained within that body part. Herein lies the key to the relationship of tissue thickness and radiographic density: quantity. In general, the thicker the body part, the larger the organs contained within the part and the greater the quantity of tissues that must be traversed before the radiation can reach the film-screen recording system in order to produce a specific radiographic density. Therefore, the thicker the body part, the greater the quantity of radiation needed to produce a desired radiographic density. If an x-ray beam is directed at two different thicknesses of the same material, the number (quantity) of x-rays passing through the thinner part will be greater than the number passing through the thicker part. Therefore, without compensation, the radiographic density produced in the recorded image of a thinner body part would be greater than the radiographic density of a thicker body part because a greater quantity of radiation will have passed through the thinner part and reacted with the film-screen recording system. In the thicker body part, a greater quantity of radiation will have been absorbed by the part and a lesser amount of radiation would be available to react with the film-screen recording system, reducing the radiographic density of the recorded image. Obviously, a thicker body part containing greater quantities of tissue would require an appropriate increase in the intensity of the x-ray beam in order to produce the same radiographic density as was produced in the thinner part using a lesser amount of radiation.

In solving problems related to the quantity of tissues, or thickness of the body part to be examined, the technologist must consider the quantity (intensity) of radiation required to produce and maintain the proper radiographic density of the recorded image. There are a number of factors that affect the thickness of the body part and/or the quantity of tissue contained therein and therefore have an influence on the control and maintenance of radiographic density.

Physique

The patient's overall physique (body habitus) and general body characteristics will affect the thickness of the body part and/or the quantity of tissue contained therein. The patient's physique will also determine the size, shape, position, and location of the internal organs contained within the chest and abdominal cavities. Patients are frequently categorized as having a specific type of physique: asthenic, sthenic, hyposthenic, or hypersthenic. This categorization is really an oversimplification of a more complex association. The technologist must recognize that such categories are never absolute and that these same general patterns of physique frequently overlap one another. Some patients may be categorized as having a sthenic physique at 6 feet and 190 pounds, whereas others may be identified as possessing a sthenic physique at 5 feet 4 inches and 130 pounds. The following general characteristics are present within the four major categories of body habitus:

1. **Sthenic.** In general and for the purpose of our evaluation, the sthenic patient would be identified as someone of normal stature (height and weight relationship), fairly angular in appearance, healthy and strong, and with good muscle tone. The sthenic-type patient represents approximately 50% of the population.
2. **Hypersthenic.** The hypersthenic patient includes the overweight, obese patient with a tendency toward roundness, especially around the abdomen. Hypersthenic patients have the general appearance of possessing less muscle tone than sthenic patients and represent approximately 5% of the population. The chest of the hypersthenic patient would tend to be broad, deep, and barrel-

shaped. The lungs and heart would have the appearance of less height and would be much wider than in the sthenic patient, and it is not uncommon for the technologist to have to examine the chest of the hypersthenic-type patient with the 14" by 17" (36 cm × 43 cm) cassette placed widthwise in the cassette holder. In the abdomen, the stomach and gallbladder would lie higher and lie in a more transverse direction compared with the sthenic-type patient; the gallbladder would lie under the rib cage and further from the midline of the body.

3. **Hyposthenic.** The hyposthenic-type patient, representing approximately 35% of the population, would be identified as the patient who appears thin and lean, perhaps underweight, and possesses less than the normal musculature present in the sthenic patient. Hyposthenic patients may be tall and lanky or short and willowy. The technologist may find it difficult to demonstrate such a patient's entire chest cavity on the length of the 14" by 17" (36 cm × 43 cm) cassette. The stomach is in a more longitudinal direction (j-shaped), and frequently its lower border will lie within the upper portions of the pelvic cavity. The gallbladder lies lower within the abdomen, frequently below the level of the rib cage, and lies more toward the midline of the body.
4. **Asthenic.** The asthenic-type patient represents approximately 10% of the population. The asthenic patient would be observed as having more than just a normal thin appearance. Asthenic patients may appear undernourished, with the presence of protruding bony processes both in the extremities and within the torso of the body. Their muscles may appear stringy, and they may lack the muscle tone associated with the sthenic patient. The asthenic appearance of the patient may result from disease processes, pathological conditions, and/or physical disabilities. Like the hyposthenic-type patient, the chest cavity will appear long and narrow, and the image of the heart will appear long and slender. The stomach and gallbladder will lie considerably lower and more toward the midline of the body, frequently dipping into the pelvic cavity when compared with that of the sthenic patient. The transverse section of the colon may appear completely within the pelvic cavity, whereas in the sthenic patient, it would be located above the crest of the ilium.

As you can see, a definitive description of a patient's physique is difficult to specify, but about 50% of all patients fall within the range of the sthenic-type body habitus for the purpose of radiological evaluation. It is also important to note that the locations and associations of the internal organs will change related to the patient's position and respiration. Depending on whether the patient is erect or supine, prone or supine, lateral or Trendelenburg, the position of the various organs contained within the body cavities will change. The position of the diaphragm and therefore the location and appearance of the structures contained within the chest and abdominal cavities will differ from full inspiration to full expiration. Chest radiography is normally performed on full inspiration, abdominal radiography, on full expiration. As you can see, the different categories of possible patient physique (body habitus) can significantly influence both the thickness of the body parts as well as the distribution of the opacity of the tissues (bone, muscle, soft tissue, and fat) contained therein.

Age

The age of the patient will be a major influence contributing to different thicknesses of the body part or the structures of interest to be examined. For example, the thigh of an infant may measure only 1.5" (4 cm) in diameter; a 7-year-old's thigh may measure 4" (10 cm); and a young, healthy adult's thigh may measure 10" (25 cm) or more in diameter. Overall, the same type and ratio of differing tissue opacities (bone, muscle, soft tissue, and fat) are contained within the thigh of the infant, the child, and the adult. Therefore, the technical factor changes required to maintain radiographic density between the three examples are primarily caused by the variation in the quantity of tissue (thickness of the part) being examined. A somewhat similar influence would be seen in thicker body parts, such as the abdomen. However, in other areas of the body, differences in the age of the patient would not result in as wide a variance of part thickness. As an example, the measurement of a lateral skull on a 7-year-old child differs very little from that of a 37-year-old adult, and, therefore, the exposure factors employed with both examinations, all other factors remaining the same, would be similar.

Development

The developmental status of the body part or structures of interest is also a major factor influencing the thickness of the part. The thickness of the shoulder of a 27-year-old professional weightlifter will frequently be considerably different from that of a 27-year-old office worker. The thickness of the thigh of an active, 12-year-old junior high school student is frequently considerably greater than that of a 12-year-old paraplegic. The developmental status of the body part can frequently relate to the muscle tone and development of the body part or structures of interest. The ratio of bone to muscle, soft tissue, and fat in the weightlifter or active student will frequently be quite different from that of the office worker or the paraplegic patient. In this case, the different thicknesses of the part caused by the differences in the ratio of the different opacities of the tissues contained within the body part or structures of interest are readily observed and can be measured, and the required technical factor adjustments can be performed to maintain the radiographic density.

A more difficult assessment related to the development of the body part or structures of interest are those cases in which the overall thickness of the part is similar between two patients, but the ratio of bone to muscle, soft tissue, and fat differs. For example, two patients are being prepared for an examination of the femur. A measurement of the thickness of the part in each patient indicates the thicknesses are identical. However, one patient is a professional football player, and the other is a professional mattress tester. In this instance, although the actual thickness of the body parts are the same, the types and ratios of the different tissues contained within the body part are of more importance than the actual thickness of the part. The quantity (intensity) of radiation may require but minor changes, if any, but the quality (penetrating ability) of the beam must be altered in order to maintain the proper radiographic density and thus the appropriate visibility of the recorded detail.

Pathology

In raising the question of how pathology, pathological conditions, and disease processes can affect the thickness of the body part or the structures of interest, we are virtually opening up the proverbial Pandora's box. Changes in the thickness of the body part are common, and depending on the cause and the progress, development, or stage of the disease or condition, the technologist may have to change the quantity (intensity) of radiation, the quality (penetrating ability) of the x-ray beam, or both.

Some diseases or conditions produce a histolytic effect on the tissues and organs of the body, and the thickness and opacity of the body part or structures of interest will decrease. Less quantity and/or penetrating ability of the x-ray beam is required in these instances to produce the same radiographic density. Conditions such as emphysema of the lungs or osteoporosis of the bones would be included in this category. Other disease processes or conditions produce an increase in the thickness and/or opacity of the tissues and organs in the body. Greater amounts of radiation and/or an increased penetrating ability of the x-ray beam is required in these instances to produce the same radiographic density. Most tumors, benign and carcinogenic, and conditions such as Paget's disease would be included in this category.

This is why all requisitions for radiographic examinations must include pertinent clinical information. The technologist must be familiar with the more common conditions and pathology that will influence the status of the thickness and opacity of the body part. Additive and destructive pathology can produce a hypertrophic state or produce an atrophic condition in the structures of interest, which are examples of tissue thickness and/or tissue opacity changes associated with pathological conditions.

Tissue Opacity

The tissue opacity (density) of the body part and structures of interest is an altogether different consideration as a patient factor affecting radiographic density. We have learned that similar thickness of body parts may still require the different application of exposure factors because of the development of the part: the variances in the ratio of different tissue densities (bone to muscle, soft tissue, and fat) that make up the structures of interest. It is a somewhat more difficult assessment to evaluate the development of the body part than simply to measure the overall thickness of the part.

The greater the density (mass per unit volume) of a tissue, the greater its ability to absorb radiation. The tissues of the body possessing greater tissue opacity (density) will absorb greater quantities of radiation and transmit less radiation toward the film-screen recording system. The recorded image of such tissue would have less radiographic density. Herein lies the key to an understanding of the relationship of tissue opacity and radiographic density: **penetrating ability**. As the tissue density (mass per unit volume of the tissue) increases, a greater percentage of the overall beam of radiation is absorbed by the part, and the radiographic density of the recorded image associated with those tissues is reduced. The x-ray beam emerging from the x-ray tube consists of a beam of radiation composed of a wide range of wavelengths (penetrating ability). The radiation being absorbed by tissues of greater opacity is that of longer wavelength, which is less capable of penetration through the more dense tissue. Therefore, the more radiopaque (dense) the tissues of the body part are, the greater the penetrating ability of the beam necessary to produce and maintain a desired radiographic density.

In solving for problems related to the opacity of the different body tissues to be examined, think in terms of radiation quality (penetrating ability) of the x-ray beam.

There are a number of factors that affect the opacity of body tissue and therefore affect the selection of exposure factors required to produce the desired radiographic density.

Cellular Composition

Basic to the consideration of tissue opacity is the actual molecular composition of the cells making up the tissues. The chemical compounds within the molecular structure of a muscle cell differ considerably from those contained within a bone, nerve, or fat cell. The higher the atomic number of the elements and the more complex the molecular structure of the elements making up the cell, the greater the opacity of the tissues that contain those cells. The approximate atomic numbers of bone, muscle, and fat cells are 13.8, 7.4, and 5.9, respectively. Obviously, if there is a higher ratio of high atomic number elements in the composition of the structures of interest, the tissue opacity of those structures compared with adjacent structures will be greater.

Compactness of the Cells

The compactness of the cells and tissues making up the various body organs is also a consideration of their mass per unit volume. Muscle tissue is relatively dense not only because of its complex molecular structure, but also as a result of the compact arrangement of its cells within the muscle tissue. Evidence of muscles is clearly visualized within the radiographic image. The psoas muscles that connect the spine with the femur are readily seen within an examination of the abdomen. The presence of these muscles helps to make the radiographic image of the size, shape, and location of the kidneys, which are surrounded by fat and therefore composed of different substances, more discernible.

On the other hand, red blood cells, although highly complex in their molecular composition, lack compactness; they are diluted within the liquid medium, the plasma, in which they travel, as well as being spread throughout the entire cardiovascular system. However, within the heart and major vessels, there are greater quantities of blood and, thus, a greater total volume in these specific areas. Therefore, the radiographic image of the major vessels entering and leaving the heart and the heart itself are partially discernible due to the presence of large quantities of blood flowing through these structures.

Spaces between the Cells

Not all of the tissues or organs of the body are in close proximity to one another. There are many spaces and cavities between different tissues and organs. This is also true at the cellular level. The tissue opacity can differ as a result of these spaces between the cells and tissues as well as the types of materials that are located within these spaces. The molecular composition of muscle cells is greater in density than bone cells, and the compactness of the cells contained in the muscle comprise a greater mass per unit volume than those of bone tissue. Why then do bones absorb more radiation and produce a recorded image of less radiographic density than do muscles? Bone cells are not closely joined or compacted to one another, but between the bone cells are various chemical compounds of organic and inorganic matter, such as calcium phosphate. It is this dense material filling in the spaces between the bone cells that provides the added density to the bones and enables them to absorb the x-rays more readily than other tissues of the body. Of all the naturally occurring tissue opacities of the body, bones are the greatest.

Status of Hollow Organs

Many organs within the body are hollow and have lumens within them. These lumens may be as small as the lumen of a capillary blood vessel that allows for blood cells to pass through in a single file, or they may have the expansion capacity and space to allow for the ingestion of a quart of milk as in the case of the stomach. The hollow spaces within a large number of organs contained within the body can result in considerable changes in the overall radiopacity of the organ. The organs of the digestive system are a good example. At times, these organs are empty, and they literally collapse on themselves making their overall tissue opacity more compact and radiopaque. At other times, the hollowed spaces and lumens are expanded and filled with air. The air, being a very low density, tends to decrease the overall opacity to radiation. On still other occasions, the hollowed spaces and lumens are filled with various substances, ranging from steak and potatoes to ice cream and cake, and, depending on the density of the substances filling the hollow organs, the radiopacity of the structure may significantly change. In fact, it is the very presence of the hollowed spaces and lumens within the organs of the digestive system that permits their radiological examination. The multiple structures of the digestive system are similar in their molecular composition and the compactness of their cells, and, as such, it is difficult to differentiate between the individual structures of the stomach, small intestines, and colon. However, the use of a highly radiopaque substance such as barium sulfate ingested by the patient will outline and mimic the size and shape of the various organs that contain it. Thus, a radiological examination of the esophagus, stomach, and small bowel is easily performed. The barium sulfate increases the overall radiopacity of the structures of interest so they are easily identified and visualized as separate, individual structures and can be examined and evaluated. For the colon, the barium sulfate is administered by means of an enema and performs the same function to produce an increased radiopacity of the structures of interest.

Many areas of the body, including the paranasal sinuses, will differ in radiopacity depending on the person's state of health. Normally filled with air, the sinuses fill with fluid in certain inflammatory diseases and pathological conditions and become quite dense due to the presence of fluid in the normally air-filled cavities.

Patient Factors Summary

It should be obvious from our investigation of the patient factors that the factors of tissue thickness and tissue opacity and their influences are vital considerations when it comes to the selection of exposure factors to produce a desired radiographic density. The adjustment of technical factors to maintain the proper radiographic density when the tissue thickness and/or tissue opacity differ or are altered by disease or trauma is also an important consideration for the technologist. This is why a thorough knowledge and understanding of human anatomy and physiology is basic to the success of the radiologic technologist. The patient factors of tissue thickness and tissue opacity are primary influences in every radiographic procedure. Failure to recognize and understand these influences begins to make the performance of your procedure a "technical guessing game." The complexity of the patient factors that influence the radiographic density of the film can be identified as the real unknown technical factor in your x-ray procedure, since each patient you examine will present a different set of influences to be considered.

There are many instances when the structures of interest will not possess sufficient differences in tissue thickness or tissue opacity to record them as a clear, visible radiographic image. Many procedures and techniques have been developed that enable the technologist to alter the tissue thickness or tissue opacity of the structures of interest so that a definitive radiological examination can be performed. Examples of these procedures and/or techniques include the use of an inorganic substance, barium sulfate, in the examination of the organs of the digestive system, previously introduced in the discussion of the status of hollow organs in the tissue opacity section. Additional examples include the use of organic compounds referred to as *contrast media* or *contrast agents* injected into the body in order to visualize previously unidentifiable structures or functions. The intravenous pyelogram to demonstrate the structures and functions of the urinary excretory system and the cerebral angiogram to demonstrate the structure of the blood vessels and the circulatory functions of the brain are examples of the use of contrast agents for this purpose.

Milliamperage

Milliamperage (mA) is a measurement of the quantity of the electron stream flowing within the electrical current as it travels through the x-ray tube from cathode to anode (see Figure 4-2). Increases in the tube current milliamperage (mA) increase the number of electrons traveling through the x-ray tube to strike the tube target on the anode and produce x-radiation. The greater the tube current milliamperage, the greater the number of x-rays produced. In fact, the quantity of x-rays produced increases or decreases in direct proportion to the quantity of milliamperage employed. If a selected tube current of 100 mA produced 100 units of x-radiation, then a tube current of 200 mA would produce 200 units of x-radiation. Therefore, it would be appropriate to identify the milliamperage as the **controlling factor** of radiographic density. A specific quantity of x-radiation milliamperage produces a recorded image possessing a specific quantity of radiographic density. It is important to associate changes in milliamperage directly to the influence it will have on the radiographic density of the recorded image. (**Perform Experiment 20.**)

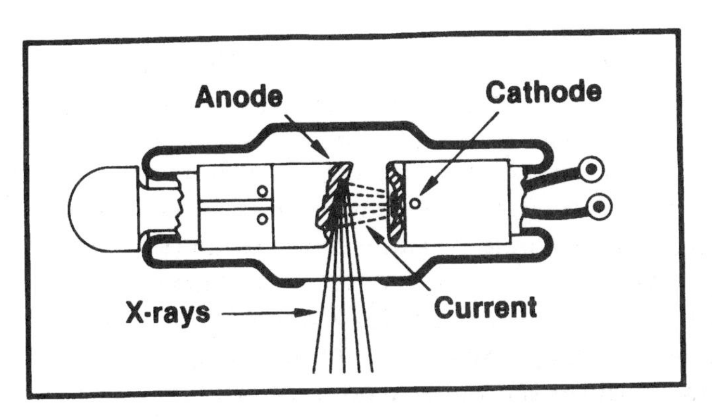

Figure 4-2
Diagram of X-ray Tube Depicting Production of X-rays

Time of Exposure

The time of exposure also represents a controlling factor of radiographic density. Time, of course, is a measure of the duration of events. In radiography, the time of exposure indicates the actual period or duration when x-rays are produced. The time of radiation exposure is generally measured in seconds or fractions thereof. The longer the time of exposure selected, the greater the total number of x-rays produced. In fact, the quantity of x-rays produced increases or decreases in direct proportion to the time of exposure employed. If a 1-sec time of exposure produces 100 units of x-radiation, then a 2-sec time of exposure would produce 200 units of x-radiation. A specific time of exposure (seconds) produces a recorded image possessing a specific quantity of radiographic density. The technologist should associate changes in the time of the exposure directly to the influence it will have on the radiographic density of the recorded image. (**Perform Experiment 21.**)

Milliampere-Seconds (mAs)

A relationship between the factors of milliamperage and the time of exposure and their influence on the production and control of radiographic density can be established. Since each factor directly controls the quantity of x-radiation produced by the x-ray tube, it is obvious that the total quantity of radiation emitted by the x-ray tube can be determined by finding the product of the milliamperage and the time of exposure (mAs). Knowing the mAs, the technologist also knows the total quantity of radiation that is going to be produced. In the manipulation of exposure factors in order to maintain the radiographic density, if any two of the factors in this relationship are known, the third factor can be determined from the following equations. You should memorize these equations, as they will be used frequently in the routine performance of your duties as a radiologic technologist.

$$\text{mA} \times \text{Time} = \text{mAs}$$
$$\text{mAs} \div \text{mA} = \text{Time}$$
$$\text{mAs} \div \text{Time} = \text{mA}$$

The radiographic density of the recorded image is directly affected by the total quantity of x-radiation produced. In the production and control of proper radiographic density, the technologist must understand the interrelationship of these two factors. When applied to the control and maintenance of radiographic density, the milliamperage and the time of exposure (mAs) are inversely proportional. For example, the selection of an exposure using 100 mA for a 0.l-sec exposure time produces a specific quantity of radiation, represented by the product of these two factors, 10 mAs (Figure 4-3). A 10-mAs exposure will produce a specific amount of radiographic density within a recorded image. If the milliamperage was increased to 200 mA, and the time of exposure remained the same, the radiographic density of the recorded image would increase, since the total quantity of radiation has also increased. In this case, the quantity of radiation will have increased by a factor of 2 times (100%). However, if you want to control and maintain the radiographic density of the recorded image using the new milliamperage of 200 mA, you would have to reduce the time of exposure (seconds) in direct proportion to the increase in milliamperage. In this example, a reduction of the exposure time from 0.1 sec to 0.05 sec will maintain the total quantity of radiation at 10 mAs, and therefore the radiographic density of the image will also be maintained. Remember, as the milliamperage is increased, the time of exposure must be decreased by the same proportion. If the milliamperage is decreased, the time of exposure must be increased by the same proportion. These two factors that control the intensity (quantity) of the x-ray beam and ultimately the radiographic density of the recorded image are inversely proportional to one another. (**Perform Experiment 22.**)

In fact, the adjustment of the milliamperage (mA) and time of exposure (seconds) relationship in order to control the mAs and to maintain the radiographic density of the recorded image is the most basic and important exposure factor control at the technologist's disposal. (Table 4-1 can be used to find the total exposure mAs over a wide range of milliamperages and exposure times.) Using Table 4-1, identify the milliamperage to be employed from the list at the top of the table, choose the time of exposure from the list to the left of the table, and examine the body of the table to determine where the two lines of milliamperage and time intersect. This is the mAs. For example, the selection of 400 mA with the application of a .0166- (1/60-) sec exposure will produce 6.66 mAs.

There are many instances where the relationship between the milliamperage and time of exposure and its appropriate manipulation will enable the technologist to reduce significantly the time of exposure while maintaining the radiographic density of the recorded image and effectively prevent image unsharpness due to motion. As with any exposure factor manipulation, the technologist must consider how that change will affect the total radiographic image. The reduction of the time of exposure requires a comparable increase in the milliamperage in order to maintain the total quantity of radiation (mAs) produced. Increases in the milliamperage produce an increase in the number of electrons available to produce x-radiation and a spreading out of the stream of electrons causing them to strike a much larger area on the tube

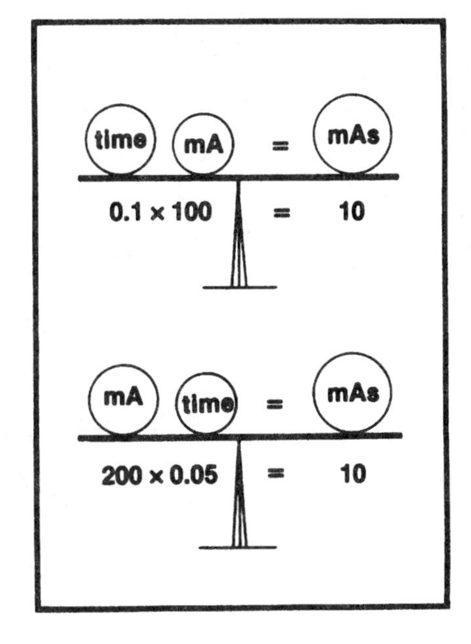

Figure 4-3
Proportional Relationship of Exposure Factors of Time and Milliamperage

Table 4-1 Milliampere–Seconds (mAs) Chart

Time of Exposure			Milliamperes										
Milliseconds	Decimal	Fractions	50	100	150	200	300	400	500	600	800	1000	1200
1	0.001	1/1000	0.05	0.1	0.15	0.2	0.3	0.4	0.5	0.6	0.8	1	1.2
2	0.002	1/500	0.1	0.2	0.3	0.4	0.6	0.8	1	1.2	1.6	2	2.4
3	0.003	3/1000	0.15	0.3	0.45	0.6	0.9	1.2	1.5	1.8	2.4	3	3.6
4	0.004	1/200	0.2	0.4	0.6	0.8	1.2	1.6	2	2.4	3.2	4	4.8
5	0.005	1/250	0.25	0.5	0.75	1	1.5	2	2.5	3	4	5	6
6	0.006	3/500	0.3	0.6	0.9	1.2	1.8	2.4	3	3.6	4.8	6	7.2
7	0.007	7/1000	0.35	0.7	1.05	1.4	2.1	2.8	3.5	4.2	5.6	7	8.4
8	0.008	1/125	0.4	0.8	1.2	1.6	2.4	3.2	4	4.8	6.4	8	9.6
8.3	0.0083	1/120	0.414	0.828	1.24	1.66	2.48	3.33	4.16	5	6.66	8.3	10
10	0.01	1/100	0.5	1	1.5	2	3	4	5	6	8	10	12
12	0.012	3/250	0.6	1.2	1.8	2.4	3.6	4.8	6	7.2	9.6	12	14.4
16.6	0.0166	1/60	0.83	1.66	2.49	3.33	5	6.66	8.3	10	13.3	16.6	20
18	0.018	9/500	0.9	1.8	2.7	3.6	5.4	7.2	9	10.8	14.4	18	21.6
20	0.02	1/50	1	2	3	4	6	8	10	12	16	20	24
24	0.024	3/125	1.2	2.4	3.6	4.8	7.2	9.6	12	14.4	19.2	24	28.8
25	0.025	1/40	1.25	2.5	3.75	5	7.5	10	12.5	15	20	25	30
30	0.03	3/100	1.5	3	4.5	6	9	12	15	18	24	30	36
33.3	0.033	1/30	1.66	3.33	5	6.66	10	13.3	16.6	20	26.6	33.3	40
40	0.04	1/25	2	4	6	8	12	16	20	24	32	40	48
50	0.05	1/20	2.5	5	7.5	10	15	20	25	30	40	50	60
66.6	0.066	1/15	3.33	6.66	10	13.3	20	26.6	33.3	40	53.2	66.6	80
100	0.1	1/10	5	10	15	20	30	40	50	60	80	100	120
120	0.12	1/8	6	12	18	24	36	48	60	72	96	120	144
130	0.13	13/100	6.5	13	19.5	26	39	52	63	78	104	130	156
160	0.16	4/25	8	16	24	32	48	64	80	96	128	160	192
200	0.2	1/5	10	20	30	40	60	80	100	120	160	200	240
250	0.25	1/4	12.5	25	37.5	50	75	100	125	150	180	240	300
300	0.3	3/10	15	30	45	60	90	120	150	180	240	300	360
400	0.4	2/5	20	40	60	80	120	160	200	240	320	400	480
500	0.5	1/2	25	50	75	100	150	200	250	300	400	500	600
600	0.6	6/10	30	60	90	120	180	240	300	360	480	600	720
700	0.7	7/10	35	70	105	140	210	280	350	420	560	700	840
750	0.75	3/4	37.5	75	112.5	150	225	300	375	450	600	750	900
800	0.8	8/10	40	80	120	160	240	320	400	480	640	800	960
1000	1	1	50	100	150	200	300	400	500	600	800	1000	1200

target. Significant increases in the milliamperage can create a "blooming" effect on the size of focus although the small tube focus is still employed or may even require the selection of the large tube focus if the higher milliamperage cannot be utilized with the small tube focus. In either case, the blooming effect of the focus size will produce an increased geometric unsharpness in the recorded image. Remember that the size of focus directly influences the geometric unsharpness of the recorded detail. This loss of image sharpness can be tolerated when the reduced time of exposure eliminates the greater loss of sharpness attributed to motion.

Quality Assurance

It is important that the technologist be confident that the equipment being utilized is properly calibrated and routinely evaluated. A quality assurance program includes an evaluation of the operation and output of the equipment. With properly calibrated equipment, if a desired radiographic density is achieved using a specific milliamperage and time of exposure to produce the desired mAs selection, the same x-ray output and therefore radiographic density should be achieved regardless of the manipulation of the relationship of the actual milliamperage and time of exposure selected, as long as the mAs utilized remains the same. This is especially critical in an x-ray department with several examination rooms and different pieces of equipment operated by different x-ray generators. With any piece of modern radiographic equipment, the control panel has available a multiplicity of milliamperage and time-of-exposure selections to produce the required mAs and achieve the desired radiographic density. If adjusting the relationship of the milliamperage and time of exposure while maintaining the same mAs produces a significant change in radiographic density from the anticipated level, it would be appropriate to report this to the radiology department's quality control personnel. It may indicate that the calibration of the equipment is inaccurate.

Milliamperage Quality Assurance

A simple check of the tube current milliamperage can be made on most pieces of x-ray equipment. Most control panels have a milliamperage meter that will register and indicate the quantity of tube current milliamperage flowing through the x-ray tube for a given exposure. Select the small focus and the smallest milliamperage value available on your control panel. Choose a time of exposure that will be of sufficient length, 0.5 (1/2) sec, so that the milliamperage meter will register and allow you to evaluate its operation. Use a low kilovoltage for these tests. Make an exposure and record the milliamperage value. Increase the milliamperage, selecting each of the available milliamperage stations that can be utilized with the small focus, repeat the test, and record the actual tube current milliamperage. A similar series of exposures should be performed after selecting the large focus. The same testing process should be performed for each unit of x-ray equipment operating within the department. The milliamperage meter should register the same milliamperage value as the milliamperage station on the control panel selected. The observation of any differences in the tube current between the selected and the actual milliamperage value registered by the milliamperage meter is a sign that the x-ray output does not coincide with the selected milliamperage on your control panel, and your x-ray unit is not calibrated or accurate. (**Perform Experiment 23.**)

A more accurate assessment of the operation of the milliamperage for a series of x-ray units would be to record the exposures of each milliamperage station selection for the small and large tube focus on a piece of radiographic film and to evaluate the radiographic density produced using a densitometer for each x-ray unit. If the radiographic densities produced for the same milliamperage stations between different x-ray units are different, then the x-ray outputs between the units are not calibrated or accurate. (**Perform Experiment 24.**)

You can see how this can be a time-consuming activity. Most x-ray departments designate a specific person among their staff as the quality assurance technologist. This person should have the training and experience necessary to perform the many required tests associated with a comprehensive quality assurance program. Larger departments hire specific quality assurance personnel that may include a radiation health physicist and an electrical engineer or technician. Quality assurance associated with the x-ray output of the equipment is a basic control related to the production of radiographic quality and is a major example of the application of the principles of using radiation levels that are "as low as reasonably achievable" (ALARA) to help eliminate unnecessary repeat exposures. The ability to maintain radiographic density and to produce consistent radiographic quality using different control panel selections and applying the same exposure factors to different pieces of x-ray equipment is essential to the development of standardized procedures and exposure technique charts. The production of high-quality radiographic images cannot be a hit-or-miss operation. After evaluating the tube current milliamperage output of a specific x-ray unit and comparing it with the output of other x-ray units, the quality assurance technologist can have the equipment adjusted and calibrated within identified parameters by the appropriate x-ray service technicians. The x-ray service personnel will adjust the tube output for the different milliamperage stations so that the tube current of each will comply with the specified tube output and will adjust these values to the other x-ray units so that the tube current milliamperage is accurate throughout the entire department.

Time-of-Exposure Quality Assurance

The second factor in the mAs equation is the time of exposure. It is essential that the timing controls on x-ray units be accurate and function properly. X-rays are generated by several different methods. Half-wave-rectified equipment produces 60 pulses of energy/sec; full-wave-rectified equipment produce 120 pulses of energy/sec. The timing mechanisms of these systems of x-ray generation can be accurately evaluated using a spinning-top device. Half-wave-rectified equipment is seldom employed in modern radiography, so we will not refer to it in this evaluation. Many modern x-ray units producing higher amounts of energy use three-phase generating equipment or, in the case of portable (mobile) x-ray units, capacitor discharge after storing electrical energy. These units produce a near-constant output of energy, and as a result of their lack of pulsation, their timers cannot be evaluated using the simple spinning top.

The spinning top is a simple device that can be used to evaluate the operation of x-ray timers. It consists of a heavy metal circular disc measuring about 25 cm in diameter. The disc has a single, small hole drilled into it near the perimeter. The disc is placed on a platform designed to hold it that allows the disc to be rotated (spun) during its testing operation.

The spinning top is placed on the surface of a cassette, and the x-ray beam is directed to its midpoint using a perpendicular beam. Several exposures utilizing different exposure times can be made on a single radiographic film by dividing and blocking the cassette into sections. Using a low milliamperage station (50 or 100 mA) and a low kilovoltage (40–60 kVp), an exposure of 0.1 (1/10) sec is selected. The spinning top is started spinning using a twist of the fingers, and an exposure of the film is made while the spinning top is moving. The resultant radiographic image will demonstrate a series of dots that correspond to the pulses of energy emitted by the x-ray tube. The speed of the rotation of the spinning top does not change the number of dots recorded or the accuracy of the results of the test. A more rapid spinning top will simply spread the dots over a larger area of the recorded image so that they will appear as elongated dots or appear as dashes within the recorded image, but the same number of dots/dashes will be demonstrated. Since full-wave-rectified equipment produces 120 pulses/sec, a 0.1- (1/10-) sec exposure should produce a total of 12 dots upon the radiographic film, (0.10×120 pulses/sec = 12 pulses). (**Perform Experiment 25.**) If the radiographic image of the spinning-top test demonstrates any number (higher or lower) other than 12 dots within the recorded image, the timing device of your x-ray unit is not operating properly. Inaccurate or malfunctioning times will result in repeat examinations, unnecessary patient exposure, and poor-quality radiographic images.

Although the timing mechanisms of three-phase x-ray-generating equipment are more properly evaluated using an oscilloscope, a modification of the spinning-top device can be helpful in assessing all but the shortest of exposure times. In this case, the spinning top has been adapted so that it spins at a constant rotation rate of 1 revolution/sec using a synchronous motor. Selecting an exposure time of 0.5 (1/2) sec, the spinning top will travel in a half a circle during the exposure, recording the image of an arc representing a half a circle or 180°. Shorter exposure times, such as 0.25 (1/4) sec, will produce a 90° recorded arc; 0.125 (1/8) sec will produce a 45° recorded arc; and an exposure of 0.083 (1/12) sec will produce a 30° recorded arc. It is obvious that as the exposure time gets shorter, the recorded arc also becomes smaller and more difficult to assess. Three-phase generating equipment has the advantage of using high-intensity exposures of 1,000 mA or greater for very short exposure times, as short as 1 msec, 0.001 (1/1,000) sec. Therefore, the assessment of the accuracy of the timing mechanism using a synchronous spinning top is limited. (**Perform Experiment 26.**)

In either instance, the assessment of the timer mechanism accuracy is another example of the application of a quality assurance program and a concern for the principles of ALARA.

Producing Proper Radiographic Density

Frequently, the problem confronting the technologist is how to produce the proper radiographic density in the first place. Remember that proper radiographic density will enable you to visualize all of the structures of interest. The proper radiographic density for a given part, however, may differ depending on the purpose of the examination and the structures of interest contained therein. As we learned earlier on, the proper radiographic density for lung radiography is considerably different from the requirements of a rib examination, even though the same part of the body is being examined. It is essential to correlate your understanding of the thickness and opacity of the tissues of the part with your knowledge of the anatomy and physiology of the body, if you are to achieve a radiographic density that enables visualization of all of the structures of interest. A thorough evaluation of the purpose of the examination and a determination of the structures of interest within the part are also vital to the selection of the basic mAs to be employed.

While analyzing the radiographic image, ask yourself this question: Am I able to see all of the structures of interest that should be visible within the image? If the answer to your inquiry is affirmative, then the proper radiographic density has been achieved. There are instances when the exposure factors selected produce an inadequate image. The radiographic density of the image is either insufficient or excessive. The same basic problem exists regardless of whether the image is too light or too dark. There has been a loss in the ability to visualize the structures of interest. **(Perform Experiment 27.)**

When this occurs, you have to analyze the radiographic image before attempting to repeat the exposure. The basic concern is to determine how much lighter or darker the radiographic image is relative to the proper and desired radiographic density. A common suggested correction for insufficient density is to increase your time of exposure by a "step in time." For example, 100 mAs is selected for an exposure of the abdomen using a 0.5- (1/2-) sec exposure time and 200 mA. In your analysis of the radiographic image, you have determined that the radiographic density is insufficient. With many control panels, the next available time of exposure will be 0.6 (3/5) sec. Therefore, an increase by a step in time requires you to select an exposure time of 0.6 sec. The 120 mAs produced by this adjustment represents an overall exposure increase of 20%.

A forearm examination is performed using a 0.1- (1/10-) sec exposure time and 100 mA for a total exposure of 10 mAs. Again, your radiographic image is determined to lack sufficient radiographic density. In this instance, the next available time of exposure may be 0.133 (13/100) sec. The 13.3 mAs produced by this increase represents an overall exposure increase of 33%.

A "step in time" 20% vs. 33%

Abdomen:	**old**	**200 mA × 0.500 sec = 100 mAs**
	new	**200 mA × 0.600 sec = 120 mAs (20% increase)**
Forearm:	**old**	**100 mA × 0.100 sec = 10.0 mAs**
	new	**100 mA × 0.134 sec = 13.3 mAs (33% increase)**

The fallacy of simply increasing or decreasing the exposure by a step in time as a correction for problems of insufficient or excessive radiographic density is apparent when you consider the wide variety of different radiographic equipment and the differences in exposure times available on different control panels. The required change may also differ with the specifics of the examination being performed. It is obvious from this discussion that an increase or decrease by a step in time is actually a step in the direction of the technical guessing game. Is it logical to assume that a 20% increase would be proper in one instance, whereas a 33% increase would be necessary in another, to produce a similar increase in radiographic density? How much exposure increase would be necessary to produce a visible change in the radiographic density of the recorded image? Would 20% be sufficient? Would 33%

be too much? In analyzing this situation you will find that to produce a visible change in the radiographic density of the recorded image requires a minimal exposure adjustment of at least 30%. (**Perform Experiment 28.**)

To Demonstrate a Radiographic Density Change:
A MINIMUM mAs ADJUSTMENT OF 30% IS REQUIRED

Depending upon your analysis of the improper radiographic density of the structures of interest in your recorded image, the increase or decrease of exposure required to produce the proper radiographic density may constitute a 50% or even a 100% change from your original technique. The analysis of an improperly exposed image can be a challenging assignment. The experience garnered by the performance of multiple experiments and your participation in a number of radiographic procedures under the supervision of a qualified radiologic technologist will soon enable you to analyze the situation related to the achievement of the proper radiographic density. However, keeping in mind the fact that a minimal change of 30% is necessary to produce perceptible differences in the radiographic density of the recorded image provides you with a starting point toward the achievement of producing proper radiographic density.

Automated Exposure Control

A significant percentage of the radiographic equipment utilized in the modern radiology department includes equipment with automated or automatic exposure control. Basically, automated exposure control devices are designed to interrupt and terminate the exposure automatically when a desired amount of radiation, and, therefore, a specific radiographic density of the recorded image, has been achieved. Automated exposure control devices have been used in radiography and during fluoroscopic procedures for more than 40 years. Their design has improved, and the accuracy of their exposure control has increased to the point where the majority of routine radiographic procedures can be performed using these devices. The use of automated exposure control for routine radiography is a positive example of the application of the principles of ALARA since the application of automated exposure devices can frequently reduce the number of repeat examinations that result from the inappropriate selection of exposure factors. However, there are some precautions and considerations in the proper application of automated exposure control devices. There are also a number of limitations imposed on the use of automated exposure control. Automated exposure cannot be utilized for tabletop (non-Bucky) procedures because the mechanisms required for the operation of these devices must be placed either between the patient and the cassette or below the cassette itself. Automated exposure control devices fall into two main categories: (1) phototimers and (2) ionization chambers.

Phototimers

Phototiming devices have been employed in radiography for over 40 years. Like the name of the device implies, phototimers are devices that employ a light-sensitive system to control the exposure time. The phototiming mechanism is placed below the surface of the tabletop and beneath the Bucky apparatus. Radiation must pass through the patient, the cassette, and the Bucky tray in order to trigger the photosensitive mechanism. The Bucky tray must be designed in such a way that a hole is located in the center of the tray to enable radiation to pass through it to activate the exposure terminating photosensitive apparatus. In addition, the cassettes used with phototiming techniques have been specially designed for that purpose. General purpose cassettes have a thin layer of lead at the back of the cassette to absorb secondary and scattered radiations that could potentially bounce back toward the intensifying screens and produce additional unwanted and uncontrolled radiation. Lead backing within phototiming cassettes would be contraindicated since the radiation must be able to penetrate through the cassette to reach the phototiming device located beneath the Bucky tray.

The phototiming unit consists of a highly sensitive photomultiplier tube that reacts to very low levels of light. Adjacent to the photomultiplier tube is a small fluoroscopic screen that emits light in proportion to the amount of radiation it receives. The phototimer controls permit the technologist to preset a specific amount of exposure that will trip the phototimer tube to terminate the exposure when the light given off by the fluoroscopic screen has reached the predetermined level. The preset exposure is based on the type of procedure being performed and the overall body type of the patient. The technologist can select a control panel button that represents either a small, average, or large patient and place a predetermined exposure value into the system. Once the amount of radiation required for the desired radiographic density of the film has been received, the exposure will be automatically terminated. That determination is controlled by the amount of light given off by the fluoroscopic screen. When sufficient light has been received by the photomultiplier tube, the preset capacitance of the tube has been reached, and the exposure is terminated automatically.

Ionization Chambers

The second type of automatic exposure control uses an ionization chamber, not unlike the general characteristics of the ionization chamber associated with the Geiger-Muller counter used for the detection of the presence of ionizing radiation, including x-rays. The ionization chamber is usually a thin, radiolucent, sealed chamber placed between the patient and the image recording system. It would be located below the tabletop surface and above the cassette placed in the Bucky tray. Ionization chambers contain gas that when exposed to x-radiation releases electrons and produces electrical charges (ions) within the chamber. Like the phototiming device, the ionization chamber automatically terminates the exposure when a preselected exposure applicable to the charge released within the chamber has been reached. The control panel of x-ray units having ionization chambers also has a number of automated exposure control buttons that allow the technologist to select predetermined levels of radiographic density associated with small, average, and large patients. The ionization chamber devices have the added advantage of usually consisting of multiple chambers, so that depending on the examination to be performed and/or the positioning and centering of the patient, the termination of the exposure will coincide with the appropriate radiographic density for the procedure being performed. As an example, for an examination of the thoracic spine, the technologist may choose the middle ionization chamber so that the preselected exposure will reflect the exposure required to produce sufficient radiographic density to visualize the thoracic spine. The adjacent structures outside of the mediastinal area would be overexposed and have excessive radiographic density. However, if the technologist wanted to demonstrate a proper radiographic density of the lungs, he or she would select a different ionization chamber that would coincide with the right side of the thorax so that the chamber would preselect the appropriate time of exposure to produce the proper radiographic density for the lung cavity of the chest. In this instance, the midline structures of the thoracic spine would be recorded with insufficient radiographic density to visualize those structures. The use of multichambered ionization chambers is a significant improvement over the use of the simple phototimer device in the accuracy of the operation of automated exposure control devices.

Application of Automated Exposure Control

An important consideration to the proper application of automated exposure control devices is that the kilovoltage necessary to penetrate the part properly must be selected in advance. The technologist can then select the milliamperage desired for the procedure. The higher the milliamperage selected, the more rapidly the predetermined exposure value selected will be reached, and a shorter time of exposure will result. The automated exposure control device will simply adjust the total time of exposure required to produce and maintain the required radiographic density. The more sophisticated devices will have a variety of control panel selections related to specific areas of the body or specific radiographic procedures. When performing an examination of the shoulder, the technologist would simply select the shoulder button on the control panel.

The positioning of the patient related to the centering of the structures of interest to the midline of the table and to the center of the film is essential for the proper utilization of automated exposure control systems. Inaccurate positioning and centering of the structures of interest or body part over the phototiming mechanism or the ionization chamber can result in an over- or underexposed radiographic image. If the part to be examined is not properly positioned or centered to the automated exposure device, the apparatus will receive an inaccurate reading and may cause the exposure to terminate prematurely, producing a radiographic image with insufficient radiographic density. On the other hand, the exposure could continue for a period of time longer than required for the procedure or the part that is being examined, producing a radiographic image with excessive radiographic density.

Automated exposure control devices limit the technologist's ability to manipulate or adjust the exposure factors in order to change (shorten or lengthen) the radiographic contrast scale of the recorded image. Therefore, even with the sophisticated devices available today, the technologist still has to recognize and understand all of the factors that influence the radiographic density of the recorded image and be able to adjust and manipulate those factors to advantage when the specific circumstances of the patient or the procedure require it.

Source-Image Distance

As the distance between the focus of the tube and the image receptor (source-image distance [SID]) is changed, you will notice significant differences in the radiographic density of the recorded image. As the beam of radiation emitted from the x-ray tube travels further from its source of origin, its area of total coverage (influence) increases. The same quantity of radiation is in the x-ray beam at 20" (51 cm) as at 40" (102 cm), but the total area covered by and influenced by the x-ray beam will be considerably greater as the distance from its source of origin increases. What this means is that the measurement of the quantity of radiation in a specific area covered by the beam of radiation will be less when the distance is increased, and the total quantity of radiation in the beam is spread out over a larger total area. The overall quantity (intensity) of the beam is spread over a larger area as the source-image distance increases, and, therefore, the influence on the radiographic film located within any specific portion of the total beam will be less. The radiation intercepted by the film will be less, and the radiographic density of the recorded image will be less (Figure 4-4). The reverse of this concept is also true. As the SID is decreased, the total area influenced by the beam will be less. The total quantity (intensity) of the beam will be concentrated within a much smaller total area, and, therefore, the amount of radiation intercepted by the film will be more and the radiographic density of the recorded image will be greater.

This relationship can be demonstrated by an application of the inverse square law. The inverse square law states that the intensity of radiation varies inversely to the square of the distance. In radiography, this applies to the SID.

Using a 20" (51 cm) SID, a beam of radiation covering a 2" (5 cm) square (length and width) measures an intensity of 100 roentgens (R)/min. The total area affected by the radiation is 4 (25.8) square in/cm. If the SID is increased to 40", the spreading of radiation coverage would increase in all directions, so that it would now affect a 4" (10 cm) square. The total area affected would be 16 (100 cm) square in/cm. The same amount of radiation would now be spread over an area four times as great. Therefore, the intensity of radiation measured at any point within this larger area using the 40" (102 cm) SID would only be one fourth as great as the intensity of radiation measured at 20" (51 cm). (**Perform Experiment 29.**)

Unless compensation for the loss of radiation intensity is provided, when the SID is increased, the overall radiographic density of the recorded image will decrease according to the principles established by the inverse square law. (**Perform Experiment 30.**) A decrease in the SID will increase the radiographic density of the recorded image according to the same principle.

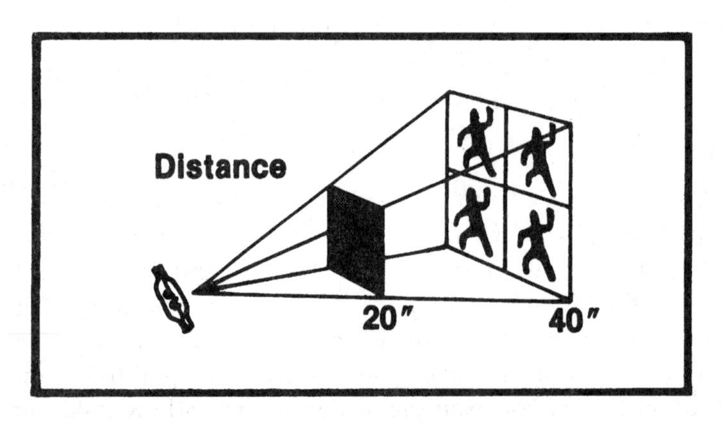

Figure 4-4
Diagram of the Relationship Between SID and Radiographic Density

In radiography, an understanding of the principle of the inverse square law is used to advantage in order to apply the necessary adjustment of exposure factors that will enable you to maintain the proper radiographic density when a change in the SID is suggested or required. A practical working formula has been developed for this purpose:

$$\text{New mAs} = \frac{\text{Old mAs} \times \text{New Distance}^2}{\text{Old Distance}^2}$$

For example, a lateral radiograph of the cervical spine was taken at a 36" (91 cm) SID using 30 mAs. The radiographic density of the recorded image was satisfactory. However, you have been requested to repeat the radiograph using a 72" (183 cm) SID in order to reduce the magnification and geometric unsharpness of the image caused by the increased object-image distance (OID) that occurs with this projection. What exposure must you employ at this new distance in order to maintain the radiographic density of the recorded image?

Old mAs = 30 Old Distance = 36"
New mAs = x New Distance = 72"

$$\frac{30 \times 72 \times 72}{36 \times 36} = 120 \text{ mAs}$$

Applying the density maintenance formula, you discover that the necessary exposure to maintain the radiographic density of the recorded image at the new SID would be 120 mAs. Therefore, in performing the radiograph of the cervical spine at the new SID of 72" (183 cm), the technologist would select 120 mAs, and the radiographic density would be the same as in the original radiograph taken at 36" (91 cm) SID and using 30 mAs.

If all SID changes required the technologist to change the distance to exactly double or one half of the original, it would be easy to calculate the exposure required by the new distance mentally. Unfortunately, such SID changes are often the exception rather than the rule. More frequently, the adjustment requires a change from an erect chest radiograph using 72" (183 cm) SID to a supine chest radiograph using 48" (122 cm) SID, or something similar. With routine distance changes, figuring out the exposure factors in your head will prove rather difficult, and simply estimating the necessary exposure adjustment will produce errors from examination to examination. There is no need to estimate the required exposure factor change when a simple formula enables you to control your exposure accurately and maintain your radiographic density. (**Perform Experiment 31.**)

Radiographic density can be controlled and maintained regardless of changes in the SID by the application of the principles established by the inverse square law. Table 4-2 identifies the multiplication factor to be used in order to maintain the radiographic density of the recorded image when the SID is changed. Find your original SID in the column to the left of the table and the new SID at the top of the chart. Within the table, find the point of intersection between these two distances. Use this multiplication factor to determine the new mAs by using the formula provided at the top of the table. The new mAs is equal to the old mAs times the multiplication factor. For example, if the original SID was 40", and a satisfactory

Table 4-2 Multiplication Factors for Changes in SID (Old mAs × Multiplication Factor = New mAs)

Original SID in inches	New SID in inches: 30	36	40	48	60	72
30	1.00	1.44	1.77	2.56	4.00	5.73
36	.69	1.00	1.23	1.77	2.77	4.00
40	.56	.81	1.00	1.44	2.25	3.24
48	.39	.56	.69	1.00	1.56	2.25
60	.25	.36	.44	.64	1.00	1.44
72	.17	.25	.31	.44	.69	1.00

radiographic image was produced using 50 mAs, the mAs required if your SID has been changed to 60" would be 112.5 mAs (50 × 2.25 = 112.5 mAs).

Kilovoltage

Kilovoltage is often referred to as the quality factor of the x-ray beam because it has a major influence on the radiographic contrast of the image and because it controls the penetrating ability of the beam. These influences are examined in detail in Chapter 5, which addresses radiographic contrast. Kilovoltage does, however, have a significant influence on the radiographic density of the recorded image, although this influence is somewhat indirect.

We have learned that the quantity of radiation produced is directly controlled by the number of electrons flowing through the x-ray tube and the period of time that the x-ray tube is energized—in other words, the mAs. An increase in the kilovoltage does not increase the number of electrons flowing through the tube, so it does not directly influence an increase in the production of x-rays. However, it increases the speed of the electron flow and, therefore, increases the energy and penetrating ability of the x-rays produced when the higher speed electrons strike the target material of the anode. The penetrating ability of x-rays is related to the wavelength of the radiation produced. The shorter the wavelength of the radiation produced, the higher the penetrating ability of the x-rays. Since higher kilovoltage increases the energy and penetrating ability of the x-rays produced, it is obvious that increases in kilovoltage generate radiation of shorter wavelength. Shorter wavelength radiation has greater penetrating ability.

Every x-ray beam is made up of a wide range of wavelengths from very short to very long, regardless of the kilovoltage applied. An increase in the kilovoltage changes the shortest possible wavelength and therefore the most penetrating radiation that can be produced within the x-ray beam. There may not be significant amounts of radiation produced at this new, shorter wavelength, but the maximum penetrating ability of the overall beam will have increased as a result of using a higher kilovoltage. Higher kilovoltage also produces a change in the overall spectral distribution of the energy levels within the beam. In addition to producing some radiation possessing a greater penetrating ability due to the shorter wavelength, there is an overall shift in the percentage of other higher energy radiation within the beam. The end result of these processes is that a greater quantity of the radiation produced in the x-ray tube is able to penetrate through the inherent and added filtration of the tube and its housing, emerge from the tube, and reach the patient. Thus, a measurement of the quantity of radiation within a beam of radiation produced using an 80 kVp beam compared with the quantity of radiation produced using a 60 kVp beam will actually demonstrate an increase in the intensity (quantity) of radiation even though the mAs selected for both exposures was the same.

Additionally, the shorter wavelength radiation and the overall higher percentage of greater penetrating radiation at higher kilovoltage permits a greater quantity of the radiation to pass through the body part and reach the film, producing an increased radiographic density within the recorded image. The tissues of the body interact with the radiation passing through it and produce secondary and scattered radiation as a result of these interactions. Changes in the energy distribution within the x-ray beam will alter the patterns of absorption, including the energies of the secondary radiation and scattered radiation produced as a result of that absorption. Ultimately, it will alter the quantity of radiation that can traverse a body part and emerge from the body as image-forming radiation. This change in the pattern of energy absorption by the body tissue and the greater quantity of radiation that can exit the body at higher kilovoltage will ultimately increase the radiographic density of the recorded image.

The wide range of wavelengths present within a given beam of radiation and, therefore, the energy distribution within the beam are not constant over the range of kilovoltage employed in diagnostic radiography. The influence of increased kilovoltage on the radiographic density of the image is not proportional as it was with the milliamperage and the time of exposure. As you increase the milliamperage or the time of exposure, the radiographic density of the image is increased proportionally. However, if you were to double the kilovoltage from 60 to 120 kVp, the effect on the quantity of radiation reaching the film and therefore the radiographic density of the image would be considerably greater than double. (**Perform Experiment 32.**) Another important factor to remember is that the effect of kilovoltage on the radiographic density of the recorded image is not constant throughout the entire range of kilovoltage applied. When operating in the lower kilovoltage range (40–50 kVp), an increase of only 2 kVp will produce a noticeable increase in the radiographic density. However, at a higher kilovoltage range (70–80 kVp), an increase of at least 6 kVp is required to produce a similar radiographic density change. (**Perform Experiment 33.**) As the kilovoltage selected increases to 100 kVp and above, the kilovoltage required to produce a noticeable increase in radiographic density continues to increase. Above 100 kVp, an increase of at least 10 kVp is required to demonstrate a radiographic density change in the recorded image. Additionally, the influence of kilovoltage on radiographic density depends on the materials selected to record the image. With direct exposure (nonscreen) radiography using screen-type film in cardboard holders, the visible change in radiographic density resulting from increases in the kilovoltage is far less than the change that occurs when you are employing intensifying screens. This is because kilovoltage is one of the external factors affecting the speed of the intensifying screens. (**Perform Experiment 34.**)

Thus far, you have learned that kilovoltage can and does affect radiographic density, but for different reasons and in a different manner than milliamperage or the time of exposure. Is there a specific application of kilovoltage related to the production and maintenance of proper radiographic density? Since kilovoltage controls the penetrating ability of the beam, it should properly be employed to produce a satisfactory contrast scale within the recorded image rather than be employed to produce a desired radiographic density. The selection of a specific kilovoltage or the manipulation of the kilovoltage would not be the ideal factor to choose in order to produce or control the desired radiographic density of the recorded image. The kilovoltage necessary to penetrate the part must be identified and must be used in every radiographic procedure. Having selected the required kilovoltage to penetrate the structures of interest properly, we can then select the appropriate mAs in order to produce the desired radiographic density. Using insufficient kilovoltage to penetrate the part would require a significant increase in the mAs required to produce the desired radiographic density, but the overall visibility of the structures of interest would still suffer. Using greater quantities of radiation than necessary to produce the recorded image would also be a poor example of the principles of ALARA. In fact, no practical amount of mAs increase can compensate and produce a satisfactory recorded image if the kilovoltage used does not provide for adequate penetration of the part. **(Perform Experiment 35.)**

kVp/mAs Relationship

If the kilovoltage necessary to penetrate the part has been selected, a basic relationship between the kVp and the mAs does exist, and the radiographic density of a recorded image can be maintained by appropriate manipulations of this relationship. If the basic criterion of penetration has been met, then the kVp/mAs relationship can be adjusted, not to produce a desired density, but to maintain the overall radiographic density when circumstances require or suggest this adjustment. It must be borne in mind that, although an adjustment of the kVp/mAs relationship enables you to maintain the radiographic density, it also alters the absorption characteristics and ratios of the structures of interest and, therefore, changes the contrast scale produced within the recorded image. In general, an increase in the kilovoltage together with a reduction in the mAs to maintain the radiographic density of the image will produce a longer contrast scale within the recorded image. There are circumstances when an alteration of the contrast scale is acceptable and even purposeful; however, there are other occasions when a change in the contrast scale will impair the visibility of the recorded image. Thus, the use of the kVp/mAs relationship related to the maintenance of radiographic density is a principle that is employed infrequently in the practice of radiologic technology.

The relationship between kilovoltage and mAs is embodied in the 15% rule. This rule states that you can maintain the radiographic density of an image by increasing the kilovoltage by 15% and reducing the mAs by 50% of its original value, provided that the kilovoltage originally selected was sufficient for penetration.

DENSITY: kVp/mAs (15% Rule)
15% kVp Increase = 50% mAs Decrease

The application of this rule can be especially useful in the maintenance of radiographic density when a reduction in the time of exposure is needed or when a change in the overall contrast scale of the image is desired. **(Perform Experiment 36.)**

Additional consideration of kilovoltage and its influences on the photographic properties of the recorded image is contained in Chapter 5. The influence of kilovoltage on scattered radiation and the effect of radiographic fog on radiographic density and image contrast are examined there.

Intensifying Screens

Intensifying screens are devices that absorb the invisible, high-energy, short wavelengths of x-radiation and transform them to a lower energy, longer wavelength pattern of visible light. The active layer of intensifying screens is made up of various types of chemical compounds (phosphors) that have the ability to luminesce when exposed to high-energy x-radiation.

In our investigation of the radiographic quality factor of recorded detail, we found that the use of intensifying screens was responsible for the greatest amount of image unsharpness outside of actual motion. However, we also discovered that intensifying screens are used in most radiographic procedures because they allow for a significant reduction in exposure factors as compared to those required for direct exposure (nonscreen) radiography, which represents a positive application of the principles of ALARA. The use of intensifying screens even with the material unsharpness produced by them can actually improve recorded detail by eliminating motion unsharpness. They can also significantly reduce the overall radiation exposure to the patient. There are two major factors to consider when employing intensifying screens in order to maintain radiographic density: (1) the screen's intensification factor and (2) the speed of the screen. Each factor affects radiographic density in a slightly different manner.

Screen Intensification Factor

The screen intensification factor represents a comparison of the exposure required without intensification screens (direct exposure) to produce a desired radiographic density to the exposure required when intensifying screens are employed. It is basically a multiplication factor representing how much the screen procedure intensifies the action of the x-rays. A consideration of the screen intensification factor is necessary when changing from direct exposure radiography to intensifying screen radiography or vice versa. Although direct exposure radiography is not a recommended procedure in radiologic technology today, we will investigate the value of the intensification factor from the historical perspective in order to appreciate the improvement that intensifying screen radiography has made to the reduction of patient exposure. The intensification of the screen is affected by a number of intrinsic and extrinsic factors.

The major extrinsic factor influencing the intensification of the screen is kilovoltage. As the kilovoltage increases, resulting in a more penetrating, higher energy beam, the intensification factor of the screen also increases. However, this increase in the intensification factor is not constant throughout the range of kilovoltage applied in diagnostic radiology. Kilovoltage affects the quantity of visible light emitted by the screen's phosphor material, as well as the spectral distribution of the light. Radiographic film, which is manufactured to be sensitive to visible light and to the specific spectrum of light emitted by intensifying screens, is also affected by changes in the kilovoltage related to the production of radiographic density. **(Perform Experiment 37.)**

The intensification factor specifies the increased light emission properties of the intensifying screen at different levels of kilovoltage. It contrasts this property of the intensifying screens to radiography performed without screens by comparing the exposure required without the use of screens to the exposure required when screens are employed in order to maintain the radiographic density of the recorded image. Since the intensification factor is influenced by the kilovoltage, and different levels of kilovoltage are employed in diagnostic radiology, a number of different intensification factor values corresponding to the usual kilovoltage applied for the various radiographic procedures are usually specified.

The intensification factors of calcium tungstate ($CaWO_4$) screens have been thoroughly investigated and standardized for a wide range of kilovoltage levels. With different phosphor materials, the intensification factor may differ considerably. Information supplied by screen manufacturers indicate that most of the recently developed phosphor materials, including the rare earth phosphors, possess a response to kilovoltage similar to that of the calcium tungstate screens. With some of the newer phosphor materials, however, the influence of kilovoltage on the intensification factor of the screen appears to be far less. This is especially true in the higher kilovoltage range above 80 kVp. Given that direct exposure radiography, when applied, was generally limited in its application to tabletop procedures of small body parts, the kilovoltage selected for these procedures was seldom more than 60 kVp. Therefore, the different intensification responses of some of the newer phosphor materials observed at higher kilovoltage are not a practical consideration when changing from a direct exposure procedure to one employing the use of intensifying screens or in our investigation of the intensification factor. The intensification factor formula and the intensification factors applicable to a 100 speed calcium tungstate intensifying screen recording system are provided here.

INTENSIFICATION FACTOR FORMULA

$$\text{Intensification Factor} = \frac{\text{Exposure without Screens}}{\text{Exposure with Screens}}$$

SCREEN INTENSIFICATION FACTOR
At 50 kVp the intensification factor is 60
At 60 kVp the intensification factor is 80
At 70 kVp the intensification factor is 100

If a different phosphor material is employed, a review of the screen manufacturer's specifications and a suitable testing program would have to be conducted in order to determine the specific intensification factor of the new phosphor material using different kilovoltage. Different film emulsion responses also affect the intensification factor. The above list identifies the intensification factor for medium speed film emulsion responding within the blue and blue/violet color spectrum. Additionally, if the change from direct exposure radiography involves a screen speed other than a 100 speed system, an additional adjustment in the exposure must be performed in order to compensate for the radiographic density change that would be attributable to the differences in screen speeds utilized for the examination.

Applying the intensification factor appropriate to the kilovoltage selected, we can maintain the radiographic density of the film when changing from a direct exposure procedure to an intensifying screen procedure. For example, if a direct exposure procedure using 50 kVp requires 400 mAs to produce a satisfactory radiographic density, you could perform the procedure more effectively using a 100 speed film-screen system. The intensification factor formula could be used to maintain the radiographic density of the recorded image when the intensifying screen system is employed. The use of a film-screen system will enable you to reduce the radiation exposure to your patient and eliminate the possibility of motion.

The intensification factor for a 100 speed $CaWO_4$ film-screen system at 50 kVp is 60. If 400 mAs was required for the direct exposure procedure, you can solve for the new exposure to be employed using the 100 speed film-screen system by applying the intensification factor.

New Exposure = 400 mAs ÷ 60
New Exposure = 6.66 mAs

In this example, at 50 kVp, the same radiographic density can be produced using a 100 speed film-screen system at 6.66 mAs that required an exposure of 400 mAs as a direct exposure procedure. At 60 kVp, with an intensification factor of 80, the new exposure would be 5 mAs; using 70 kVp, with an intensification factor of 100, the new exposure would be 4 mAs. The reduction of patient exposure is significant, as is the reduction in the time of exposure in order to eliminate motion. At 50 kVp, if the direct exposure procedure utilized 400 mA × 1 sec to produce 400 mAs, a 100 film-screen system utilizing 400 milliamperes would only require an exposure time of 1/60 (0.0166) sec to produce 6.66 mAs. At 60 kVp, the film-screen system procedure would require an exposure time of 1/80 (0.0125) sec to produce 5 mAs; at 70 kVp, the film-screen system procedure would require an exposure time of 1/100 (0.01) sec to produce 4 mAs.

Since the intensification factor relates to a comparison between direct exposure radiography and intensifying screen radiography, and direct exposure radiography is no longer advocated in radiology, it is not an applicable factor for the technologist to use in the performance of today's radiographic procedures. However, this review of the intensification factor principle enables us to understand why direct exposure radiography has been discouraged and how much the use of intensifying screens has improved the performance of radiographic procedures. (**Perform Experiment 38.**)

Intensifying Screen Speeds

Intensifying screens used in radiography are utilized with appropriate light-sensitive film in a film-screen recording system available in a variety of different speeds. Knowing the speed of the film-screen system enables the technologist to select the proper exposure factors to produce or maintain a desired radiographic density. We examined the factor of intensifying screen speed in Chapter 2 in our discussion of material unsharpness. A review of this section will assist you in your consideration of the influence of intensifying screen speeds on radiographic density. We discovered that many of the factors influencing the speed of intensifying screens are inherent factors produced during the manufacture of the screen. The major factors affecting the speed of intensifying screens include the following:

- The reaction of the specific phosphor material used
- The size of the crystals of the phosphor
- The thickness of the active phosphor layer, which includes a consideration of the quantity of phosphor materials contained therein and the compactness or distribution of the phosphor material within the active layer
- The reflectance of the phosphor layer backing material

As the speed of the screen increases, a greater quantity of light is emitted by the phosphor layer of the screen for the same relative exposure, and, therefore, the radiographic density of the recorded image will increase. (**Perform Experiment 39.**) Therefore, the radiographic density of the recorded image can be maintained by reducing the exposure when a more rapid film-screen recording system is employed. From the very slow speed system to the most rapid speed system, each system provides a useful purpose for a specific radiographic application. However, it is important to remember that even the slowest speed system produces a measured loss of image sharpness. As a general rule, as the speed of the screen increases, the loss of image sharpness also increases. However, with the application of some of the rare earth recording systems, we have discovered that a higher speed rare earth system may have similar or better image resolution properties than a slower speed calcium tungstate system. The application of different film-screen recording systems in radiography must consider the influence of material unsharpness on the recorded detail of the radiographic image. Where recorded detail is important, the use of systems that provide for greater resolution properties should be employed even though their overall speed may be less than optimum. On the other hand, when minimal radiation exposure, such as abdominal radiography of a pregnant patient, or possible motion is the prime concern, the use of rapid imaging systems with the fastest film-screen recording speed may be more appropriate.

The standard film-screen recording system speed identified by the industry is a 100 speed system and is favorably compared with the old par-speed $CaWO_4$ intensifying screen system. The 100 speed system and its speed value is fairly comparable from manufacturer to manufacturer. All other quoted speed values supplied by manufacturers are based on and compared to this standard. The 100 speed system has been given a relative speed value of 1 (see Table 4-3). As the speed of the system increases, the exposure required to maintain radiographic density decreases. Table 4-3 provides a comparison between the relative speed and exposure value of a number of film-screen recording systems when a comparison exposure value of 100 mAs was chosen.

It is important to identify the speed value of the film-screen recording system you have chosen for your procedure. Once it is identified, you can calculate the exposure required to maintain the radiographic density of the recorded image regardless of changes in the system speed required by the procedure. Table 4-4 compares the exposure required to maintain the radiographic density for different film-screen recording systems using 60 kVp with an intensification factor of 80. The exposure required to maintain the radiographic density of the image in the examples provided in Table 4-4 could be anywhere from 800 mAs for a nonscreen procedure, to 10 mAs for a 100 speed recording system, to 0.83 mAs for a procedure employing the most rapid rare earth film-screen recording systems. (**Perform Experiment 40.**)

However, drastic changes in the application of different speed recording systems would not be accomplished without significant compromises in the radiographic

Table 4-3 Intensifying Screen Speeds and Exposures

Screen system speed	Speed value	Comparison exposure value (mAs)
50 Speed	1/2	200
100 Speed	1	100
200 Speed	2	50
300 Speed	3	33
400 Speed	4	25
600 Speed	6	16.6
800 Speed	8	12.5
1000 Speed	10	10
1200 Speed	12	8.3

quality of the image. Rapid imaging systems frequently accomplish their increased speed and exposure reduction at the expense of the image-resolving capabilities of the system. Manufacturers claim that the resolution properties of many of their rare earth screens are similar to the resolution properties of the 100 or 200 speed $CaWO_4$ recording systems. If a suitable testing program confirms this, the use of rare earth phosphors could be of considerable value in procedures where the lower speed systems would normally be employed. The rare earth screen systems may enable the technologist to eliminate the problems associated with motion and to reduce significantly the overall exposure to the patient. Use of higher speed recording systems is a positive example of the application of the principles of ALARA. Rapid imaging systems, however, are often associated with quantum mottle, which can become a major consideration in the overall quality of the recorded image. Review the section on radiographic noise in Chapter 2 in order to recognize more fully the problems associated with quantum mottle and the radiographic quality of the recorded image.

Table 4-4 Exposure Required To Maintain Radiographic Density with Different Film-Screen Recording Systems (Tabletop Radiographic Procedure) Using 60 kVp with an Intensification Factor of 80

Screen system speed	Speed value	Comparison exposure value (mAs)
Nonscreen		800
50 Speed	1/2	20
100 Speed	1	10
200 Speed	2	5
300 Speed	3	3.3
400 Speed	4	2.5
600 Speed	6	1.6
800 Speed	8	1.25
1000 Speed	10	1
1200 Speed	12	0.83

Central Ray Angle/SID Relationship

Most radiographic procedures are performed using a 40" (102 cm) SID. The technologist will adjust the distance from the tabletop or Bucky to the x-ray tube according to preset measurements located upon the ceiling- or the vertical-wall-mounted tube stand. The tube stand when adjusted to a certain height above the tabletop will provide the 40" (102 cm) distance utilized for most radiographic procedures. The majority of procedures utilize a central ray that is adjusted to produce a perpendicular (90°) relationship with the surface of the x-ray table or cassette holder. There are, however, a number of procedures that require varying degrees of angulation of the central ray from the perpendicular. In most instances, the central ray angulation is provided to prevent the superimposition of other structures that lie in the path of the structures of interest or to demonstrate properly the anatomy of the structures of interest when they do not lie parallel with the plane of the film. We introduced these concepts in Chapter 3 in our investigation of shape distortion. The use of an appropriate angle of the central ray can help to eliminate or minimize shape distortion in the structures of interest.

An interesting phenomenon occurs when you manipulate the central ray from the perpendicular. As the angle of the central ray increases, without any adjustment in the selected height of the tube stand, the distance from the x-ray tube to the surface of the film also increases. This increase in the SID occurs without your actually adjusting or manually changing the selected SID on your tube stand. The SID identified on your ceiling-mounted or wall-mounted tube stand will still read 40" (102 cm), but the actual distance that the radiation has to travel to reach the film will have increased due to the angulation of the central ray. Unless this phenomenon is compensated for, it will reduce the radiographic density of the recorded image. At minor angles, the increase in the SID is minimal and the radiographic density change may not be noticeable. However, as the angle of the central ray increases, the radiographic density of the recorded image will noticeably decrease.

There are two ways to adjust for the increased SID caused by the angulation of the central ray: (1) leave the height of the tube stand at its original position, and increase the exposure (mAs) by the amount necessary due to the phenomenon associated with the various tube angles; or (2) reduce the height of the tube stand in order to re-establish an actual 40" (102 cm) SID and maintain the original exposure. Either method will enable you to maintain the radiographic density of the recorded image. Table 4-5 illustrates the influence of this phenomenon. The column to the far left indicates a number of possible tube angles from 5° to 45°. The second column indicates the new mAs needed to maintain the radiographic density. The third column indicates the percentage of exposure increase required to maintain the radiographic density. The fourth column provides the new vertical tube stand height that must be used in order to maintain the radiographic density if you choose not to adjust the mAs. The final column indicates the specific distance the tube stand must be changed from its original vertical height in order to maintain the radiographic density.

Let's consider the application of Table 4-5 when using a 30° angle of the central ray. The technologist uses a 30° cephalic angle of the central ray in order to demonstrate the lumbosacral articulation in the anteroposterior projection. As a result of the phenomenon associated with the use of this angle, the original exposure of 100 mAs

Table 4-5 The Effect of X-ray Tube (Central Ray) Angulation on Radiographic Density When a 40" (102 cm) SID Has Been Utilized at 100 mA

	Maintenance of radiographic density by adjusting the original mAs		Maintenance of radiographic density by adjusting the vertical tube height	
Central Ray Angle	New mAs to maintain density	mAs % increase	New vertical tube stand distance	Reduction from vertical distance
5°	101.00	1.00	39.80"	.20"
10°	103.02	3.02	39.40"	.60"
15°	107.12	7.12	38.60"	1.40"
20°	112.36	12.36	37.60"	2.40"
25°	121.00	21.00	36.00"	4.00"
30°	128.82	28.82	34.60"	5.40"
35°	139.24	39.24	32.80"	7.20"
40°	154.69	54.69	30.25"	9.75"
45°	167.38	67.38	28.25"	11.75"

will have the same influence upon your film as an exposure of 77.63 mAs would if the central ray were perpendicular. If you choose to leave the tube stand in its original position, you would have to increase your exposure to 128.82 mAs in order to maintain the radiographic density of the recorded image. If, however, you decided to maintain your 100 mAs exposure, you could adjust the vertical height of your tube stand so that it reads 34.6" (88 cm), a reduction of 5.4" (14 cm) from its original height. This will enable you to re-establish an actual SID of 40" (102 cm) and maintain your radiographic density without having to adjust your technical exposure factors. Most technologists find it easier to adjust the vertical tube height in those procedures that require an angulation of the central ray.

Consider the number of examinations that employ various central ray angles. Unless the actual 40" (102 cm) SID is maintained by either increasing the mAs or reducing the vertical height of the tube stand, the radiographic density of the recorded image will be reduced. A review of Table 4-5 indicates that the change in the mAs or the adjustment in the vertical tube stand distance when the angle of the central ray is between 5° and 15° is minor. However, to be accurate in the performance of your procedure, the recommended changes should be made. At a central ray angulation of 20° and above, the recommended changes in Table 4-5 must be made if the radiographic density of the recorded image is to be maintained. Note that the mAs increases or the adjustments in the vertical tube stand distance are not directly proportional to the central angle changes. A 5° central ray change from a 10° to a 15° angle will require an increased mAs change of only 4.1% or a vertical distance change of .80" (2 cm) to maintain the radiographic density of the recorded image. The same 5° central ray change from a 30° to a 35° angle will require an increased mAs change of 10.42% or a vertical distance change of 1.8" (4.57 cm) to maintain the radiographic density of the recorded image. (**Perform Experiment 41.**)

Darkroom Procedures

Darkroom procedures include the care, maintenance, and storage of the radiographic film, film holders, safelights, and other materials utilized for the processing of radiographic films. Film processing will be introduced separately, although there will be many factors related to these total activities that will crossover between this section and film processing.

The Darkroom

The darkroom or film processing room is a specially designed and constructed room for the proper storage and maintenance of radiographic films, as well as for the handling of films in the process of loading and unloading cassettes and the processing of the exposed radiographic film. Ideally, the darkroom facility is centrally located within the radiology department and usually has several passboxes built into its walls that allow for the passage of cassettes into and out of the darkroom directly to the x-ray examination rooms. Because of its sensitivity to visible light and to x-radiation, radiographic film must be protected from both sources of exposure. In fact, exposed radiographic film has even greater sensitivity to additional exposure than an unexposed film.

During construction, darkroom walls are lead-lined to protect the stored film from sources of radiation. Darkrooms are also specially designed to prevent the entrance of extraneous light sources. Radiographic fog, an unwanted form of radiographic density, can occur from either light sources or radiation sources reaching the film. Therefore, it is important that all areas, joints, ceiling tiles, door jams, electrical fixtures and outlets, plumbing connections, the various film processing equipment, and so forth, and especially the darkroom entrance, be light- and radiation-proof. Since the construction of the darkroom facility requires that the walls be lead-lined, potential radiation leaks into the darkroom are rare. However, light entering the darkroom from the outside is a major source of radiographic fog. Suitable testing of the light-proof security of the darkroom should be conducted prior to its utilization for the storage or processing of radiographic film and at regular intervals thereafter.

Testing for Light Leaks

One of the best methods to test the light-proof security of a darkroom is to adapt fully to the darkened environment of the darkroom facility (a minimum of 15–20 min) and to turn out all of the darkroom lighting including the safelights. You should now be in a totally dark environment. Carefully look for light leaks starting with the ceiling, especially at the joints where the ceiling meets with the walls and the corners of the room. Next, look at the electrical outlets and fixtures for any light leaks where they are connected to the walls or ceiling. Check the location of the plumbing connections for any light leaks. Continue your investigation by checking for light leaks around the perimeters of the passboxes where they emerge through the walls, as well as around the doors of the passboxes and where the film processing units join with the darkroom wall. Finally, check around the darkroom entrance area for any light leaks. (**Perform Experiment 42.**) If any sources of extraneous light are discovered, report it to your supervisor and make arrangements to correct the light leak immediately. The most common places for extraneous light leaks into the darkroom are around the passboxes, the perimeter areas around the film processing units, and especially around the darkroom entrances.

Darkroom Entrances

In larger departments, a labyrinth entrance may be used. A properly designed labyrinth eliminates the need for darkroom doors and, constructed with a 180° turn from the outside of the darkroom, this type of entrance also eliminates the possibility of extraneous light leaks. This type of entrance, however, takes up considerable space and increases the costs of construction and, therefore, may not be the entrance of choice for many departments. Additional elaborate darkroom entrances include the construction of a revolving-door entrance that effectively prevents light leaks into the darkroom. Less costly entrances include a double-door entrance in which access to the darkroom is accomplished by opening an outer door enabling you to enter a darkroom anteroom. An electrical system of locks prevents you from opening the inner door leading into the darkroom itself until the outer door is completely closed. The locking devices are arranged to prevent both doors from being opened at the same time. In smaller departments, a single-door entrance is frequently employed. This type of entrance, although the least expensive and requiring minimal space, is also the greatest potential source of light leaks into the darkroom. The various materials used around the door frame where it meets with the door itself are a constant, potential source of light leaks as the materials deteriorate with age, wear out with usage, or change from their original position due to abuse and produce small gaps between them and the door frame, so that they no longer provide a snug, light-proof fit. The light source security of darkrooms should be checked routinely as part of the department's overall quality assurance program.

Darkroom Ventilation

With the use of chemical agents, it is important that a good ventilation system is provided for the darkroom. The ventilation system also provides the climate environment most favorable to the storage and maintenance of radiographic films. Radiographic films should be stored between 50°F and 70°F and in a 30% to 50% humidity range. Most darkrooms are provided with a positive pressure ventilation system. That is, the air flow and air pressure within the darkroom is greater than the air pressure outside the darkroom. This helps to keep the darkroom clean by reducing the amount of dirt, dust, and extraneous matter that can enter the darkroom every time the door is opened. Reducing the amount of dirt and dust entering the darkroom helps to reduce the potential for film artifacts associated with extraneous material entering into the cassettes during loading and unloading procedures.

Unexposed film is kept in large, light-proof film storage bins that hold up to several hundred films of different sizes. Many film storage bins have the added safety feature of an electrical connection with the darkroom door. These film storage bins cannot be opened unless the darkroom door is closed and any lights other than safelights are turned off. Additional radiographic film boxes should be stored on their sides and not placed on top of one another. Unexposed boxes of films placed on top of one another can produce pressure artifacts from the weight of the successive boxes lying on top of them. The expiration date, located on every box of film, should be considered when storing films. Films with the earliest expiration date should be used first. The proper care and storage of radiographic film and prevention of it being exposed to x-radiation and visible light are essential since the cost of radiographic film represents a considerable portion of the radiology department's budget.

Safelights

Actually, the term **darkroom** is not entirely correct. Procedures performed within the darkroom do not have to be done in complete darkness. There is a considerable amount of light allowed within the darkroom to assist personnel in the performance of their duties. However, because the radiographic film is sensitive to light, the illumination employed within the darkroom must meet certain requirements. As a group, the illumination devices used during procedures when unexposed or exposed film is being handled are referred to as **safelights.** Exposed films being removed from cassettes and being placed into the feeding trays of film processing units are up to five times more sensitive to light than is unexposed film. Darkroom lamps or safelights are employed to provide a maximum of "safe" light so that operators can see to perform their job with safety. Safelights must be employed in all procedures performed in the darkroom when radiographic film is being handled.

Safelights are fitted with specific, selective filters for use with different light-sensitive radiographic films. Blue-sensitive film employs an amber (Wratten 6-B) filter. This filter allows light having wavelengths longer than the spectral response of the blue-sensitive film to emerge from the filter to provide light for personnel to perform their procedures. Therefore, the radiographic film will not be affected by that light. Green-sensitive film should not be exposed to an amber (Wratten 6-B) filter. A filter tending more toward the red or reddish-brown spectrum of color is employed with green-sensitive film. This type of filter can also be used with blue-sensitive film and, therefore, is the most frequently employed filter used in darkrooms today.

Safelights are designed to be used with a specific wattage of light bulb. If the wattage of the light bulb used with the safelight exceeds the manufacturer's recommendations, the light emitted by the safelight may be excessive and cause fogging of the film. Safelights are also designed to be used at a specific distance from the tabletop working surface where the films are being handled. Safelights are designed as either direct or indirect illumination devices. The illumination from ceiling-mounted safelights is indirect. The light emitted by this type of safelight is directed toward the ceiling, bounces off the ceiling, and spreads out and diffuses before reaching the working surface of the darkroom. Direct illumination safelights are usually employed directly over the work area to provide a maximum, safe amount of light to assist the operator in performing the tasks of opening and closing cassettes, loading and unloading film, using film identification devices, and handling all materials in a safe and proper manner. Because the intensity of light is directed toward the work area surface, direct illumination safelights are designed to be used at a minimum of 3' from the work area surface. Mounting of direct illumination safelights closer than the manufacturer's recommended distance of 3' may cause fogging of the film. The safelight housing should be checked for light leaks, and the filter should be checked for any cracking caused by age or heat. Safelight devices found with any of these problems must be replaced.

Safelight Testing

The ultimate safety of the illumination of the work area surface of the darkroom and the evaluation of the safelight system employed requires that the system be tested. The best place to test the safelight system is the film-handling area where the cassettes are loaded and unloaded. This is the area that would be under the direct safelight illumination devices and would receive the greatest intensity of light during the darkroom procedures.

Expose a film to a small exposure in order to sensitize the film. Remember, exposed film is approximately five times more sensitive to light than unexposed film. In the darkroom, open the cassette, and lay the exposed film in the middle of the work-area surface. Divide the film's length into five equal sections. Cover all but one fifth of the surface of the film with a firm piece of opaque material; a lead film divider or even a thick piece of cardboard can be used. With a pencil, marker, or lead numeral mark the surface of the exposed portion of the film with a number five. Using the darkroom timer, expose the film portion for 1 min to the safelight. At the end of the minute, move the opaque material in order to expose another one fifth of the film surface while covering the remainder of the film. A total of two fifths of the film surface is now being exposed to the safelight. Expose these sections for another minute, marking this new film section with a number four. Continue this process until all five sections of the film have been exposed to the safelight. Process the film and evaluate it on a radiographic illuminator. The radiographic film should have the numbers one through five written on it or be potentially recorded as an image if a lead number marker was used for the test. The numbers correspond to the time that the section was exposed to the safelight. The section marked with a 5 was exposed to the safelight for a period of 5 min; the section marked with a four was exposed for 4 min, and so forth. If the overall radiographic density of the film is constant throughout its entire length, it indicates that your safelight system is safe for at least a 5-min operational period. However, if the film demonstrates any radiographic density differences in any of the 5 film sections, you can determine exactly when the change occurred, and identify the period of time when the unsafe condition began.

A further evaluation of the safety of the safelight system is required for a practical application of this test. If no radiographic density change is apparent until the fourth or fifth minute, you must determine whether any darkroom procedure in which an exposed film will be under the safelight illumination will require 4 or 5 min to complete. In most instances, if there are no changes in the radiographic density of the tested film for up to 4 min or more, the overall safety of the safelight illumination system can be considered acceptable. If, however, the radiographic density of the film changes within the first 3 min of the test, the safelight illumination system cannot be considered safe and must be adjusted. Perhaps the total illumination provided in the darkroom is excessive, too many safelights are being used, and the removal of some of the devices will correct the problem. Perhaps the adjustment may require the simple reduction of the wattage of the light bulb used in the safelight device or an increase in the distance that the safelight is placed above the work-area surface. **(Perform Experiment 43.)**

Radiographic Film Processing

The most modern radiographic equipment and materials, ideal motion control, perfect positioning of the patient, and the selection of proper exposure factors ensure only an excellent latent image upon the radiographic film. All of these factors have no ultimate value if the methods employed to process the film and produce a permanent, visible image are carelessly or improperly performed. The radiographic quality of the recorded image ultimately owes its very existence to the standardization and quality control of the film processing procedure. The ultimate radiographic density of the recorded image is controlled not only by the amount of radiation reaching the film, since this only produces the latent image within the film emulsion, but also by the procedures utilized to process the latent image into a permanent, visible image. When properly exposed, radiographic films will provide for maximum visibility of the structures of interest only when they are properly processed using a standardized procedure that includes a regulated control of the time and temperature of the processing solutions. This declaration holds true whether the processing of the radiographic film is performed using manual or automated film processing procedures.

Manual Film Processing

The manual processing of radiographic film is introduced here primarily to present the historical value of the procedure and to contrast it with the quality controls present in the modern, automatic procedures for processing of radiographic film. Modern radiographic procedures and the workload associated with most radiology departments or physicians' offices could not be accomplished with the application of manual film processing. The student technologist may not be introduced to manual film processing during training and in all likelihood will not encounter it within the current practice of radiologic technology. Prior to semiautomated film processing introduced in the mid-1950s, manual processing was the only method available to produce the permanent radiographic image.

Manual processing used metal film frames appropriate to the film sizes to hold the film while it was manually advanced through the various processes of development, acid stop bath, fixing, fixer neutralizer, washing, and drying. Open tanks of chemicals within the darkroom often led to chemical spills, noxious chemical smells, and higher humidity levels. Quality control using manual film processing was difficult to achieve and frequently absent. There were so many variables in manual processing that the technologist found it hard to maintain a definitive degree of quality control. With manual processing, the simple but essential regulation of the temperature of the developer at 68°F and adherence to the recommended development time of 3 min were frequently beset with problems. Add to this the problems inherent in the control and maintenance of the various solutions' chemical activity, and manual processing often became a procedure of trial and error.

Some quality control did exist, and many radiology departments performed numerous tests to ensure the proper control of the developer temperature/time relationship and the examination of the various chemical solutions for chemical activity and exhaustion. However, a dedicated system of standardization and control was lacking in many radiology departments. By and large, manual processing proved to be a less than effective method for processing large quantities of films.

Adding to this lack of standardization and control was the frequently employed manual processing method referred to as **sight development.** Sight development was a method employed by unsure technologists when they suspected that their radiographic exposure was improper, and it was a prime example of the technical guessing game employed by many. Instead of controlling their film development using a standardized method of control (regulation of the developer temperature and the time of development), they would frequently pull the developing film from the developer tank and attempt to determine the radiographic density by sight by bringing the radiographic image close to the safelight. This process not only created radiographic fog from the close approximation of the film to the safelight but also created various stains upon the radiographic image caused by the exhaustion of developing chemicals adhering to the film surface when the film was taken out of the developer tank for extended periods of time. Overexposed films rapidly achieved great radiographic density and were pulled from the developer tank and placed into the acid stop bath long before the proper time of development, 3 min, had elapsed. This underdevelopment of the overexposed film produced a radiographic image lacking in appropriate radiographic contrast and demonstrating inadequate visualization of the structures of interest. Underexposed films were left in the developing solution for longer than the proper time of development in an attempt to increase the overall radiographic density of the image. Increases in the radiographic density of the image could potentially be achieved by overdevelopment, but again, the radiographic image suffered as the increased radiographic density resulted from a form of chemical (developer) fog. In either case, sight development eliminated what quality controls manual film processing possessed and did not produce recorded images of radiographic quality.

With manual film processing, a number of additional human errors associated with the mishandling of the film when placing it on the film frames or improper handling of the film frames when placing them into the various chemical solution tanks produced a number of film artifacts within the recorded image. The subject of film artifacts is introduced later in this chapter. (**Perform Experiment 44.** Experiment 44 can only be performed at facilities that maintain manual film processing equipment.)

Automatic Film Processing

Semiautomated film processing was introduced in the mid-1950s. These early devices still used metal frame film holders, but after the film was placed on a frame by the operator, the frame was automatically advanced through the film processing system. The time required to process the film fully to a dry radiograph ready to be removed from the film frame holder was approximately 50 min. This was quite a delay, especially in those instances when the examination involved an emergency or required the patient to wait until the examination could be reviewed to determine if any additional procedures would have to be performed. Therefore, the semiautomated processing unit was mainly used for the processing of routine examinations. Manual film processing was still employed for radiographs that required an immediate "wet reading."

In the early 1960s, a fully automated system for the processing of radiographic film was introduced using a roller feed system. The early models of these fully automated systems required as little as 15 min to complete the entire film processing procedure and was a significant advance over the semiautomated systems or the use of manual film processing. In addition, these systems provided for considerable standardization of the control of the solution temperatures and the activity of the chemical solutions, as well as strict adherence to a regulated control of developing time required for the processing of the films. The use of automated systems for the processing of films also significantly reduced the human errors associated with the production of film artifacts by eliminating the need to use metal film frame holders and the need to advance the film manually through the various solutions and processes associated with the film processing procedure. The radiographic film was simply fed into a "feeding tray," and the film was automatically advanced by a series of roller systems through the various chemical solutions and processes until the dry, fully processed radiograph emerged from the processor. An additional advantage of the automated system was that all of the chemicals and most of the plumbing and controls for the chemical solutions were located outside of the darkroom. This reduced the chemical smell and humidity associated with the processing chemical solutions located within the darkroom. The only portion of the automated processor located inside the darkroom was the film loading tray. The location of the processor outside of the darkroom made for greater ease in the cleaning and maintenance of the system and also created additional working space within the darkroom.

In the early 1970s, processors that could perform the film processing procedure in as little as 7.5 min were designed. By the mid-1970s, improvements in the design and operation of these processors enabled films to be processed in as little as 3 min, and more recently, processors that can perform the entire film processing procedure in as little as 90 sec have replaced most of the previous models. The initial introduction of automated film processing was limited to large radiology departments with budgets sufficient to purchase these expensive processors. Over the years, the processors became smaller and faster, and the cost became more affordable. In addition, the regulation of the various processes required for the film processing procedure has become better controlled.

There are many distinct advantages to automated film processing, not the least of which is the accurate and consistent control and regulation of the temperatures of the solutions. The need for a more accurate control of the temperatures of the solutions in automated film processing becomes obvious when one recognizes that manual film processing recommended a developer temperature of 68°F, and automatic processors recommend a developer temperature in the range of 92°F to 96°F. Most chemical activity is affected by higher temperatures, and the increased temperatures associated with automated film processing are no exceptions. The impact on the development process and the radiographic density of the recorded image with the difference of a few degrees of temperature in a 68°F solution when the total time of development is 3 min compared with the impact produced by the difference of a few degrees of temperature in a 94°F solution when the time of development is measured in seconds is obvious. The accurate control of the temperature of the chemical solutions in an automatic processor is critical to the consistent quality of the processed radiograph. The regulation of the development time when the total time of development is so short is also critical and in the automatic processor is accurately controlled by the roller transport system, which rotates at a specific speed and advances the film through the solutions within a very narrow, defined period of time. The chemical activity level of the processing solutions is also regulated by the automatic processor. The processor automatically replenishes the chemical solutions to maintain the required level of chemical activity for each processing solution. The human error so prevalent with manual film processing and the mishandling of the film during the various processing steps leading to film artifacts are virtually eliminated since the film is automatically advanced through the processor, and the processing steps occur within the machine and are not accessible to interference or mishandling.

The automatic processor enables the technologist to standardize film processing and to develop and implement a quality assurance program of processor control that allows for an examination and evaluation of past and current performance, as well as for a prediction of future performance related to a specified standard of operation. Implementation of a quality assurance program for the processing of films enables the technologist to depend on the processing system and to anticipate and/or eliminate problems as they occur or even to take steps to prevent their occurrence. A quality assurance program associated with the automatic processing of films can bring about a standardization of control and evaluation of the system that was not possible with previous film processing systems. The automated processing of films and the quality control of these processes removes one very large variable that the technologist had to consider related to the production of a recorded image of radiographic quality. (**Perform Experiment 45.** Experiment 45 can only be performed if automatic processing equipment is available that students can adjust and use without interfering with the normal operation of the radiology department or adversely affecting its quality assurance program.)

Basic Automatic Processor Control

Even with the best of the automated systems, the human element is never completely eliminated, so the importance of establishing standards of operation and the implementation of a quality assurance program for the processing of films cannot be overemphasized. Without detailing the many elements of a specific quality assurance program, the following basic factors are essential to the standardization of your film processing system. A more detailed analysis of quality control is introduced in Chapter 6.

1. Follow the processor manufacturer's recommendations related to the following:
   - Daily/weekly/periodic schedule of routine clean-up of the roller systems and crossover rollers
   - Selection of the proper solutions for the processor
   - Adherence to the routine, regular maintenance schedule and the upkeep of accurate maintenance records
2. Maintain the proper concentration of chemicals:
   - Follow the instructions on the solution container.
   - Mix or dilute concentrated solutions accurately.
   - Store chemicals in a cool, darkened environment.

- Bleed the solution tubing to eliminate air bubbles when changing or replacing solutions.
- Check the replenishment solutions and adjust the replenishment rates according to the established standards.

3. Maintain strict temperature control by evaluating and regulating the internal tank solution temperature, as well as observing and recording the temperatures identified by the external processor gauges.
4. Check for processor operation:
   - Keep a supply of parts that frequently require replacement on hand and available.
   - Use the proper processor parts for repairs or replacement. It is false economy and may cause processor damage to attempt to interchange parts that are not specifically intended or designed for that purpose.
   - All seals within the processor should be checked periodically for leaks or wear and replaced as needed.
   - All compartments and processor cover seals should be checked periodically for wear or damage. Failure of any of these seals can create light leaks, which would be disastrous to the processing of films.

Many radiology departments and physicians' offices maintain contracts with film processor maintenance companies to clean, service, and maintain their processing equipment. It is important that the service contract include elements of quality control and that the activity and services provided by the maintenance company comply with and adhere to the strict quality assurance standards that you have established for your department. A quality assurance program related to the standardization, control, and evaluation of the radiographic film processing is another example of the application of the principles of ALARA. Properly handled and processed films will eliminate the need for repeat examinations and, therefore, ultimately will result in a reduction of patient exposure.

Radiographic Artifacts

Radiographic artifacts frequently influence the radiographic density of the image and are often associated with the handling and processing of the radiographic film. This is why this subject is introduced in this chapter. An artifact is basically a defect in the film itself. It is the unwanted and unexpected appearance of something upon or within the radiographic image that was not intended, expected, or designed. Radiographic artifacts can be produced through human error, chemical errors, or equipment or processor errors. We will review a list of the more common radiographic artifacts, describing them, identifying their causes, and recommending methods to eliminate or prevent their occurrence. Remember that the exposed radiographic film is five times more sensitive to radiation exposure, external light sources, and mishandling than is unexposed film. Many film artifacts that were so frequently observed with manual film processing have been eliminated by the use of automated film processing procedures and the application of quality assurance programs that include film processing controls.

Artifacts Caused by Humans

Crinkle/Crescent Marks

Crinkle marks are crescent-shaped artifacts located within the radiographic image and are produced by the mishandling of the film when it is placed into or removed from the film storage bin or cassette, or when it is being placed upon the film feeding tray of the automatic processor. These artifacts are caused by the bending or pinching of the film during handling, are more common in the larger film sizes, and more frequently occur near the edges of the film than in the center of the film. Technologists must be careful to grasp, handle, and move the film with caution and avoid crimping it or pinching it between the fingers.

Smudge/Fingerprint Marks

Smudges and/or fingerprint marks will be produced on the film if the film is handled when the hands are wet or when they have an oily or chemical residue on them. Films should only be handled when the hands are clean (free from oils or chemical residues) and dry.

Static Electricity Discharges

Static electricity discharges on the film appear as various patterns, including tree-like streaks, crown-like streaks, or odd-shaped smudges frequently circular in appearance. These discharges are caused by low humidity in the darkroom and the build-up of static charges on the film during handling. Taking the film out of the film storage bin or the cassette too rapidly and in a sliding manner where it rubs across other surfaces can cause a build-up of static electricity, which, unless grounded, will discharge upon the film, creating a pattern of artifacts. Methods of reducing static electricity discharges upon the film include maintaining the recommended humidity level within the darkroom, grounding the countertop surfaces in the work areas, and designing the floor of the darkroom with antistatic materials or embedding a wire mesh during the construction of the darkroom floor in order to ground the operator during the performance of processing procedures. The routine maintenance of intensifying screens includes their cleaning with a recommended cleaning solution. These cleaning solutions have antistatic agents in them to assist in the elimination of static electricity discharge upon the film.

Artifacts Caused by Chemicals

The majority of artifacts caused by chemicals have been eliminated or decreased as a result of automated processing procedures. However, a few of the chemical artifacts may still occur when films are improperly processed or when the quality controls of the automated system have broken down. Many of these artifacts appear as stains or chemical fogging, which adds to the radiographic density of the radiographic image.

Brown Stains

Brown stains throughout the radiographic image will result from the presence of exhausted developer in the processor. The concentrations of the developer replenishing solution or the control of the developer replenishment rate must be adjusted so that the chemical activity of the developer solution is at the level required for the proper development of the radiographic image.

Yellow Stains

Yellow stains throughout the radiographic image will result from the presence of exhausted fixer in the processor. The concentrations of the fixer-replenishing solution or the control of the fixer replenishment rate must be adjusted so that the chemical activity of the fixer solution is at the level required for the proper fixation of the radiographic image.

Milky-white Stains

A milky-white appearance over the radiographic image will result from an inadequate rinsing and washing off of the processing chemicals during the final stages of the processing cycle. The rate of water exchange and its proper circulation within the wash tank must be checked and adjusted in order to eliminate this problem.

Artifacts Caused by Equipment or Processors

Streaks/Streakiness

A general appearance of streaks or streakiness within the radiographic image can be caused by the poor circulation of the processing solutions within the various tanks of the processor. Adjusting the replenishment rates of the solutions and checking the roller transport systems through each tank to see that they are operating smoothly and consistently with the requirements of their operations will help to eliminate this problem.

Hesitation Marks

Small lines of increased radiographic density across the film may be caused by slight hesitations or stoppage of the film roller transport system. The hesitation of the movement of the film through the film roller transport system may also be due to uneven pressures between the rollers. A check of the roller transport system of each section of the processor and the adjustment of the rollers or, if necessary, replacement of the rollers may be needed.

Guide Shoe Marks

These marks are frequently associated with scratches upon the film surface in longitudinal patterns. They are caused by the film guides in the crossover rollers being misaligned and causing excess pressure on the film surface when guiding the film into the next section of the film processing system. This can be corrected by adjusting the film guides so that they do not cause excessive pressure upon the surface of the film. In severe cases of misalignment, the entire crossover roller system may have to be replaced.

White Spots/Rough Surface

White spots or a rough, uneven surface of the film may be caused by a processor that has not been cleaned or maintained according to the maintenance schedule. Small particles of dust or dirt can get into the system and adhere to the film as it is transported through the system. If occurring within the developing section, the extraneous material adhering to the film surface will prevent the chemicals from coming into contact with those areas of the film causing a white, spotty appearance throughout the radiographic image. The rough, uneven surface of the film is created when these extraneous materials remain on the film and are embedded into its surface when the film goes through the washing and drying cycle. Most of these problems can be eliminated by the conscientious care and maintenance of the processor and the routine cleaning of the internal sections according to a daily/weekly schedule. White spots throughout the radiographic image can also be caused by dirt and dust on the surface of the intensifying screens or by the pitting of the surface of the intensifying screen as a result of age and abuse. This problem can be corrected by cleaning the intensifying screens; however, if the surface of the screen is pitted or cracked, it will be necessary to replace the screen.

Radiographic Fog

Film artifacts are also caused by radiographic fog. The undesired and uncontrolled increase in radiographic density produced by fog is a major problem hampering the control of radiographic quality. Radiographic fog is caused by a number of factors that include the following: exposure to extraneous light (cassette leaks, film storage-bin leaks, darkroom light leaks, improper safelights), exposure to unwanted radiation, storage of the film at high temperatures, and pressure fog created by storing boxes of film on top of one another.

Anode Heel Effect

Although not a factor related to the actual production of radiographic density, the influence of the **anode heel effect** on the radiographic image should be included in this analysis because it is a factor that affects the overall uniformity of the intensity of the x-ray beam emerging from the x-ray tube. Unlike the other factors you have investigated, the anode heel effect is not a factor that you can control or regulate. However, by recognizing its potential influence on the radiographic density of the recorded image, you can plan your procedures so that its influence is minimized when it is detrimental to radiographic quality. When its influence can be used to improve the recorded image, the technologist can apply the appropriate techniques to accomplish this objective.

The rate of emission (intensity) of the x-ray beam is not uniform as it emerges from the x-ray tube along its longitudinal axis from the cathode to the anode end of the tube. As the x-ray beam spreads further as it travels between the x-ray tube and the patient, the uneven intensity of the emission rate that occurs between the cathode and anode ends of the beam becomes greater. If you were to measure the rate of emission (intensity) of the radiation using a 40" (102 cm) SID at the extremities of the beam coverage when the beam limitation device is opened to its widest coverage, you would discover a difference in the intensity of the radiation from one edge of the beam to the other. Depending on the type and characteristics of the x-ray tube producing the x-rays, this difference can be minimal or significant. The anode heel effect is an inherent characteristic associated with the design of the tube occurring during manufacture. The anode heel effect differs from manufacturer to manufacturer and between different x-ray tubes from the same manufacturer and depends on the size of the focus and the actual angle of the target of the x-ray tube's anode. The angle of the target area of the x-ray tube's anode is designed by the manufacturer to produce a smaller effective focus in order to improve the recorded detail of the recorded image. Different target angles produce different anode heel effect influences.

As the electron stream strikes the target area of the anode, x-rays are produced at every point along the length and width of the surface of the anode area struck by the electrons. The x-rays produced by this process are emitted from the target area of the anode and begin to radiate in all directions from that point. The majority of the x-rays produced in this manner are absorbed by the lead shielding located in the x-ray tube housing. At the lower end of the tube housing is a small area referred to as the **tube window.** The tube window, aligned in a direction toward where the patient will be located, allows for a limited, defined beam of radiation to emerge from the x-ray tube. The x-rays emerging from the tube window form the useful beam employed for the production of the radiographic image.

A measurement of the x-ray beam intensity within the beam of radiation emerging from the x-ray tube will demonstrate that the quantity of radiation measured at different locations within the beam is not equal. If the length of the beam's coverage at the surface of a radiographic film measures 17" (44 cm), you would find very little difference in the beam's intensity from the central ray to the measurements taken within the beam located several inches on either side of the central ray. However, if you were to measure the two extremes of the beam's coverage at the surface of the film from one end of the film to the other, you would notice an obvious and, depending on the x-ray tube design, a potentially substantial difference in the beam's intensity from one edge of the beam to the other.

This influence occurs because, as the angle formed between the surface (plane) of the x-ray tube target and the beam of radiation produced at that point increases, the intensity of the radiation produced also increases. In x-ray tubes, the angle of emission of the x-ray beam becomes greater for those beams that travel out of the tube in a direction toward the cathode end of the tube. Therefore, the intensity of the radiation emitted from the tube traveling in a direction toward the cathode end of the tube will be greater than the intensity of the radiation projected toward the anode end of the tube (Figure 4-5). In Figure 4-5, you can see that the central ray of the x-ray beam produces a 20° angle of emission with the surface (plane) of the target. The portion of the beam traveling toward the anode end of the x-ray tube produces an emission angle of only 4°. In the portion of the beam traveling toward the cathode end of the x-ray tube the angle of emission formed with the target increases to an emission angle of 36°. The intensity of the x-ray beam in Figure 4-5 demonstrates a significant variation in the intensity of radiation within the length of the beam: from a 56% intensity at the anode end of the tube to a 102% intensity at the cathode end of the tube. Checking your tube housing, you can easily determine which side of the tube is the cathode end. The cathode side of the x-ray tube will be identified with a minus (–) symbol marked on the tube housing; the anode side of the x-ray tube will be marked with a plus (+) symbol.

In those procedures in which the anode heel effect is recorded on the radiographic film, one end of the film receives a greater amount of radiation and will record a greater radiographic density. As a result of the anode heel effect, the end of the film adjacent to the cathode end of the x-ray tube would receive the greater intensity of radiation. The anode end of the x-ray tube would receive less intensity of radiation, and, therefore, the radiographic density of the recorded image located at this end of the film would be less. (**Perform Experiment 46.**)

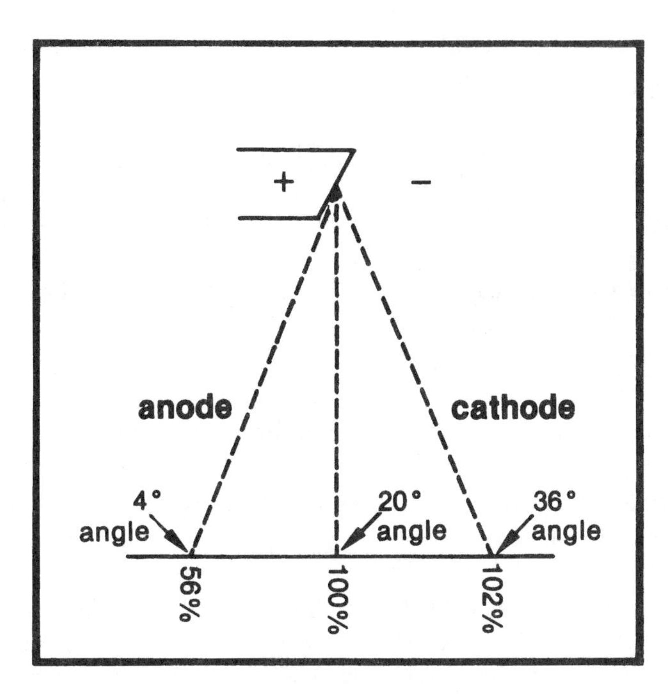

Figure 4-5
Diagram of the Anode Heel Effect

This effect is more severe and noticeable when the total length of the available beam of radiation is used during a procedure. It is less noticeable when the total area covered by the emission of radiation is limited to only the central axis of the beam. It must be noted that the anode heel effect is not apparent or visible in the majority of radiographic examinations. This is because the actual size of the beam of radiation used for all examinations is limited to the area covering the structures of interest. In many procedures, the actual size of the beam of radiation utilized represents only a small portion of the total potential beam size. Therefore, the differences in the intensity of the radiation represented by the extreme ends of the beam (cathode to anode) are not recorded within the area included on the radiographic film.

There are several examinations, however, where the anode heel effect will be visible on the radiograph, and its influence may be undesirable. In those examinations where it is determined that the anode heel effect may influence the overall radiographic density of the recorded image, the technologist can utilize certain techniques to minimize its influence and, in specific instances, can even utilize its influence to advantage. In either case, the technologist is simply attempting to produce a recorded image that possesses a similar radiographic density throughout the entire radiographic film. The following considerations would apply to the anode heel effect in order to minimize its influence within the recorded image or to utilize its influence to advantage.

Application of the Anode Heel Effect

1. Remember that it is the *cathode* end of the x-ray tube that possesses the greater intensity of radiation.
2. The anode heel effect is only an influence on the radiographic density of your recorded image if portions of the beam recording the image include the peripheral portions (extreme ends) of the x-ray beam.

- At a routine 40" (102 cm) SID, the anode heel effect is imperceptible on radiographic films measuring 10" by 12" (25 cm × 30 cm) or less. If you use larger film sizes, such as a 14" by 17" (36 cm × 43 cm) film, at this distance you must, however, consider this influence, since a significant amount of the peripheral beam will be included in the recorded image.
- At an increased SID of 72" (183 cm), the influence will not be perceptible even on the larger film. At increased SID, considerable collimation (beam limitation) is employed, and only the central portion of the beam will be required to record the image on even a 14" by 17" (36 cm × 43 cm) film.
- At a reduced SID below 40" (102 cm), the influence may be perceptible on a smaller film size such as a 10" by 12" (25 cm × 30 cm) film, since at the reduced distance, you must open the collimator more in order to record the image, thereby using more of the peripheral beam.

3. The anode heel effect can be used to advantage only in those procedures that examine structures or body parts that exhibit appreciable differences in thickness and/or tissue opacity along their longitudinal axis from one end of the part to the other. The following list identifies some of the examinations that can generally take advantage of the anode heel effect when performed on adult patients. Note that all the procedures represent examinations that would require a large film size for the examination so that the beam of radiation included within the recorded image will include the extremities of the beam and, therefore, the significant differences in the intensity of the beam:
   - Examination of the femur (thicker at the hip)
   - Examination of the leg (thicker at the knee)
   - Examination of the humerus (thicker at the shoulder)
   - Examination of the forearm (thicker at the elbow)
   - Examination of the thoracic spine (thicker toward the abdomen)
   - Examination of the lumbosacral spine (thicker toward the sacrum)
4. To use the influence of the anode heel effect to advantage, identify the cathode end of the x-ray tube, and position the structure of interest or part to be examined so that the thicker or more opaque portion is placed toward the direction of the cathode end of the tube.

In those procedures where the anode heel effect is recorded on the film, applying the above principles enables you to distribute more uniformly the intensity of radiation and produce an image possessing a more uniform radiographic density throughout the recorded image. Improperly applying the influence of the anode heel effect or ignoring its influence may produce an image with excessive radiographic density at one end and inadequate radiographic density at the opposite end. (**Perform Experiment 47.**)

Orthopedic Cast Material

In the examination of a large number of skeletal injuries, the technologist is faced with the problem of examining the same body part following the application of a thick plaster cast. It is obvious that the selection of exposure factors for this postreduction examination will have to differ from that of the original examination, but in what manner and by what amounts?

Many of the answers related to the problem of performing a radiographic examination through a recently applied plaster cast come from a review of some of the principles related to the production and maintenance of radiographic density examined earlier in this chapter. The key to dealing with this problem is to remember that you need to demonstrate the same structures of interest and produce the same radiographic density and contrast (visibility of the image) on the postreduction examination as you did on your original examination. Ideally, you want to produce an image possessing a radiographic density and scale of contrast similar to the original film taken without the cast. A review of the **patient factors** of tissue thickness and tissue opacity will help you to understand the solution.

When we studied the factor of tissue thickness, we learned that tissue thickness is equated with the quantity of tissue to be examined. We discovered that a thicker body part (a greater quantity of tissue) requires a greater quantity of radiation to produce a desired radiographic density than does a thinner body part. We compared the thickness of the thigh of a 7-year-old, measuring 4" (10 cm) in diameter, with that of a young adult's, measuring 10" (25 cm) in diameter, and indicated that an increase in quantity of exposure (mAs) would be necessary in order to maintain the radiographic density of the image in the adult examination because of the increase in the quantity of tissue to be examined.

A similar situation exists with the postreduction examination. The actual measurement of the thickness of the part to be examined has increased because of the addition of the plaster cast material. This should automatically indicate to you that an increase in the quantity of radiation is necessary. Depending on the new total thickness of the part including the cast, compared with the original thickness of the part, it will be necessary to increase the quantity (mAs) of the radiation in order to maintain the radiographic density of the recorded image. Another consideration related to a change in the thickness of the part may result from the new position of the part within the cast compared with the original position of the part. In the reduction of fractures, the physician frequently has to place the structures into positions of different angles or directions in order to correct for the injury. The injured part is often cast in this new position in order to maintain the reduction of the fracture. When examining the part as a postreduction procedure, it may be impossible to achieve the same positions for the examination, and it may be necessary to pass through much greater thickness as a result of the different position of the structures within the newly applied cast. It is important for the technologist to identify the greatest thickness of the part that he or she has to examine in order to determine what increase in mAs is required to maintain the radiographic density of the image.

However, we have only addressed half of the problem and identified part of the solution. It is also necessary to determine if any changes have occurred in the tissue opacity of the part. In our investigation of the patient factors, we discovered that the higher the tissue opacity, the greater the tissue's ability to absorb radiation. As the tissue opacity of the part or structures of interest increases, we determined that it is necessary to increase the penetrating ability of the beam. With the application of the plaster cast, not only has the total thickness of the part been increased, but additional opacities to the beam of radiation, consisting of water, gauze and cotton binding, and plaster materials made of an inorganic mixture of lime (including calcium carbonate), have been added to the part. The tissue opacity of the part to be examined has been increased by the additional opacities of the casting materials, including the large quantities of water used in the preparation of the cast. The technologist must also compensate for this increase. Since the material of the cast has added to the opacity of the structure, it is necessary to increase the penetrating ability of the beam in order to examine the structure properly and to maintain the radiographic contrast necessary for the proper visibility of the image.

Therefore, in addition to the mAs increase required as a result of the increased thickness of the part, an increase in the kilovoltage to compensate for the increased opacity of the part is also necessary. You will find that increasing the mAs without an appropriate increase in the kilovoltage produces a less than satisfactory result. Remember, it is your goal not only to produce a similar radiographic density, but also to maintain a similar contrast scale within the recorded image. A formula to compensate for the application of an orthopedic cast has been developed. The new exposure factors required after the application of an orthopedic cast are as follows:

New Exposure = 2 × mAs + 10% ↑ kVp

This formula is intended to be used with the postreduction "wet" cast. The doubling of the mAs is used to compensate for the increased thickness of the part appropriate to an average cast thickness. The 10% increase in the kilovoltage represents the compensation required by the increased opacity of the part. Note that the kilovoltage increase is a percentage of the original and not a specific number. At 60 kVp, an increase of 10% would represent 6 kVp, while at 80 kVp, an increase of 10% would require an increase of 8 kVp. You should recall in our investigation of kilovoltage, we learned that the necessary increase in kilovoltage to produce a visible change within the recorded image differs depending on the kilovoltage selected. At the lower kilovoltage range, a smaller increase in the kilovoltage will produce a visible change. At the higher kilovoltage range, a greater increase is necessary to produce a visible change. Increases in the kilovoltage produce the necessary increases in the penetrating ability of the beam and help to compensate for the screen intensification differences that contribute to the maintenance of the radiographic density when the kilovoltage is increased. (**Perform Experiment 48.**)

For follow-up examinations of orthopedic injuries with "dry" casts, you will find that you have to adjust the formula and the exposure compensation slightly. In drying, the cast loses some of its opacity (mass per unit volume) primarily by the evaporation of most of the water used in the preparation and application of the cast. Therefore, you can normally examine a dry cast with an increase in the mAs alone. Examination of a dry cast can frequently be performed with a doubling of the mAs.

For abnormally large or thick casts or casts that incorporate additional layering of wood or metal to help maintain the position of the part, it would be necessary to adjust your formula appropriately. In these examinations, it may be necessary to triple or even quadruple the original mAs in order to maintain the radiographic density of the image.

Beam Filtration

Filtration in radiography is used primarily to "harden" the beam of radiation. Hardening the beam will significantly reduce the radiation dosage to the patient's skin and superficial tissues. The proper use of beam filtration is a positive example of the principles of ALARA. Filtration performs this function by attenuating (absorbing) a portion of the radiation intensity emitted from the x-ray tube. The filtration of the beam of radiation performs this attenuation process in a very selective and useful manner.

As we discovered in our investigation of the kilovoltage factor, x-rays are produced in a wide range of wavelengths and energy levels. Although x-rays are produced in a wide range of wavelengths, from 0.01 angstroms (Å) to more than 1,000 Å, the useful diagnostic range of x-rays are limited to the wavelength range from approximately 0.1 Å to 0.5 Å. The length of an angstrom unit is equal to one ten-billionth of a meter (10^{-10}). The shorter the wavelength, the greater the penetrating ability of the radiation. The wavelength of the radiation and therefore the penetrating ability of the beam are controlled by the kilovoltage applied to the x-ray tube during exposure. Regardless of the kilovoltage selected, the emerging beam of radiation will contain a wide range of wavelengths, including a percentage of long wavelengths having less penetrating ability. Much of the longer wavelength radiation produced by this process has insufficient energy to penetrate the body part or structures of interest and is therefore unable to contribute to the process of recording an image upon a radiographic film. Such radiation is useless to the production of a radiographic image but is biologically ionizing and, therefore, potentially harmful to living tissue.

Our analysis of beam filtration will deal with the investigation of the total filtration in the beam. Total filtration includes the inherent and added filtration in the beam. **Inherent filtration** refers to the filtration of the beam prior to its emergence from the tube housing. **Added filtration** includes those materials, usually thin aluminum metal sheets, placed in the path of the x-ray beam after it emerges from the x-ray tube housing but before it reaches the patient.

There are several reasons why beam filtration is positively applied in diagnostic radiology. It would be inappropriate to subject the patient's body tissues to low energy levels of radiation that would not contribute to the radiographic process. Only those energy levels of radiation potentially capable of participating in the process of producing a radiographic image should be allowed to reach the patient. Low-energy radiation directed toward the patient can be absorbed by the patient's skin and may be capable of penetrating to depths of more than a centimeter of tissue, yet have insufficient energy to travel through the patient's body and produce a radiographic image. These radiations are absorbed by the soft tissues of the body including the skin and are capable of producing potentially serious biological effects.

Inherent Filtration

All x-ray tubes have a level of inherent filtration related to the glass envelope surrounding the tube components. In general-purpose x-ray tubes, the inherent filtration level approximates the filtration equivalent of 0.5 mm aluminum. With age and extended tube use, the inherent filtration tends to increase due to the deposits of tungsten elements vaporized from the filament and tube target that are deposited upon the inside of the glass envelope. Some special-purpose x-ray tubes, including those operating at low kilovoltage levels such as mammography tubes, have special thin tube windows often made of beryllium, which reduce the inherent filtration to only 0.1 mm aluminum.

Added Filtration

In diagnostic radiography, it is important to filter the beam of radiation in order to absorb the less penetrating, longer wavelength radiation that will not participate in or contribute to the quality of the radiographic image but could be absorbed by the patient's skin and soft tissues. Filters made of varying thickness of aluminum are placed within the beam of radiation emerging from the tube housing to accomplish that objective and are referred to as **added filtration.** Filtration used in radiography has the positive potential of selective absorption when interacting with a beam of radiation. That is, depending on the filters' thickness and the material they are made of, they will filter out only those wavelengths of radiation that are incapable of penetrating through them. The energy of this longer wavelength radiation represents the lower end of the energy spectrum occurring within the beam of radiation. At the same time, these added filters will permit the higher energy, more penetrating radiation to pass through them. This is referred to as **hardening** of the beam. By selectively absorbing the longer wavelength radiation and permitting the shorter wavelength, higher energy radiation to emerge within the beam of radiation directed toward the patient, the technologist has effectively reduced the patient's overall exposure to harmful radiation by a significant level. Those energy levels of radiation that can potentially penetrate through the patient's body and contribute to the formation of the radiographic image are allowed to emerge through the added filtration placed in the path of the x-ray beam.

In radiographic procedures employing less than 70 kVp, it is recommended that the total filtration (inherent and added filtration) in the beam be no less than the equivalent of 1.5 mm of aluminum. Above 70 kVp, the total filtration recommended should be no less than the equivalent of 2.5 mm of aluminum. Since radiographic equipment can operate over a wide range of kilovoltage, above and below 70 kVp, it would be impractical to adjust or change the total filtration every time a different examination is performed. Therefore, most radiographic installations maintain a constant total filtration in the beam of 2.5 mm aluminum equivalent.

There are few occasions when the technologist will find it necessary to adjust or change the filtration in the beam. However, it must be understood that since beam filtration attenuates the beam of radiation produced, it has an influence on the quantity (intensity) of the radiation emerging from the x-ray tube. As the filtration thickness or the atomic number of the filtration material used increases, the intensity of the emerging radiation directed toward the patient decreases. As a result, it is important to recognize that beam filtration does have an effect on radiographic density. By attenuating the intensity of the exposure and hardening the beam of radiation, filters actually reduce the exposure rate measured in the beam and, depending on the structures of interest being examined, could reduce the radiographic density of the recorded image. (**Perform Experiment 49.**)

Total filtration above 2.5 mm aluminum placed in the path of the x-ray beam during routine radiographic procedures is not recommended. As the thickness of the beam filtration increases, it eventually reaches a point of diminishing returns. That is, its usefulness compared to its potential negative effects on the quality of the radiographic image diminishes. Filtration above 2.5 mm aluminum will begin to attenuate or harden the beam too much. It will begin to absorb some of the wavelengths of radiation necessary for the production of the contrast scale so important to the visibility of the radiographic image.

Compensating Filters

Another application of filters in selected radiographic procedures or projections is the use of compensating filters. Compensating filters are specially designed filters made of aluminum or other materials having different designs depending on the purpose of their application. The most commonly employed design of a compensating filter is a wedge-shaped filter (thicker at one end, thinner at the other end). The filter gradually decreases its thickness from one end to the other.

Compensating filters operate on the same principle of hardening the beam, but for a different practical purpose. These filters are added to the normal total filtration present within the beam of radiation. Since there is an additional attenuation of the beam, the overall intensity of radiation measured within the beam would be less. When employing wedge-shaped or other forms of compensating filters, the technologist will have to increase the exposure factors in order to produce and maintain the same radiographic density of the recorded image produced without the use of the special filter. There are many areas of the body that differ greatly in thickness or density from one end to the other, and it is difficult to demonstrate all of the structures of interest with proper radiographic density. If sufficient radiographic exposure is selected to demonstrate properly the thickest portion of the body part, those structures contained within the less thick portions of the part will frequently be "burned out"—possess so much radiographic density that it is impossible to visualize portions of the structures of interest. Consider the examination of the foot of an adult patient in the anteroposterior projection. A measurement of the thickness of the distal portions of the foot adjacent to the area of the toes would be considerably less thick than a measurement of the thickness of the foot in the area of the instep. As a result of this considerable difference in the thickness of the part from one end of the structure to the other, it is a difficult examination to perform related to the achievement of an image of radiographic quality. If the appropriate exposure factors are selected to demonstrate properly the distal end of the foot in order to visualize the distal portions of the metatarsal bones and the toes, the thicker areas of the foot adjacent to the instep would be recorded with insufficient radiographic density. On the other hand, if appropriate exposure factors are selected to demonstrate properly the proximal end of the foot adjacent to the instep in order to visualize the tarsal bones and the proximal metatarsals, the thinner areas of the foot adjacent to the toes would be recorded with excessive radiographic density.

Compensating filters can be used to compensate for the differences of tissue thickness or opacity within the body part in order to produce a recorded image possessing a more uniform radiographic density throughout. A wedge-shaped compensating filter can be effectively used in the anteroposterior projection of the foot to produce a more desirable radiographic image.

In practice, the wedge-shaped compensating filter is attached to the collimator or beam limitation device so that it lies parallel between the x-ray beam and the patient. Many collimators and other types of beam limitation devices have slots or grooves built into their lower edges to accommodate a variety of compensating filters. The proper placement of the compensating filter within the x-ray beam is essential to the success of the procedure. The thicker section of the filter must be placed in the direction of the thinner section of the body part to be examined. This will place the thinner section of the filter in the direction of the thicker section of the body part. The foot is an example of a structure that is examined more successfully with the use of a compensating filter. For a foot examination, the thicker portion of the wedge-shaped filter is placed in the direction of the toes and the thinner portion of the filter in the direction of the instep of the foot. The result of using the wedge-shaped compensating filter would be the production of a radiographic image of the foot with a more uniform radiographic density throughout its entire structural length. (**Perform Experiment 50.**)

The overall intensity of exposure when using a compensating filter would not have to be increased. However, it would be necessary to increase the kilovoltage by 6 kVp to 10 kVp in order to compensate for the radiation absorbed by the filter. Compensating filters can be purchased in a variety of shapes that are designed for specific body parts or examinations. The most commonly employed compensating filter is the wedge-shaped filter. Other examinations that could utilize a wedge-shaped compensating filter include the following:

- Anteroposterior and lateral projections of the cervicothoracic spine
- Anteroposterior and lateral projections of the thoracic spine
- Anteroposterior and lateral projections of the femur, leg, or humerus
- Anteroposterior and lateral projections of the sacrum and coccyx
- Axial (tangential) projections of the os calcis

Radiographic Density Summary

We have now investigated a number of factors contributing to and influencing the radiographic density of the recorded image. The major factors that control and influence radiographic density have been identified. We have examined the concept of proper density related to the photographic properties of the recorded image and discovered that proper radiographic density will demonstrate all of the structures of interest.

The emphasis in this chapter has been on correctly evaluating the radiographic image for a proper radiographic density and on investigating the factors that will enable you to achieve this objective. Additionally, we examined the many methods used to manipulate and adjust these factors in order to produce and maintain the proper radiographic density. It is important to understand that the manipulation and adjustment of the technical exposure factors should not be performed without a reason or purpose. That purpose should be to maintain or improve the overall radiographic quality of the recorded image. You should also recognize that the manipulation of exposure factors does not have to be a technical guessing game. A number of excellent concepts, principles, formulas, and materials have been introduced to assist you in selecting the proper exposure factors for any given set of circumstances.

By understanding the factors and principles presented in this chapter on radiographic density, the technologist will be able to select the appropriate manipulation of the exposure factors. However, it is only the experience gained by the application of these principles within a radiological setting that provides the self-confidence necessary to produce recorded images that are consistently of diagnostic and/or radiographic quality. As experience increases, the selection of materials and the adjustments in exposure factors necessary to produce successful radiographic images become easier to perform.

Other factors that influence radiographic density were not included in this chapter, as their overall influence is greater on the radiographic contrast of the recorded image. These include radiographic grids, beam limitation devices, and other methods designed to control and reduce the level of scattered radiation reaching your film. Scattered radiation produces an undesirable radiographic density known as **fog.** These factors are analyzed in detail in the next chapter, which introduces the other factor related to the photographic quality of the recorded image: radiographic contrast. The factors that we have investigated in this chapter are summarized below.

Patient Factors. The patient factors are perhaps the most important influence of all and the factors over which the technologist has the least control. It must be the first influence evaluated related to the production of proper radiographic density. A careful assessment of the structures of interest related to the thickness of the part and the composition of the tissues (tissue opacity) of the part must be performed before the selection of any radiographic materials or exposure factors.

Remember: For tissue thickness differences, adjust the quantity of radiation (mAs); for tissue opacity differences, adjust the penetrating ability of the radiation (kilovoltage).

Milliamperage. The milliamperage is a measurement of tube current, that is, the quantity of electrons flowing through the tube and striking the anode that are capable of producing x-rays.

Remember: The quantity of x-rays produced will be directly proportional to the milliamperage used.

Exposure Time. The time of exposure indicates the total period of time, measured in seconds, that x-rays are being produced.

Remember: The quantity of x-rays is directly proportional to the time of exposure selected.

Milliampere-Seconds. The product of milliamperage and time is mAs. Milliampere-seconds represents the total intensity of the beam and is the controlling factor of radiographic density.

Remember: The relationship of milliamperage and the time to mAs and the maintenance of radiographic density is inversely proportional. An increase in the milliamperage must be accompanied by a proportional decrease in the time of exposure if the radiographic density of the recorded image is to remain constant.

Source-Image Distance. The relationship of SID to radiographic density is an inverse relationship. That is, as the SID increases, the radiographic density of the recorded image will decrease.

Remember: Use the principles of the inverse square law formula or SID published tables when you change your SID from your original distance in order to determine the exposure factors to be utilized at the new distance. In this manner, you can be sure to maintain the radiographic density of the recorded image.

Kilovoltage. Although kilovoltage is better identified with the production and control of radio-graphic contrast, it also affects the radiographic density of the recorded image. As the kilovoltage increases, your radiographic density will also increase. However, the relationship between kilovoltage and radiographic density is not a proportional one.

Remember: The relationship expressed by the kVp/mAs 15% rule can be used to maintain your radiographic density when motion is a major consideration and when you want to reduce your time of exposure, or to produce a longer contrast scale within your recorded image.

Intensifying Screens. The intensification factor and screen speed of intensifying screens also influence the production, control, and maintenance of radiographic density.

Remember: The ability to select film-screen recording systems that possess different intensification factors and screen speed factors will enable you to employ a wide range of imaging materials to advantage. Higher speed systems may be advocated in specific examinations, such as those on infants where motion is of major concern. Slower speed systems may be advocated in examinations where the sharpness and resolution of the recorded details are of greatest interest.

Central Ray Angle/SID Relationship. The use of various degrees of central angulation is necessary in many radiographic procedures. As the central ray angle from the perpendicular increases, the SID will increase unless the vertical height of the tube stand is adjusted to maintain the original distance. Radiographic density decreases as the angle of the central ray is increased, and either the true SID distance must be maintained or the exposure factors must be adjusted to maintain the radiographic density.

Remember: A general rule of thumb suggests a reduction in the position of the vertical height of the tube stand by 1" (2.54 cm) for each 5° of central ray angulation. However, a more accurate control to maintain the radiographic density would be to utilize the factors identified in Table 4-5.

Darkroom Procedures. Radiographic fog caused by light leaks in the cassettes or within the darkroom during the handling and processing of the radiographic film will produce significant changes in the radiographic density of the recorded image.

Remember: Routine checking of the darkroom and the radiographic film holders for light leaks as well as testing of the safety of the safelight system within the darkroom will help to reduce and eliminate unwanted radiographic density resulting from the exposure of the film to extraneous light sources. The exposed film is up to five times more sensitive to light or radiation than is an unexposed film and, therefore, should be handled with that much more care and caution.

Film Processing. Proper processing of the radiographic film is essential to the radiographic quality of the visible image. The standardization of the processing procedures and the implementation of a quality assurance program for processor control are essential to the maintenance of radiographic quality.

Remember: Quality control begins with the maintenance of proper time and temperature processing controls and continues with the identification of processing standards and a conscientious testing program to evaluate every element of the process.

Radiographic Artifacts. The elimination of radiographic artifacts must have a high priority in the production of every radiographic image. Radiographic density can be significantly influenced by the presence of film artifacts.

Remember: Careful handling of the radiographic films, attention to details, and the implementation of a quality assurance program that includes processor control and equipment and material testing will help to eliminate most of the common radiographic artifacts.

Anode Heel Effect. The anode heel effect, although not directly related to the production of radiographic density, does have an influence on the uniformity of the overall radiographic density of the image. In specific applications, the anode heel effect can even be employed to advantage.

Remember: To take advantage of the anode heel effect, the radiographic image must include the peripheral portions of the x-ray beam. You must be examining an area that possesses significant differences in thickness and/or opacity from one end of the part to the other. Finally, you must place the thicker and/or more dense area in the direction of the cathode end of the x-ray tube.

Orthopedic Cast Material. The application of orthopedic cast material adds both thickness and opacity to the part, and you must compensate for both. The radiographic density of the recorded image will be severely affected by the addition of an orthopedic cast.

Remember: Double the mAs for increased part thickness; increase the kilovoltage by 10% for increased part opacity with a wet cast. With a dry cast, double the mAs. This compensation is for the average cast thickness. Large, thick casts may require triple or even quadruple the mAs used for the original examination.

Beam Filtration. Added filtration above 2.5 mm of aluminum is not recommended since it would be necessary to increase your exposure factors significantly to maintain the radiographic density. Less than this recommended filtration, however, exposes your patient to unnecessary radiation.

Remember: Beam filtration is utilized to reduce patient exposure to radiation. It is a positive example of the application of the principles of ALARA. Filters should never be removed during routine radiographic procedures.

Compensating Filters. The use of compensating filters designed in specific shapes and thicknesses can be used to produce a more uniform radiographic density throughout a body part possessing considerable variance in the thickness and/or opacity from one end of the part to the other.

Remember: The wedge-shaped compensating filter is the most commonly employed. The thicker portion of the filter's wedge should be placed adjacent to the part of the body that is thinner or that possesses less tissue opacity. Generally, an increase of 6 kVp to 10 kVp will compensate for the use of a wedge-shaped compensating filter.

Radiographic Density Review Questions

Name ____ Date ____________________

1. Identify the two major factors influencing the photographic properties of the recorded image.

2. Of all the major factors influencing the radiographic quality of the recorded image, __________ is considered the basic and most important property.

   a. recorded detail
   b. radiographic density
   c. radiographic contrast
   d. shape distortion

3. An important goal in the manipulation of exposure factors is to produce a minimal loss of recorded detail but always to maintain the __________ of the recorded image.

   a. image sharpness
   b. mA/time relationship
   c. radiographic density
   d. radiographic contrast

4. Comparing a slightly overexposed radiographic image with a slightly underexposed radiographic image, which would be of greater critical concern? Why?

5. What is meant by this statement: Proper radiographic density will demonstrate a complete image of the structures of interest?

6. Discuss the significance of the patient's condition/diagnosis and clinical information/reason for the examination to the production of proper radiographic density.

7. The thicker the structures of interest or body part, the (*circle one*) more/less radiation absorbed by that part. What effect would this have on the radiographic density of the recorded image, and how would you compensate for this influence?

8. Differentiate between the terms **sthenic, hypersthenic, hyposthenic,** and **asthenic** related to the characteristics of the patient's physique and the location of the internal organs.

9. Comparing the exposure factors required for an anteroposterior projection of the femur of a 7-year-old child with those required for a 25-year-old adult, you would find it necessary to increase the __________.

   a. kilovoltage
   b. SID
   c. mAs
   d. none of the above

10. The measurement of the thickness of the part for an anteroposterior projection of the shoulder of a 27-year-old trapeze artist with that of a 27-year-old quadriplegic indicates that the thickness of the part is similar. What change, if any, would be required in the exposure factors between the two examinations? __________

    a. increased kilovoltage
    b. decreased SID
    c. increased mAs
    d. none of the above

    Give reasons for your choice.

11. In most instances, tumors would be considered a (*circle one*) additive/destructive type of pathology. How would you adjust your exposure factors for most tumor conditions?

12. As the intensity of radiation reaching the film increases, the radiographic density of the recorded image (*circle one*) increases/decreases.

13. Areas of low tissue opacity within the body are visible as a (*circle one*) greater/lesser radiographic density when compared with tissues of higher opacity.

14. Describe how the following factors would affect the radiopacity of the tissues/structures of interest:

    Cellular composition

    Compactness of the cells

    Spaces between the cells

    Status of hollow organs

15. The exposure factor related to the intensity of the radiation produced is __________.

    a. kilovoltage
    b. time
    c. milliamperage
    d. SID

16. Write the formulas for finding the mAs, the milliamperage, and the time of exposure when the other two factors are known:

    mAs =

    mA =

    time =

17. As the milliamperage applied to the x-ray tube increases, the radiographic density of the film (*circle one*) increases/decreases. Why?

18. The relationship between the milliamperage and time of exposure to the mAs in order to maintain the radiographic density of the recorded image is __________.

    a. nonproportional
    b. directly proportional
    c. inversely proportional
    d. indirectly proportional

19. If 25 mAs are required to produce a desired radiographic density, identify the appropriate time of exposure using the milliamperage values indicated below:

a. 50 mA __________ time of exposure

b. 100 mA __________ time of exposure

c. 200 mA __________ time of exposure

d. 300 mA __________ time of exposure

e. 500 mA __________ time of exposure

f. 800 mA __________ time of exposure

20. If 15 mAs are required to produce a desired radiographic density, identify the appropriate milliamperage using the time of exposure indicated below:

a. .020 seconds __________ mA

b. .033 seconds __________ mA

c. .050 seconds __________ mA

d. .066 seconds __________ mA

e. .20 seconds __________ mA

f. .40 seconds __________ mA

21. Using the following exposure factors, indicate the total quantity of exposure delivered:

a. 50 mA for .012 seconds __________ mAs

b. 100 mA for .075 seconds __________ mAs

c. 150 mA for .033 seconds __________ mAs

d. 400 mA for .018 seconds __________ mAs

e. 600 mA for .008 seconds __________ mAs

f. 1000 mA for .006 seconds __________ mAs

22. In Question 21, which combination of factors represents the greatest quantity of x-ray exposure? __________ Which combination represents the least quantity of x-ray exposure? __________

23. To produce a visible change in the radiographic density of the recorded image, an increase or decrease of at least __________ from the original mAs is required.

a. 15%
b. 30%
c. 50%
d. one step in time

24. Calculate the new mAs required to produce a visible increase in the radiographic density of the recorded image in the following procedures:

    a. original mAs, 5 mAs new mAs ________

    b. original mAs, 16.6 mAs new mAs ________

    c. original mAs, 90 mAs new mAs ________

    d. original mAs, 150 mAs new mAs ________

25. Discuss the implementation of a quality assurance program that includes an evaluation of the tube current. Describe how you could check this factor.

26. Discuss the implementation of a quality assurance program that includes an evaluation of the timer mechanism. Describe how you could check this factor on a full-wave rectified, two-phase generation x-ray unit.

27. Indicate how you would adjust your procedure to check the timing mechanism on a full-wave rectified, three-phase generation x-ray unit.

28. Explain the purpose of automated exposure control.

29. Describe the basic phototiming device, and indicate how it performs the function of automated exposure control.

30. Can automated exposure control procedures be performed as tabletop procedures? ________. Explain your answer.

31. Identify the advantages of the ionization chamber over the use of the phototimer for automated exposure control.

32. Discuss the significance of proper positioning and centering of the part to be examined to the operation of automated exposure control devices.

33. As the distance between the tube focus and the radiographic film decreases, the radiographic density of the recorded image ________.

    a. increases
    b. decreases
    c. doubles
    d. quadruples

34. The equation governing the intensity of the x-ray beam and its relationship to the SID is known as the ________.

    a. inverse intensity law
    b. reverse intensity law
    c. intensity square law
    d. inverse square law

35. To maintain radiographic density, as the distance between the tube focus and the radiographic film increases, the quantity of the exposure must ________.

    a. increase
    b. decrease
    c. double
    d. quadruple

36. The formula for maintaining the radiographic density of the recorded image when changes in the SID are made is ________.

    a. new mAs = old distance^2 × new distance^2 ÷ old mAs

    b. new mAs : old mAs :: old distance^2 : new distance^2

    c. old mAs = new mAs × new distance^2 ÷ old distance^2

    d. new mAs = old mAs × new distance^2 ÷ old distance^2

37. You have taken an examination of the abdomen using a 50" (127 cm) SID and an exposure of 200 mAs. Another radiograph is taken, and the new distance will be 30" (76 cm). Identify the new mAs necessary to maintain the radiographic density of the recorded image. (Show all work.)

38. A chest radiograph was performed using a 72" (183 cm) SID and 7.5 mAs in the department. A similar examination is requested later in the day as a portable examination, and the patient is now unable to sit up. A 40" (102 cm) SID must be used. Identify the new mAs necessary to maintain the radiographic density of the recorded image. (Show all work.)

39. If your original mAs was 15, changing from an SID of 36" (91 cm) to 72" (183 cm) would make it necessary for you to adjust your mAs to __________.

    a. 3.75 mAs
    b. 7.5 mAs
    c. 30 mAs
    d. 60 mAs

40. The **quality factor** of the x-ray beam refers to its __________.

    a. radiation intensity
    b. penetrating ability
    c. contrast scale
    d. maximum wavelength

41. The exposure factor that controls the quality of the x-ray beam is the __________.

    a. milliamperage
    b. SID
    c. kilovoltage
    d. beam filtration

42. The wavelength of the x-rays produced determines their penetrating ability. As the wavelength of the x-rays produced is decreased, they become (*circle one*) more/less penetrating.

43. Shorter wavelengths and therefore greater penetrating ability are produced by (*circle one*) increasing/decreasing the kilovoltage.

44. The actual intensity of radiation produced by the x-ray tube (*circle one*) is/is not affected by changes in the kilovoltage. Explain your answer.

45. The total exposure to your patient is higher when the exposure factors selected include a high mAs and a low kilovoltage. *True or false?* __________

46. The overall visibility of the structures of interest is generally improved when the exposure factors selected include a low kilovoltage and a high mAs. *True or false?* __________

47. A visible change in radiographic density will be observable when you increase your kilovoltage (kVp) from 40 to 44 (*true or false*) __________; increase your kilovoltage from 80 to 84 (*true or false*) __________. Explain your answer.

48. A relationship between the kilovoltage and mAs for the maintenance of radiographic density exists, although not directly proportional. This relationship is defined by the __________.

    a. 12.4/kVp formula
    b. inverse square law
    c. 15% rule
    d. intensification factor

49. An increase in the kilovoltage together with a reduction in the mAs may be recommended in cases of motion. In the following examples, indicate the proper exposure factor adjustment to be made:

    a. Anteroposterior hip: original exposure 60 mAs at 70 kVp

       New exposure = _____ mAs at _____ kVp

    b. Lateral lumbosacral articulation: original exposure 200 mAs at 80 kVp

       New exposure = _____ mAs at _____ kVp

50. In Question 49, the kVp/mAs changes were made knowing that the original kilovoltage was sufficient to ______________ the part.

51. The unwanted radiographic density resulting from secondary and scattered radiations reaching your film would be referred to as __________.

    a. absorbed radiation
    b. remnant radiation
    c. radiographic fog
    d. film mottle

52. The amount of scattered radiation influencing the radiographic density increases as __________.

    a. the density of the part decreases and the kilovoltage decreases
    b. the thickness of the part decreases and the kilovoltage increases
    c. the thickness of the part increases and the mAs increases
    d. the thickness of the part increases and the kilovoltage decreases

53. Overall, the following factors are predictable in their control of radiographic density:

    a. mAs (*true or false*) __________
    b. kilovoltage (*true or false*) __________
    c. SID (*true or false*) __________

54. Define the **screen intensification factor.**

55. Generally, as the kilovoltage is increased, the screen intensification factor (*circle one*) decreases/increases.

56. If the screen intensification factor at 60 kVp is 80 × and at 70 kVp, 100 ×, calculate for the new mAs using 70 kVp when the original mAs at 60 kVp was 60 mAs.

57. As the speed of the film-screen recording system increases, the exposure required to maintain the radiographic density of the recorded image (*circle one*) increases/decreases.

58. If the original exposure utilizing a 100 speed film-screen recording system required 25 mAs, solve for the new mAs when the following film-screen recording systems are employed:

    a. 200 speed film-screen system ________ new mAs

    b. 300 speed film-screen system ________ new mAs

    c. 500 speed film-screen system ________ new mAs

    d. 800 speed film-screen system ________ new mAs

    e. 1200 speed film-screen system ________ new mAs

59. Since higher speed film-screen recording systems help to reduce the required exposure and, therefore, address the problems of motion, why don't we employ the highest speed system available with all radiographic procedures?

60. Describe the purpose of using a higher speed recording system in the following situations:

    a. 3-month pregnant patient for an abdomen

    b. 18-month-old child for a skull series

61. In what manner does the use of a 40° cephalic angle of the central ray influence the SID and, therefore, the radiographic density of the recorded image? Explain how you would adjust for this influence.

62. Why are darkroom walls normally lead-lined?

63. Describe how you would evaluate the complaint of a possible light leak somewhere in the darkroom that may be producing radiographic fog upon your processed films.

64. Describe the advantages and disadvantages of the following types of darkroom entrances:

    a. labyrinth entrance

    b. revolving-door entrance

    c. double-door entrance

    d. single-door entrance

65. Differentiate between the safelights employed for blue-sensitive and green-sensitive radiographic films.

66. Describe how you would test for the safety of a safelight system in the darkroom.

67. Discuss the significance of film processing to the overall success of the radiographic procedure.

68. Does manual processing normally provide for a system of adequate quality control? Explain.

69. How has the development of the automatic processor improved the overall quality of the processed radiograph?

70. Discuss the importance of a standardized program of routine cleaning and maintenance and the implementation of a quality assurance program that includes these elements to the overall performance levels of the radiology department.

71. What are radiographic artifacts?

72. Identify the following film artifacts and indicate how you would eliminate or reduce their incidence within the radiology department:

    a. crinkle/crescent marks

    b. smudge/fingerprint marks

    c. static electricity discharge

73. Describe how a quality assurance program will help to control or eliminate the appearance of chemical and equipment or processor artifacts.

74. The exposure rate of an x-ray beam varies as a result of the angle the x-rays form with the surface (plane) of the anode of the x-ray tube. This effect is known as the __________.

    a. bremsstrahlung effect
    b. minimum wavelength effect
    c. inverse square effect
    d. anode heel effect

75. Using a 40" (102 cm) SID, you would probably see this effect (Question 74) if you were employing a __________ film size. Explain the reasons for your choice.

    a. 14" × 17"
    b. 11" × 14"
    c. 10" × 12"
    d. 8" × 10"

76. Considering the influence of this effect (Question 74), respond to the following problems and situations:

    A. The body part measures 16" in length, and the thickness and opacity are uniform throughout. What influence, if any, would this effect have on the radiographic density of your recorded image?

B. If the body part measures 17" in length, and the thickness and opacity of the part differ considerably from one end to the other, could you take advantage of this influence using a 40" (102 cm) SID? Explain your answer. What if the SID was increased to 72" (183 cm)? Explain your answer.

C. If the body part measures 10" in length, and the thickness and opacity of the part differ considerably from one end to the other, could you take advantage of this influence using a 40" (102 cm) SID? Explain your answer.

D. When examining a femur on an adult patient using an SID of 40" (102 cm), explain how you would arrange your patient's body in relation to the x-ray tube in order to produce a more uniform radiographic density of the recorded image.

77. The addition of orthopedic cast materials requires you to re-evaluate your original patient factors of the ______________________ of the part and the ______________________ of the part.

78. Comparing the radiograph of the forearm of an injured patient taken prior to the application of an orthopedic cast to one taken after its application, if the same exposure factors were employed, what would you expect to happen to the radiographic quality of the image related to the following photographic properties of the recorded image?

    Radiographic density:

    Radiographic contrast:

79. Relating to the procedure described in Question 78, your goal would be to produce a radiographic image possessing similar photographic properties after the application of the orthopedic cast. Using the general formula applicable to examining plaster cast materials in their "wet" state, adjust your exposure factors in the example provided below:

    Original exposure factors: 15 mAs at 60 kVp

    New wet cast factors: _____ mAs at _____ kVp

    The next day, an additional examination is requested on the dry cast. Determine your exposure factors related to the *original* factors employed above. Explain the reason for the changes, if any, in the selection of your exposure factors.

    New dry cast factors: _____ mAs at _____ kVp

80. Describe the purpose of beam filtration in routine radiography.

81. Differentiate between the terms **inherent filtration** and **added filtration** as they relate to the total filtration of the beam.

82. A total filtration of __________ mm aluminum equivalent is recommended when operating radiographic equipment below 70 kVp. When operating above 70 kVp, __________ mm aluminum equivalent filtration is recommended.

83. Describe the purpose of compensating filters in general radiography.

84. Indicate an example of an examination that could use a compensating filter to advantage.

85. Describe the shape of the most commonly employed compensating filter used in diagnostic radiology.

86. Describe the application of this type of filter (Question 85) for the examination indicated in Question 84.

Radiographic Density Analysis Worksheet

Let's review our analysis of radiographic density with the following exercise. A satisfactory diagnostic radiograph of the pelvis was produced using the following technical factors:

100 mA
1 sec (time of exposure)
80 kVp
40" (102 cm) SID
1-mm focus size
Minimum OID
Regular screen-type film
100 speed film-screen system
8:1 ratio, 100-line moving grid
Proper collimation
2.5-mm aluminum filtration
92°F development at 90 sec
Normal thickness and opacity of the part

Without compensation, the changes listed in the table below are made one by one. Indicate the effect, if any, each change has on the radiographic density of the radiographic image.

1. If the radiographic density of the recorded image is increased, mark a plus (+) in the space provided.
2. If the radiographic density of the recorded image is decreased, mark a minus (–) in the space provided.
3. If the radiographic density of the recorded image is unchanged, mark a zero (0) in the space provided.

Radiographic Density

Proposed change	Effect	Proposed change	Effect
Use a 2.0-mm focus	_____	Change to a 5:1 ratio, 100-line moving grid	_____
Use 10" (25 cm) OID	_____	Change to a 12:1 ratio, 100-line moving grid	_____
Reduce the SID to 25" (63 cm)	_____	Remove all collimation	_____
Increase the SID to 48" (122 cm)	_____	Remove all filtration	_____
Use screen-type film in a cardboard holder	_____	Develop at 100°F for 90 sec	_____
Use a 200 speed film-screen recording system	_____	Body thickness reduced by an atrophic condition	_____
Use 200 mAs	_____	Pathological condition reduces opacity of body tissues	_____
Increase the kVp to 90	_____		
Omit the use of a grid	_____		

5

Radiographic Contrast

Another of the photographic properties of the recorded image is the radiographic contrast. It is the second factor that is related to the visibility of the image. Consider the general meaning of the word **contrast.** A dictionary definition includes the following: the noticeable differences exhibited when items are compared or set side by side and to compare or appraise in respect to differences. These definitions fit the concept of radiographic contrast perfectly (Figure 5-1). Whereas radiographic density identifies the completeness of the structures of interest within the recorded image, the radiographic contrast provides the visibility between different structures and the ability to discern minute differences between the different radiographic densities recorded within the radiographic image.

Radiographic density and radiographic contrast are two different properties of radiographic quality, yet it is impossible to separate the two factors completely because there is an interrelationship and interdependency between them. Many of the materials and factors that influence the radiographic density of the recorded image also influence the radiographic contrast, but in a different manner. Radiographic contrast only exists because there are different variations of radiographic density exhibited in the recorded image from one structure within the body part that can be compared with other adjacent body structures. A recorded image that is of radiographic quality should be able to demonstrate all of the structural differences that the body part possesses as variations of radiographic density.

Thus, when we analyze the contributions of radiographic contrast to the radiographic quality of the recorded image, we must identify radiographic contrast as both a quantity factor and a quality factor. The quantity factor of radiographic contrast refers to the total number of useful densities recorded between the minimum and maximum radiographic density visible within the recorded image. It would represent the total amount of radiographic contrast recorded. The quality factor of radiographic contrast refers to the ability to distinguish between the various recorded radiographic densities and to identify the structure within the body that they represent. The quality factor of radiographic contrast indicates that the structures of interest have been recorded within the proper range of radiographic densities required to visualize the structural differences. This range of radiographic density can more simply be referred to as the **recorded contrast scale of the recorded image.** The quality of proper contrast enables the technologist to discriminate

Figure 5-1
Radiographic Contrast Should Demonstrate the Visibility of Noticeable Radiographic Density Differences

between the multiple radiographic densities recorded within a single structure and to compare them with the radiographic densities produced in adjacent structures recorded within the radiographic image.

To better understand this quality factor of contrast, let's compare it to the relationship between two painters. One, a house painter, would cover the subject adequately by simply applying even coats of paint using a steady stroke of the brush to produce a complete coverage of the area needed to be painted. This can be compared with the overall completeness of the radiographic image associated with the production of proper radiographic density. The house painter would be performing an acceptable job within that basic criterion. The other painter, a portrait artist, would be much more inventive and discriminatory in the use of the paint and the application of his brush strokes to the canvas. The portrait artist would mix and adjust the color tones of the paints and the texture of the brush strokes in an attempt to add depth and character to the painting he or she is creating. The technologist can also perform his or her functions like the house painter and simply produce radiographs in a mechanical manner and perform an adequate job. Like the house painter, he or she would cover the subject with an overall coat of radiographic density. On the other hand, like the portrait painter, the technologist can consider and evaluate the needs of each examination performed, selecting the equipment and materials and choosing the exposure factors that will enable him or her to produce a radiographic image that possesses depth, character, and quality. Like the portrait artist, the technologist can produce a radiographic image that provides the proper radiographic contrast scale within the appropriate range of radiographic densities necessary to visualize all of the structures of interest. To achieve this, however, the technologist must have a thorough knowledge and understanding of the many factors that control and influence the production of a proper radiographic contrast scale.

Radiographic contrast frequently has been identified as one of the "gray" areas of radiographic quality. This statement can be better understood if we accept the fact that the evaluation of a painting or a radiographic image can be open to considerable subjective opinion related to its value or worth (Figure 5-2). The quality of the

Figure 5-2
The Quality of Radiographic Contrast Is Relative to the Requirement of Interpretive Evaluation

radiographic contrast within the recorded image will always represent the proper relationship of radiographic densities for that particular procedure, structures of interest or body part, and specific purpose for the requested examination. As we know, the appropriate radiographic density and contrast will be quite different for an examination of the chest for a possible rib fracture versus an examination of the chest performed for possible lung pathology. Depending on the circumstances of the procedure and the requirements needed for the proper interpretive evaluation by the physician, a certain contrast scale might be quite proper for one examination but inappropriate and unacceptable for another, even though the same body part is being examined.

It is within this somewhat subjective area of radiographic contrast that the profession of radiologic technology becomes interesting and challenging for the radiologic technologist. We soon begin to recognize that the proper radiographic density and radiographic contrast scale for one physician's evaluation may be different from the requirements of another physician. A comparison might be to evaluate the esthetic pleasure of viewing a Picasso compared with a da Vinci painting. Art critics may not always agree or may prefer one style over another, but each painting would possess the qualities of a masterpiece. It will not be suggested that it will be possible to get all of the physicians to agree with one another on all aspects of the proper radiographic contrast scale of a recorded image, but what is clear and can be defined is that regardless of their individual requirements, a radiographic image will either possess the proper radiographic contrast for the procedure being performed, or it will not. What is meant by that statement is that the radiographic image will be recorded upon the film completely or not and the structures of interest within the recorded image are either clearly **visible** as separate, individual images, or they are not. The technologist's responsibility to the physician and to the patient is to produce a radiographic image that permits the visualization of all of the structures of interest. This is an evaluation that the technologist can and must perform for each radiographic image produced.

In an analysis of the factors that control and influence radiographic contrast, it will be necessary to learn how to control both the quantity (total amount of useful radiographic density) and the quality (proper contrast scale) of the radiographic contrast of the recorded image. Basically, radiographic contrast is influenced by two major considerations: (1) subject contrast and (2) film contrast.

RADIOGRAPHIC CONTRAST

Subject Contrast	Film Contrast
Body Tissue	Emulsion (Speed/Latitude)
Volume & Type	Exposure (Screen/Nonscreen)
Opacity & Condition	Processing Procedure
Quality of Radiation	Overall Radiographic Quality

Familiarizing yourself with these major considerations and with the individual factors that influence them will enable you to produce an image with the proper quantity and quality of radiographic contrast.

Subject Contrast

The various structures and tissues of the body differ from one another in composition, thickness, and opacity. Each of these considerations affects the subject contrast produced by the patient and ultimately the radiographic contrast of the recorded image. **Subject contrast** is defined as the difference in the quantity of radiation transmitted by a particular structure or body part as a result of the different absorption characteristics of the tissues and structures making up that part (Figure 5-3). The normal differences in tissue composition, thickness, and opacity are responsible for the differences in the radiation absorption rates as the x-ray beam traverses the part. The different absorption characteristics of the x-ray beam are also affected by the quality (penetrating ability) of the x-radiation contained within the beam. The penetrating ability of the x-ray beam is controlled by the kilovoltage applied to the x-ray tube. Thus, subject contrast is affected by the normal differences in tissue composition, thickness, and opacity, as well as the quality of the beam of radiation. It should be noted that the photographic properties of the radiographic image and the visibility of the recorded image, like radiographic density, are influenced to a great extent by patient factors. Let's examine the patient factors and their influence on subject contrast and, ultimately, the radiographic contrast of the recorded image.

As the primary beam of radiation emerges from the x-ray tube, enters the body part, and traverses through it, a portion of the beam of radiation will be completely absorbed by the tissues of the part. The penetrating ability of that portion of the beam is insufficient to pass through the thickness and/or opacity of those tissues. The areas of the body represented within the radiographic image representing this absorption pattern will be recorded with the least amount of radiographic density. An example would be the general appearance of bones recorded on the radiographic film. Other portions of the beam of radiation pass through the body with little or no absorption at all. These areas possess greater quantities of radiographic density within the recorded image. An example would be the air spaces of the lungs. The majority of the beam of radiation is absorbed in differing amounts relative to the composition, thickness, and opacity of the various tissues and structures of the body part being

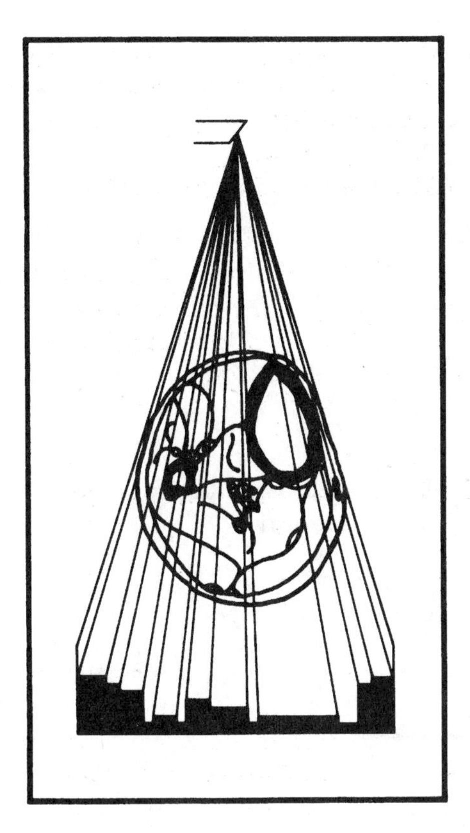

Figure 5-3
Subject Contrast Results from Different Absorption Characteristics of the Structures Making Up the Part

examined. Therefore, a multiplicity of varying amounts of radiation passing through the patient's body is capable of producing a wide range of different radiographic densities upon the radiographic film. The different radiographic densities recorded upon the film comprise the radiographic contrast of the recorded image.

The subject contrast attributed to these patient factors is something over which the technologist has little influence or control. Each patient the technologist examines will be different as to size, shape, age, and the condition of individual body parts. These are all variables that are not under the control of the technologist. However, the concept of subject contrast is not as complex as it initially appears. Within every structure of the body, there are multiple differences in tissue composition, thickness, and opacity that can be recorded within the radiographic image as separate, individual radiographic densities.

In fact, if you examine any single structure of the body, such as a femur, you will not find the same composition, thickness, or opacity throughout the entire structure (Figure 5-4). In bony structures, the outer surface of the bone below the periosteum is quite dense, but not of equal thickness throughout the entire structure. The middle of the bone shaft will possess a greater thickness of this compact, dense bone structure than will the ends (extremities) of the bone. The inner cancellous layer of the bone is less dense and is sponge-like in appearance, being filled with many mesh-like cavities. There is a greater quantity of cancellous bone in the extremities of the bone than in the middle of the shaft. The central medullary canal is far less dense than either of these two other structural layers of the bone and contains an entirely different substance with a different radiopacity: bone marrow. Therefore, within this

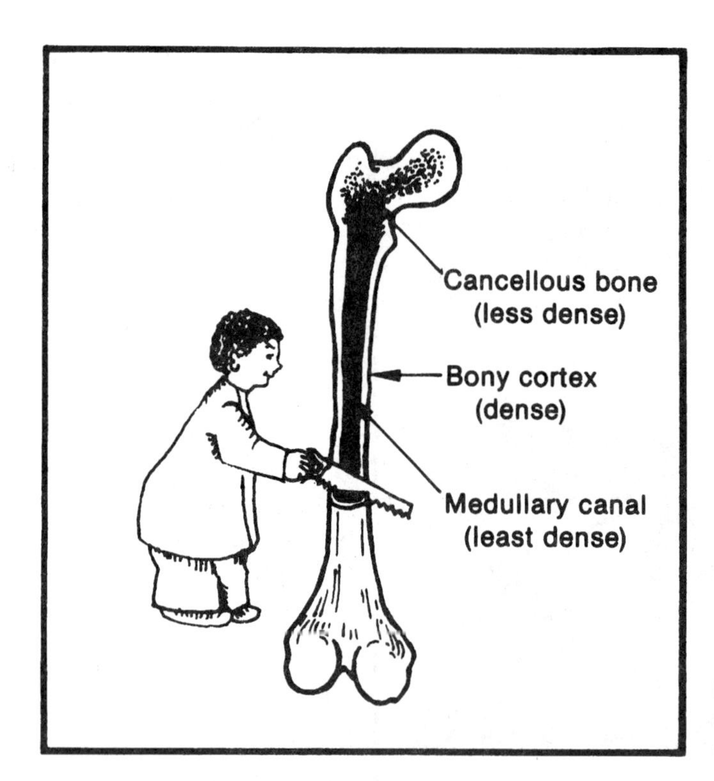

Figure 5-4
Differences in Thickness and Opacity of the Femur

single structure of the body, we have been able to identify several different tissue patterns of composition, thickness, and opacity. These differing thicknesses and opacities of the bone produce different absorption characteristics of the beam of radiation, and, therefore, different amounts of radiation will be transmitted to the film producing a wide range of radiographic density from just this single body structure. Every structure of the body contains a similar multiplicity of different subject contrasts capable of producing a visible pattern of radiographic contrast within the recorded image.

The impact of subject contrast on the radiographic contrast of the recorded image becomes more evident when you consider the multiplicity of the different body tissues, with their different thicknesses and opacities contained within any group of structures contained within the body. Some structures of the body exhibit considerable differences in these patient factors and are referred to as areas of **good subject contrast.** Examples of areas of good subject contrast are the shoulder and the hip. Radiographic examination of areas possessing good subject contrast produces a high ratio of radiographic density differences within the recorded image. Other areas of the body are composed of structures so similar in composition, thickness, and opacity that differential absorption ratios of the beam of radiation are poor. These areas of the body would be identified as possessing **poor subject contrast.** An example of an area of poor subject contrast would be the abdomen. The many soft-tissue structures of the many organs contained within the abdominal cavity are similar in composition, thickness, and opacity and are therefore difficult to demonstrate as separate, individual radiographic densities within the recorded image. The visual demonstration of a high ratio of radiographic density differences in areas of poor subject contrast can be difficult or impossible to achieve.

A thorough knowledge and understanding of the patient factors that contribute to subject contrast is essential to the technologist in order to be able to select the appropriate equipment and to choose the proper exposure factors to meet the needs required by each patient's individual examination. Using their knowledge and understanding of the patient factors that influence subject contrast, technologists can choose and manipulate the exposure factors in an effort to demonstrate all of the various differences in the absorption characteristics of the x-ray beam that the examination of the body part will permit.

Patient Factors

In studying radiographic density, we investigated a number of factors that affect the thickness and opacity of the tissues. As the thickness of the body part increases or decreases, or as the opacity of the tissues change, there will be a significant difference in the absorption characteristics of the body part. A review of the patient factors introduced in Chapter 4, "Radiographic Density," will be helpful to your understanding of their influence on radiographic contrast. A synopsis of these factors is outlined below.

Tissue Thickness

The volume of the tissue and, therefore, the thickness of the part is influenced by the following factors:

1. **Patient age: infant, child, adult, elderly**
   In general, as the patient ages from infant to child to adult, the thickness of the part increases as the patient matures and grows. However, even within different individuals of the same age classification, there will be major differences in the thickness of the various body parts. Occasionally, the elderly adult will begin to lose tissue volume. The greater the volume of tissue, the greater the absorption of radiation by the part.
2. **Patient sex: differences in tissue distribution**
   It is not uncommon to have larger deposits of fatty tissues in the area of the hips and lower pelvis in women than in men. Women with large breasts will cast additional radiographic shadows in chest radiographs compared with male chest radiographs. The greater the quantity of fat and lower density tissue, the lower the absorption of radiation by the part compared with a similar thickness of the part comprised of higher density tissue.
3. **Development: sthenic, hypersthenic, hyposthenic, asthenic**
   Differences in general body types will add to or reduce the amount of the tissue volume and, depending on the tissue, the ratio of the tissue density.
4. **Condition: pathological status of part**
   Pathological conditions can add to or reduce tissue volume, changing the thickness of the body part, and can also change the tissue density of the part, although the overall thickness may remain the same.

Each of these patient factors can and will influence the overall volume of tissue contained within the body part. Proper radiographic procedures require that all parts to be examined be measured and their values be evaluated against known measurements provided in a technique chart. Using a technique chart based on the average thickness of body parts is a positive application of the principles of using radiation levels that are "as low as reasonably achievable" (ALARA) and an avoidance of the technical guessing game. Using technique charts or guides will help to eliminate repeat examinations caused by the improper selection of exposure factors.

Tissue Opacity

Of even greater interest when appraising the absorption characteristics of body tissues related to radiographic contrast is the tissue opacity. The opacity of the tissues and therefore the absorption characteristics of the body part are influenced by the following factors:

1. **Cellular composition: chemical and molecular structure**
   As the chemical composition of the cells increases, the absorption characteristics of the tissues that comprise those cells will increase.
2. **Tissue structure: cells in close proximity or separated**
   As the cells making up the tissues are more closely compacted, the mass per unit volume of the structure increases, and, therefore, the absorption characteristics of the part will also increase.
3. **Surrounding structures: materials filling in between tissues**
   If the materials filling in between the cells or tissues are of high density, even a low-density tissue material will produce increased absorption characteristics to the beam of radiation passing through it.
4. **Status of organs: empty or filled with gas, liquid, or solid**
   Although the heart contains large cavities, these cavities are filled with large quantities of blood. Blood in large volumes presents a significant tissue density due to the high chemical composition of the cells contained within the blood. Therefore, without special radiographic techniques, the actual cavities of the heart are not visible within the recorded image of the heart on the radiograph. The absorption characteristics of the hollow organ structures of the body will change and will be dependent on whether the structure is empty or filled with a radiolucent or radiopaque substance at the time of the radiographic examination.

Natural Tissue Densities

Examples of the body's naturally occurring tissue densities listed from the least to the most dense include the following:

- Gas
- Fat
- Cartilage
- Hollow organs (empty)
- Muscles
- Solid organs
- Hollow organs (filled)
- Bone

Let's examine these naturally occurring tissue densities contained within the body as they relate to radiographic contrast.

Gas. Being the least dense substance, gas is recorded as the most dense useful shadow within the recorded image since it provides little interference with the passage of x-rays through the part containing various levels of a gaseous substance. Areas of the body that would normally show gas include the gastrointestinal tract, the respiratory passages, and the paranasal sinuses. All of these areas normally demonstrate air within their structures, and the air will provide a different absorption pattern to the beam of radiation than the structures of tissue adjacent to it. Outside of the normal patterns of gas contained within the body, certain pathological conditions such as the build-up of gas with an intestinal obstruction, the large amounts of air trapped in the air sacs of the lungs in emphysema, or the loss of the normal air-filled paranasal sinus cavities in the condition of sinusitis all alter the natural pattern of gas within these areas. An increase or decrease in the normally occurring gas patterns within the structures of the body would require a compensation in the technical factors selected in order to maintain the proper radiographic contrast and visibility of the structures of interest.

Gas can even be used to produce an artificial subject contrast in areas of poor subject contrast. Gaseous agents, including air used as an artificial contrast agent, perform this task by creating a radiolucent density around the adjacent tissue densities. Examples of gas used as an artificial subject-contrast agent include air in the double-contrast study of the colon (air-contrast barium enema), nitrous oxide in a pelvic pneumogram to demonstrate the female reproductive organs, and air within the ventricles of the brain for a pneumoencephalogram.

Fat. Fat is a naturally occurring body tissue density normally located within and around most of the tissues of the body. Therefore, it is extremely difficult to demonstrate fat as a separate tissue density. The tissue density of fat is low, and, therefore, the absorption characteristics of fatty tissue are almost nonexistent. There are instances, however, where the presence of fat actually improves the subject contrast of an area. An example is the deposit of perirenal fat surrounding the kidneys that enables you to visualize the outlines of the kidneys on a radiograph of the abdomen without the aid or need of an artificial contrast agent. However, with the hypersthenic patient who is obese and possesses considerable fatty tissue throughout the entire abdomen, it may be difficult to demonstrate the kidneys as separate structures due to the large deposits of fat contained throughout the entire abdomen that could obscure the visualization of the perirenal fatty deposits surrounding the kidneys.

Cartilage. Cartilage is abundant throughout the body, especially within the skeletal system. In fact, like fat, cartilage would float on water. Cartilage, although more dense than fat, does not provide a sufficient difference in its absorption characteristics to be visibly recorded as a separate radiographic density. The presence of the cartilage separating the bones is more obvious by the lack of radiographic density demonstrated in the spaces between the bones. The actual presence of cartilage can be determined, however, with the aid of a radiopaque contrast agent that alters the tissue density of the joint cavities thereby outlining the structure of the cartilage. The introduction of a contrast agent into a joint space, a procedure referred to as **arthrography,** enables the physician to determine the presence of cartilaginous pathology or changes in the appearance or location of the cartilage resulting from injury.

Hollow Organs. When empty and lying adjacent to one another, hollow organs exhibit very little tissue density variation. Their overall tissue density as a whole can be described as similar to that of water. (Over 65% of the body's tissues are comprised of water, and many of the hollow organ tissues have a density similar to water.) The organs of the gastrointestinal tract are mainly hollow organs and as a group possess very little subject contrast. It is extremely difficult to distinguish individual radiological structural patterns between the different organs of the gastrointestinal system within the abdominal cavity. The slight differences that may be visualized are normally due to the size, shape, and thickness differences between organs resulting from differences in the patient's physique and the development of the part.

Muscles. Because of the dense chemical composition of their cells and also due to the compactness of their structure, muscle tissues are quite dense and provide considerable absorption (attenuation) to an x-ray beam. As an example, the muscle bundles of the posterior abdomen, the psoas major and minor, are easily visualized in a radiograph of the abdomen as oblique shadows going from medial to lateral position as they lie in a superior to inferior direction. As seen through the overall water density of the tissues of the gastrointestinal organs of the abdominal cavity, they are visualized clearly as separate structures. Numerous examples of muscle shadows lying adjacent to bones or other naturally occurring body tissues are routinely demonstrated in radiographic examinations.

Solid Organs. Solid organs are frequently even more dense than their surrounding muscular attachments. The liver, a solid organ lying in the upper abdominal cavity on the right side and continuing across the midline of the abdomen, is normally clearly visible as a separate structure within the abdomen. This is due to the rich supply and large volume of highly dense blood contained within the liver, as well as the actual chemical constituents of the liver cells themselves. It also has to do with the location of the liver lying under the surface of the diaphragm that separates the thoracic and abdominal cavities. All of these factors contribute to the good subject contrast of the liver, which can usually be recorded as a very distinct radiographic image within the abdomen.

Filled Hollow Organs. We already identified the hollow organs' naturally occurring subject contrast and compared it with the contrast produced of water. However, hollow organs when filled present a rather unique consideration related to subject contrast and the ultimate visualization of their structure as a separate, individual radiographic contrast. The attenuation of the x-ray beam depends on the quantity and opacity of the material filling the hollow organ. The hollow organs of the gastrointestinal tract are classic examples of this concept of filled hollow organs. They will differ considerably in their attenuation of x-rays relative to the amount, type, and opacity of the material contained within them. In fact, the organs of the gastrointestinal tract are some of the most frequent sites of artificially induced subject contrast because of their lack of adequate subject contrast in their natural state. The most common method employed to visualize these organs is that of increasing the radiopacity of the organs as compared with adjacent structures by the introduction of a dense metallic salt, barium sulfate. Esophagrams, upper gastrointestinal series, small bowel series, and barium enemas to investigate the colon are examples of examinations of the gastrointestinal system using a radiopaque contrast agent to produce an artificial subject contrast in these organs. Visibility of the size, shape, function, and condition of these structures is easily produced in this manner.

Bone. Bone is the most dense naturally occurring tissue within the body. As a whole, bones absorb a greater percentage of radiation than does any other naturally occurring subject contrast tissue of the body. Therefore, the bones exhibit a high subject contrast compared with all other tissues of the body.

Patient Factors Summary

The recognition of these patient factors and an understanding of the principles involved in the study of subject contrast is vital to the selection of exposure factors that will record an image that possesses proper radiographic contrast. There are many factors influencing subject contrast over which the technologist has little control, but, although you have little control over the subject contrast, you can take advantage of the effect that different exposure factors will have on the absorption characteristics of the tissues that are being examined. (**Perform Experiment 51.**)

Quality of Radiation

Another factor contributing to subject contrast and the different absorption characteristics of the beam of radiation is the quality of the radiation. In fact, this factor could be considered the most important factor within the management of the technologist for the control and manipulation of subject contrast. The quality of the radiation essentially refers to the penetrating ability of the beam, and the penetrating ability of the beam is dependent on the kilovoltage selected and applied to the x-ray tube. Although it is appropriate to associate kilovoltage with the penetrating ability of the beam and to state that as the kilovoltage is increased, the penetrating ability of the beam increases, this is really an oversimplification of the relationship of kilovoltage to the quality of beam of radiation and its influence on the subject contrast.

Another way of referring to this relationship would be to state that the shorter the wavelength of the x-ray beam, the more penetrating the radiation. The wavelength of the x-ray beam is a measurement between two successive waves. It is measured in angstrom units, which are one ten-billionth of a meter (10^{-10}). As the kilovoltage is increased, the minimum wavelength produced in the beam becomes shorter. Let's compare the minimum wavelength produced using a 40-kVp beam with the minimum wavelength produced with an 80-kVp beam. The formula for determining the minimum wavelength of an x-ray beam is to divide the kilovoltage into 12.4 (12.4 is a constant).

40-kVp X-ray Beam	80-kVp X-ray Beam
λmin = 12.4/40 kVp	λmin = 12.4/80 kVp
λmin = 0.31 Angstroms	λmin = 0.15 Angstroms

It is obvious from the minimum wavelength produced by the 80-kVp x-ray beam that the 80-kVp x-ray beam would be more penetrating. However, with the increase in the kilovoltage, more than just a decrease in the minimal wavelength of the beam has taken place. The composition of any x-ray beam consists of countless numbers of different wavelengths. Indeed, within any selected beam of radiation, there will be a wide range of wavelengths. Each wavelength within the beam, from the shortest, most penetrating to the longest, least penetrating radiation, represents a certain percentage of the x-ray beam's total radiation. If you were to plot this on a chart, you would find that the energy distribution within the beam is not uniform. At 40 kVp, the minimum wavelength, 0.31 Å, may represent only 1% of the total energy within the beam. However, it must be emphasized that this does represent the most penetrating radiation contained within that particular beam of radiation. The rest of the total radiation energy is distributed throughout multiple, longer, less penetrating wavelengths. The largest percentage of the radiation within a 40-kVp beam is located in the range of 0.55 Å.

At 80 kVp, the minimum wavelength, 0.15 Å, may also represent only 1% of the total energy of the x-ray beam, but in addition to reducing the minimum wavelength, the increase in the kilovoltage also produces a significant shift in the entire energy distribution of the beam. There will be a much greater percentage of the total radiation within the beam lying within the range of shorter wavelength, higher penetrating radiation. There will be an overall lower percentage of longer wavelength, less penetrating radiation in the 80-kVp beam compared to the 40-kVp beam of radiation.

An increase in the kilovoltage, and thus an increase in the overall quality and penetrating ability of the radiation, will have a pronounced effect on the absorption characteristics of the body tissues and, therefore, on the subject contrast. An increase in the kilovoltage not only produces additional radiographic density, but also alters the patterns of tissue absorption and thus the relationship between the radiographic density of one recorded structure and another. A shift in the percentages of the various wavelengths and, therefore, the penetrating ability of the beam produced by an increase in the kilovoltage, will have a significant effect on the entire scale of radiographic contrast, and, therefore, the visibility of the recorded image is also affected. At times, adjustments in the kilovoltage are recommended in order to manipulate this relationship, and the change can be used to advantage to produce a desired radiographic contrast scale. However, there are other occasions when adjustments in the kilovoltage will produce such a photographic visibility change in the radiographic contrast scale that the recorded image is totally unacceptable. Later in this chapter, the effect of kilovoltage on the radiographic contrast scale is fully investigated.

Radiographic Film Contrast

The other major factor influencing the radiographic contrast of the recorded image is the radiographic film contrast. There are several intrinsic and extrinsic factors that affect radiographic film contrast. Intrinsically, we are referring to the film characteristics of speed and latitude inherent in the manufacturer's product. The extrinsic factors affecting the radiographic film contrast include the actual exposure factors selected and the processing methods employed to change the latent image into a permanent visible image.

Film Speed

Film speed refers to the rapidity with which a film responds to radiation. In other words, it refers to the amount of radiation required to produce a specific radiographic density. There are differences in film speed between the two types of radiographic film (screen and nonscreen) as well as between the same type of film from different manufacturers. Each manufacturer produces screen-type film with different film speeds to use in combination with intensifying screens to provide for a specific film-screen recording system speed. The different film speeds are manufactured to possess different film characteristics of not just speed, but also ability to produce a range of radiographic densities. There are reasons for these radiographic film product differences. Depending on the procedure being performed and the purpose for the examination, the equipment and materials employed, and the all-important patient factors, a slow or average speed film might be quite appropriate with one combination of factors, whereas a faster speed film might be the more appropriate choice under a different set of circumstances. In Figure 5-5, the relative speeds of two films are compared. The sensitivity response curve of Film A indicates that it will produce a radiographic density of 2.0 (measured with a densitometer) with a log relative exposure of 1.2. Film B, however, indicates that a log relative exposure of 1.5 is required to produce the same radiographic density of 2.0. *It is important to recognize that an increase in the log relative exposure of 0.3 represents a doubling of the relative exposure.* Therefore, if an exposure of 50 mAs produces a radiographic density of 2.0 using Film A, Film B would require an exposure of 100 mAs to produce the same radiographic density of 2.0. It would be correct to state that Film A is twice as fast as Film B. (**Perform Experiment 52.**)

Film Latitude

Film latitude refers to the range of log relative exposure values that will produce radiographic density in the accepted diagnostic range. The useful diagnostic range of radiographic density lies between 0.4 and 2.75, as demonstrated on a densitometer. This means that all the various structures and tissues of the body should be demonstrated between these radiographic density levels in order to be visible within the recorded radiographic image. Radiographic density levels above or below these values may record some of the tissues and structures of the body at a radiographic

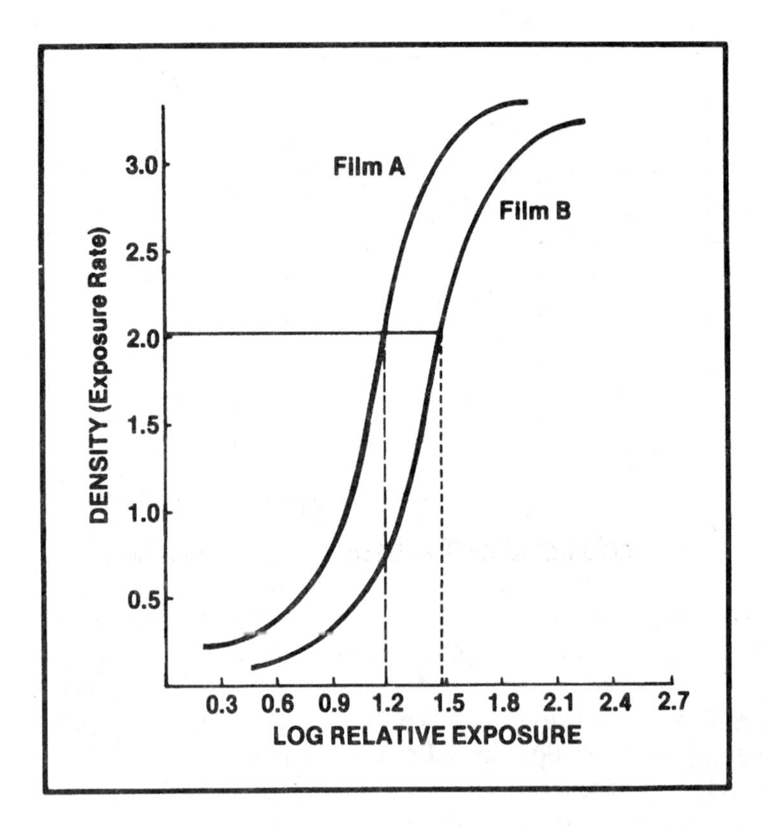

Figure 5-5
Comparison of Relative Speeds of Two Films

density where they may not be sufficiently visible for the interpretation of the physician. The manufacturer has the ability to build into the film a characteristic for either high or low contrast, and, depending on the circumstances of its application, the selection of either film will produce a satisfactory radiographic image. When used properly, the inherent property of film latitude can be used by the technologist to advantage.

Figure 5-6 compares two radiographic films: Film A is considered a film of high contrast; Film B is considered a film of low contrast. Keeping in mind the useful radiographic density range is from 0.4 to 2.75, Film A is limited to a log relative exposure range of 0.29 to 0.9 in order to produce the useful diagnostic range of radiographic densities. This means that all of the body tissues and structures will have to be recorded within the useful diagnostic range of radiographic densities utilizing this narrow range of exposure values. Film B, on the other hand, has a log relative exposure range from 0.44 to 1.84, and, therefore, possesses a much wider range of exposure values and a greater margin for exposure error while still producing radiographic densities within the useful diagnostic range. This is the practical aspect of the concept of film latitude. For the technologist, the film that possesses greater film latitude (Film B) makes the selection of exposure factors less critical to the process of achieving a film of radiographic quality. Assuming that the technologist has chosen the proper kilovoltage level for adequate penetration, he or she would have a greater margin for error in the selection of the mAs value necessary to produce a satisfactory radiographic density using a film possessing a wide latitude.

As with film speed, there are also good reasons why differing film latitudes (responses) are available. In areas of high subject contrast, such as those normally investigated and demonstrated in examinations performed for the orthopedic

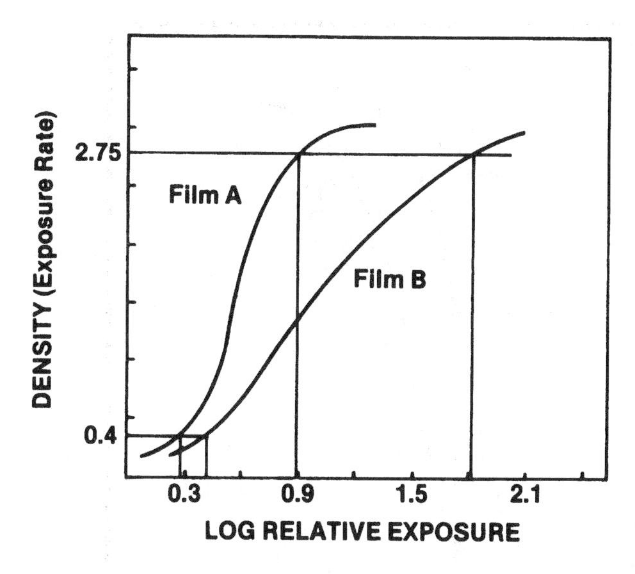

Figure 5-6
Comparison of the Latitude of Two Films

physician, Film A, having such an extremely high film contrast and short exposure latitude, may be contraindicated. In high subject contrast areas, Film A would make it more difficult to record and visualize an adequate number of different radiographic densities within the diagnostic range within your recorded image. The limitations of your film's inherent characteristics will produce and record an even higher contrast scale. Your recorded contrast scale will be more abrupt and limited in its interpretive value. In addition, the selection of your exposure factors will have to be extremely accurate. There will be little margin for exposure factor error. For orthopedic radiography, it may be more appropriate to select a film that possesses a lower film contrast and greater film latitude, such as Film B. This type of film enables you to extend the high subject contrast levels of the tissues and structures of the body recording more density differences within the diagnostic range and provides for a greater margin of exposure factor error. Overall, it enables the technologist to record and visualize more radiographic information.

On the other hand, if the procedures performed are for an internist or a urologist, the technologist would be examining areas of the body such as the abdomen that possess poor subject contrast. In these areas, it may be more advantageous to employ a higher contrast, shorter latitude film such as Film A in order to record a higher radiographic contrast within the structures being examined. By shortening the contrast scale, there is a greater chance that small, subtle differences in tissue or structural opacity can be recorded within the radiographic image. In this manner, greater information will ultimately be recorded within the diagnostic range of radiographic density.

Radiographic contrast can be said to be the result of the composite effects of subject contrast and film contrast and is influenced and determined by the selection of proper exposure factors for that particular procedure. Remember, with radiographic contrast, we are referring to the visibility of the recorded details—that is, how distinctly one structure is visible compared to an adjacent structure. Radiographic contrast is also influenced by appropriately controlled processing procedures. Properly controlled processing procedures are essential to the visualization of the radiographic contrast scale. Improper control of the film's processing can produce chemical fog, which will severely impair the radiographic contrast and visibility of the recorded image.

Radiographic Contrast Scale

A properly performed radiographic procedure will produce an image recorded within the useful diagnostic range of radiographic density. The useful diagnostic range of radiographic density recorded upon the film is visible and can be used to compare the differences between them as a ratio of densities. This ratio of compared radiographic density is called the **radiographic contrast scale.** The density ratio or contrast scale is determined by the number of useful diagnostic densities visible and the percentage of difference between them. A radiographic image is frequently described as possessing either high or low contrast.

High Contrast

In a high-contrast radiographic image, a measurement of the individual recorded densities will find that they are quite different from one another. The ratio of differences between individual radiographic densities is considerable. The total number of useful diagnostic radiographic densities, however, may be minimal. All of the many different tissue and structure thicknesses and opacities have to be recorded within this narrow range of recorded density. Another term used to describe this relationship is **short scale contrast.** This indicates that the differences in the recorded radiographic density are abrupt or shortened and that the subject contrast of the image is recorded within a minimal number of recorded and visualized radiographic densities. The radiographic density difference between two adjacent structures would be significant and would produce an image with more obvious visual differences.

As a general rule, radiographs of extremely high, short scale radiographic contrast are frequently of less diagnostic value than films that demonstrate more gradual density changes and visualize a greater total number of radiographic densities within the useful diagnostic range. High contrast, however, can be useful and used to advantage in areas of the body that exhibit poor subject contrast or procedures that require a more obvious presentation of an image or structure compared to the adjacent structures contained within the recorded image.

Low Contrast

In a low-contrast radiographic image, the individual recorded densities demonstrate more gradual differences from one another within the useful range of radiographic density. The ratio of difference from one adjacent density to another is a slight, less abrupt, and more gradual change than the radiographic contrast recorded in a high-contrast image. The total number of useful diagnostic radiographic densities would be increased. Another term used to describe this relationship of the radiographic contrast of the recorded image is **long scale contrast.** This indicates that the visible radiographic density differences are more gradual or lengthened and that the subject contrast of the image is recorded within a maximum number of useful diagnostic densities. There would actually be more information visibly recorded within a radiographic image possessing a low, long scale radiographic contrast. However, not all low-contrast images will be images of radiographic quality. It is possible to extend the scale of contrast too far, and the differences in the recorded radiographic density of the tissues and structures of the body would be so minimal that the actual information recorded and visible would be decreased. As a general rule, the extremes of either scale of contrast are contrary to radiographic quality. The technologist must be able to evaluate each procedure and examination in order to determine the proper exposure factors to select so that the recorded image possesses the appropriate radiographic contrast scale for the structures being examined and that they are demonstrated within the useful diagnostic range of radiographic density.

As an example, the useful diagnostic range of radiographic density and the radiographic contrast scale appropriate for an examination of the thorax for a rib fracture would be quite different from the proper radiographic contrast scale for a chest examination for lung pathology, even though the same structures are being recorded.

There are three major factors that influence the radiographic contrast scale:

---

1. **kVp/mAs relationship**
2. **Radiographic fog levels**
3. **Intensifying screens**

The primary factor to consider in the control of the radiographic contrast scale is the relationship established between the kilovoltage and mAs. The second major factor, no less important to the visualization of the recorded image, is the control and reduction of scattered radiations reaching the film in order to avoid producing excessive amounts of radiographic fog upon the recorded image. The third major factor influencing the radiographic contrast scale is the inherent properties of intensifying screens.

kVp/mAs Relationship

The proper application of the relationship between these two exposure factors is necessary for the production and maintenance of a proper radiographic contrast scale. When the kilovoltage selected is at the proper level for penetration of the part, the technologist has at his or her disposal the greatest amount of exposure latitude. This means that the technologist has some latitude in the selection of the mAs that will produce the diagnostic range of radiographic densities for the recorded image. The exposure latitude refers to the range of mAs from a minimum to a maximum that will still produce a diagnostic radiographic image. This allows for the maximum variance in the selection of the quantity of exposure (mAs) that can be used with the selected kilovoltage to produce an image containing sufficient radiographic density and appropriate radiographic contrast to visualize all of the structures of interest. In general, the production of a high-contrast radiographic image permits less exposure latitude than the production of a lower contrast radiographic image. **(Perform Experiment 53.)**

The technologist does exercise some control over the influence of the kVp/mAs relationship on the radiographic contrast scale recorded within the radiographic image. A shorter scale of contrast (high contrast) will result from an increase in the mAs and a reduction in the kilovoltage. When selecting this option, it is important for the technologist to recognize that the level of kilovoltage necessary for penetration must always be employed and that if the selected kilovoltage is insufficient, the overall radiographic quality of the image will suffer. A longer scale of contrast (low contrast) in the radiographic image can be produced by a decrease in the mAs and an increase in the kilovoltage. For this application of the relationship between the kilovoltage and the mAs, the 15% rule described in the chapter on radiographic density can be applied. (**Perform Experiment 54.**)

High Contrast: High mAs + Low kVp
Low Contrast: Low mAs + High kVp

By increasing the kilovoltage above the level required to penetrate the tissues and structures of interest of the body part, together with appropriately reducing the mAs (15% kVp ↑ = 50% mAs ↓), the radiographic density of the recorded image can be maintained, and the radiographic contrast scale of the recorded image will be extended. However, eventually a contrast scale will be produced that extends the radiographic contrast scale beyond a level where the proper visibility of the structures of interest can be maintained. Increases in the kilovoltage beyond the levels necessary for penetration will also produce significant levels of radiographic fog, which will degrade the radiographic quality of the recorded image. However, working within the limits of the subject contrast available within the body part to be examined and identifying the specific needs of the examination, the technologist can extend or shorten the radiographic contrast scale by manipulating the kVp/mAs relationship to produce a recorded image that possesses the total diagnostic information required by the physician to interpret and evaluate the results.

How does the technologist avoid the technical guessing game in the proper selection of exposure factors related to the production of radiographic contrast? What is the proper relationship of the kilovoltage and mAs factors? Their relationship is primarily governed by the subject contrast contained within the part to be examined and the purpose of the examination. The relationship is also influenced by the contrast capability of the film you are using and the desired radiographic contrast scale to be achieved. The technologist knows that he or she will always require the kilovoltage level necessary to penetrate the part. After the kilovoltage necessary to penetrate the part has been selected, the mAs required to produce the desired radiographic density for the examination should be selected.

---

kVp/mAs Relationship

1. **kVp: To Penetrate**
2. **mAs: To Maintain Density**

In analyzing the radiographic image, the technologist must determine whether proper penetration of the part has been achieved. If it has not, this relationship must be corrected before any additional adjustments are attempted. Insufficient kilovoltage prevents portions of the beam of radiation needed for the proper recording of the radiographic image from passing through the part. In fact, no practical increase in mAs will correct for a severely underpenetrated part. (**Perform Experiment 55.**) An analysis of the radiographic quality of the image is essential to determine the factors and influences that have contributed to the production of an unacceptable radiograph. A radiographic image possessing poor photographic properties frequently indicates that the relationship between the kilovoltage and mAs is probably unbalanced. The kilovoltage necessary for penetration of different body parts has been established (Table 5-1). Note that the kilovoltage level recommended for the skull is 80 kVp, while the paranasal sinuses recommendation is 70 kVp. This is due to the reduced opacity of the air-filled cavities of the paranasal sinuses. The kilovoltage level recommended for the gastrointestinal tract is 90 kVp, compared with 80 kVp for an abdomen. This is because a highly radiopaque substance, barium sulfate, has been added to the gastrointestinal tract in order to record it properly.

It cannot be overemphasized that when the proper kilovoltage has been selected, the technologist will be performing the procedure with the greatest possible exposure latitude. Therefore, the greatest possible margin for error between a minimal mAs and a maximal mAs that will still produce a diagnostic range of radiographic density exists. Exposure technique charts employing standardized levels of kilovoltage sufficient to penetrate the part and the selection of variable mAs levels to produce the desired radiographic density are among the most accurate and successful methods to establish quality control over the selection of exposure factors and to eliminate the problems inherent in the technical guessing game.

How does the technologist determine the proper mAs to produce the desired radiographic density for the examination? Exposure technique charts have been developed to standardize the application of exposure factors. One standardized exposure technique guide utilizes the concept of optimum kilovoltage. In this guide, once the optimum kilovoltage for penetration of the part has been selected, the mAs required to produce the desired density is chosen from a suggested table contained within the technique chart. Suggested mAs values have been developed based on the thickness of the body part and have been categorized into three main groups: (1) small, (2) average, and (3) large. To utilize the optimum kilovoltage technique chart effectively, the technologist must measure the thickness of the body part to be examined using a set of radiographic calipers. The chart will have a listing of all the various examinations and body parts together with suggested mAs values corresponding to the thickness range identified with the small, average, and large body habitus. The measurement of a small shoulder may be indicated by a thickness range

Table 5-1 Kilovoltage Necessary To Penetrate Various Body Parts

Body part	kVp
Small extremities	60
Large extremities	80
Skull	80
Paranasal sinuses	70
Abdomen, pelvis, and posterior vertebrae	80
Lateral vertebrae	80
Gastrointestinal tract	90
Chest, posteroanterior projection	80*
Chest, lateral	90
Chest, oblique	80

*Many radiology departments are using high kVp (120–150) for lung radiography using microline grids or air gap techniques.

of 6 cm to 8 cm, an average shoulder measurement by a thickness range of 10 cm to 12 cm, and a large shoulder may be identified by a thickness range of 14 cm to 16 cm on the standardized technique chart. After measuring the part and finding the mAs value corresponding to the thickness of the part, the technologist would adjust his or her control panel selection to the proper mAs identified by the technique chart. Approximately 90% of all examinations can be effectively performed using a standardized technique chart. To avoid the technical guessing game, the part to be examined must be measured accurately and the technical factors recommended by the technique chart used correctly. A detailed investigation of the application of standardized technical charts is introduced in Chapter 6. The use of standardized exposure factor technique charts is a positive application of the principles of ALARA. Using standardized exposure factor technique charts helps to reduce the number of repeat exposures to the patient.

Radiographic Fog

Radiographic contrast and radiographic fog are opposing factors and influences on the radiographic quality of the recorded image. The radiographic contrast scale deteriorates in the presence of radiographic fog. The unwanted density associated with radiographic fog can result from a number of factors (Figure 5-7). We have already investigated radiographic fog caused by light leaks, radiographic film age, and exhausted processing chemicals in Chapter 4. Our investigation in this chapter will focus on the radiographic fog resulting from secondary and scattered radiations reaching the film.

The presence of radiographic fog causes a reduction of image visibility. It degrades the entire radiographic contrast scale. Because it produces additional radiographic density, it increases the overall radiographic density of the image beyond the satisfactory, desired diagnostic levels. The increase in radiographic density resulting

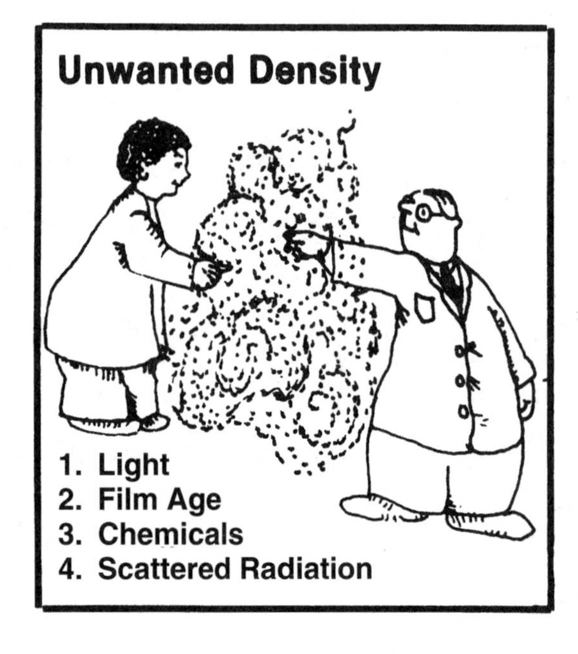

Figure 5-7
Some Causes of Radiographic Fog

from the presence of radiographic fog is not distributed equally throughout the entire recorded image. The additional radiographic density produced in areas of the body of less thickness and lower tissue opacity is frequently much greater than the additional density produced in areas of greater thickness and higher tissue opacity. Frequently, this addition of unwanted density to the recorded image creates a situation where some structures can no longer be visualized at all. Structures best recorded in the lower density levels will be increased to a radiographic density where they can no longer be demonstrated. Other structures of the body will have lost their distinctive radiographic contrast scale, making the visualization of those structures compared with adjacent structures difficult.

Radiographic fog, in and of itself, does not destroy recorded detail, since it is not a factor that controls or influences the actual recorded details of the image. Consider a comparison of radiographic fog to an early morning fog bank that has come in from the sea. As you leave your house to get in your car and drive to work, you may have difficulty seeing your car in the driveway. The car is still there, but the visibility of its location has been obliterated by the presence of the fog. The effect of radiographic fog on the visibility of the image is similar to this. The recorded details of the radiographic image are still present, but the overlying radiographic fog prevents you from visualizing or evaluating those details. By preventing the presence of radiographic fog or reducing the levels of this unwanted, uncontrolled radiographic density reaching the radiographic image, the recorded details present within the radiographic image can be visualized. The visibility of the recorded image is a photographic property of radiographic quality. The actual recorded detail of the image is a geometric property of radiographic quality. Several factors should be considered as to their influence on the production of radiographic fog.

FACTORS INFLUENCING RADIOGRAPHIC FOG

Patient Factors
1. **Tissue Volume**
2. **Tissue Thickness**
3. **Tissue Opacity**

Object-Image Distance
kVp/mAs Relationship

Patient Factors

A major factor in the production of radiographic fog is the patient factors that contribute to the subject contrast so essential to the production of radiographic contrast. Basically, as the volume of tissue increases (part thickness), a greater quantity of radiation (mAs) is necessary to produce a satisfactory radiographic density. However, as the quantity of radiation is increased in order to produce a satisfactory radiographic density in the thickest portion of the body part, other structures of less thickness may be recorded with excessive amounts of density produced by the levels of secondary and scattered radiations produced within those areas of the body part. The increase in the radiographic exposure is therefore accompanied by an increase in the amount of radiographic fog produced.

The various absorption characteristics of the beam and therefore the radiographic contrast scale also depend on the tissue opacity of the part being examined. As the opacity of the tissue increases, an increase in the kilovoltage is necessary to provide adequate penetration of the part. As a result of an increase in the kilovoltage, the absorption characteristics of the tissues and structures of the body are altered. The increase in kilovoltage necessary to penetrate the higher density tissues will produce increased radiographic fog, especially in the lower density tissues of the part. The less dense tissues of the body produce greater amounts of scattered radiation. Fluid in the tissues is one of the major factors in the production of scattered radiation. Tissues of lower density tend to have a high fluid content and, therefore, produce more scattered radiation than do tissues of higher density, which absorb greater amounts of the radiation.

Object-Image Distance

The object-image distance (OID) of the structures of interest within the part also influences the radiographic fog levels of the radiographic image. In Figure 5-8, Diagram A represents an object to be examined that has been placed directly upon the film surface. Three inches of absorbing material have been placed on top of it. The pattern and density of the radiographic image are represented below the figure. Very little scattered radiations will be produced by or emerge from the body part to influence the radiographic image. In Diagram B, the object to be examined is placed on top of the absorbing material, producing an OID of 3" (7.6 cm). The image pattern below the figure represents the increased magnification and increased radiographic density caused by the increased OID and the scattered radiations produced in the absorbing material below the structures of interest that can emerge from the body part in this arrangement. (**Perform Experiment 56**.)

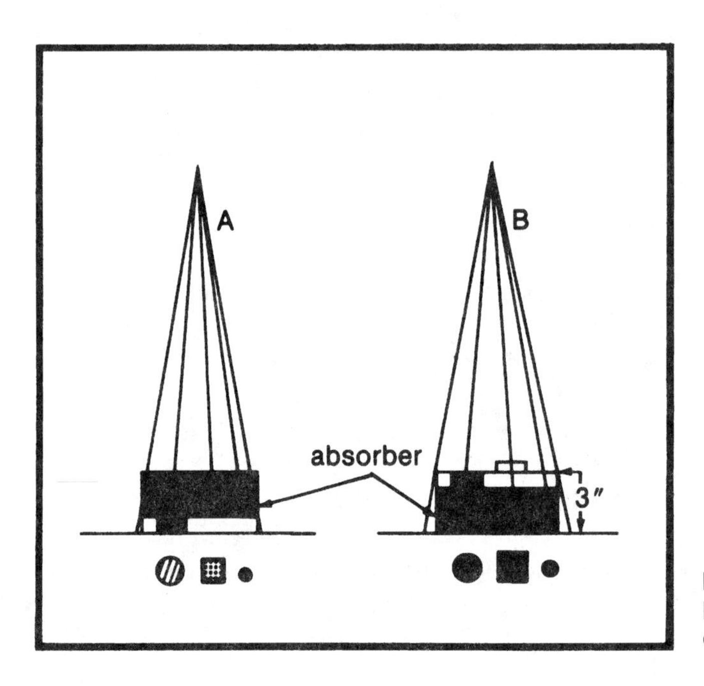

Figure 5-8
Diagram of the Effect of OID on Radiographic Fog

As the OID increases, scattered radiations produced by tissues lying closer to the film produce a layer of radiographic fog that produces an overall unwanted and uncontrolled increase in radiographic density and a reduction of the radiographic contrast in the recorded image. The greater the OID of the structures of interest, the greater the undercutting of the visibility of the image caused by scattered radiation being recorded over the primary, diagnostic level of radiographic density in the recorded image.

An interesting phenomenon related to OID and the production of radiographic fog takes place when the entire body part is placed a distance of between 6" and 10" (15 cm and 25 cm) from the film. This is known as the **air-gap principle**. Diagram A in Figure 5-9 represents an object placed on top of 3" (7.6 cm) of absorbing material. The entire structure is placed directly upon the film surface. The pattern and density of the radiographic image is represented below the figure. A significant portion of the radiographic density produced by this arrangement of the structures to be examined and the additional absorbing material results from the scattered radiations produced by the absorbing material reaching the film. In Diagram B, the object still lies on the 3" (7.6 cm) absorbing material, but the entire structure is raised 10" (25 cm) above the level of the film, creating an air gap and an OID of 10" (25 cm). The pattern of increased magnification created by the increased OID is obvious, but the overall radiographic density compared to Diagram A is decreased, and the radiographic contrast improved as a result of this air-gap principle. (**Perform Experiment 57.**)

Reviewing the diagrams in Figure 5-9, you can see that this principle may have some positive radiographic applications. Using the air-gap principle to advantage, the scattered radiations produced by the absorber are dissipated in the air gap in greater amounts than the image-forming radiations emerging from the body part because the scattered radiations are less energetic than the radiation that actually records the image. When utilizing the principle of air gap to advantage, the overall

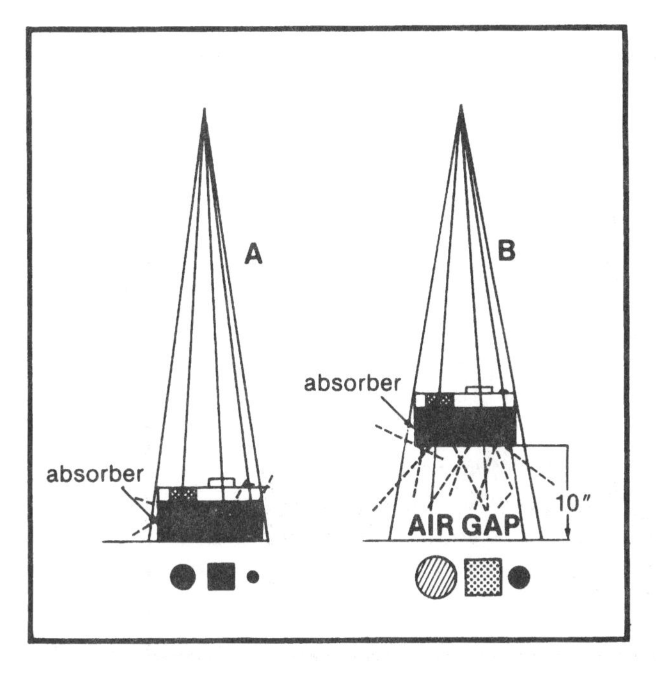

Figure 5-9
Use of the Air Gap Technique To Decrease Radiographic Fog

radiographic density of the image decreases and must be compensated for. The application of the air-gap principle can be used to advantage in specific but limited procedures. One of the most successful applications of the air-gap principle is with nongrid, high-kilovoltage chest radiography employing an increased source-image distance (SID) of 80" (203 cm). The increased SID is employed to compensate for the increased (OID) employed with the air-gap principle. The increased mAs usually necessary due to increased SID is offset by the ability to perform the radiographic procedure without the need for a radiographic grid.

kVp/mAs Relationship

As we have already discovered, the kVp/mAs relationship is one of the controlling factors that can be selected by the technologist in the production of a proper radiographic contrast scale. The relationship of the kilovoltage and mAs is also a major consideration in the production of radiographic fog. Remember, radiographic fog levels are at their minimum when the kilovoltage necessary to penetrate the structure has been selected, and the mAs has been adjusted to provide for the proper range of diagnostic radiographic density. Increases in kilovoltage above that necessary to penetrate the part will produce significant increases in scattered radiation and contribute to radiographic fog.

Reducing Radiographic Fog

The presence of radiographic fog is a major influence on the radiographic quality of the recorded image, and a number of methods and devices have been developed to reduce this unwanted, uncontrolled increase in the radiographic density of the recorded image. In fact, the reduction of radiographic fog is a major consideration in every radiographic procedure. Since radiographic fog and radiographic contrast are opposing influences on the radiographic quality of the recorded image, it would be safe to say that the reduction of radiographic fog will always increase or improve the visibility of the recorded image and the radiographic contrast. The two main methods of reducing radiographic fog are (1) the restriction of the x-ray beam and (2) the use of radiographic grid devices.

Beam Restriction

The primary method employed to reduce the radiographic fog attributed to secondary and scattered radiations is to limit the size of the primary beam as it leaves the x-ray tube. Some method of x-ray beam restriction should be employed in **all** radiographic procedures (Figure 5-10). The remnant or exit radiation emerging from the patient is the radiation that produces the radiographic image. After passing through the patient, this exit radiation consists of portions of the primary beam as well as significant amounts of secondary radiation produced within the body by the interaction of the primary beam with the body tissues and primary x-ray that has scattered and changed direction after interacting with body tissues. The energy levels of the secondary and scattered radiations emerging from the patient's body are often sufficient to produce a radiographic density upon the film.

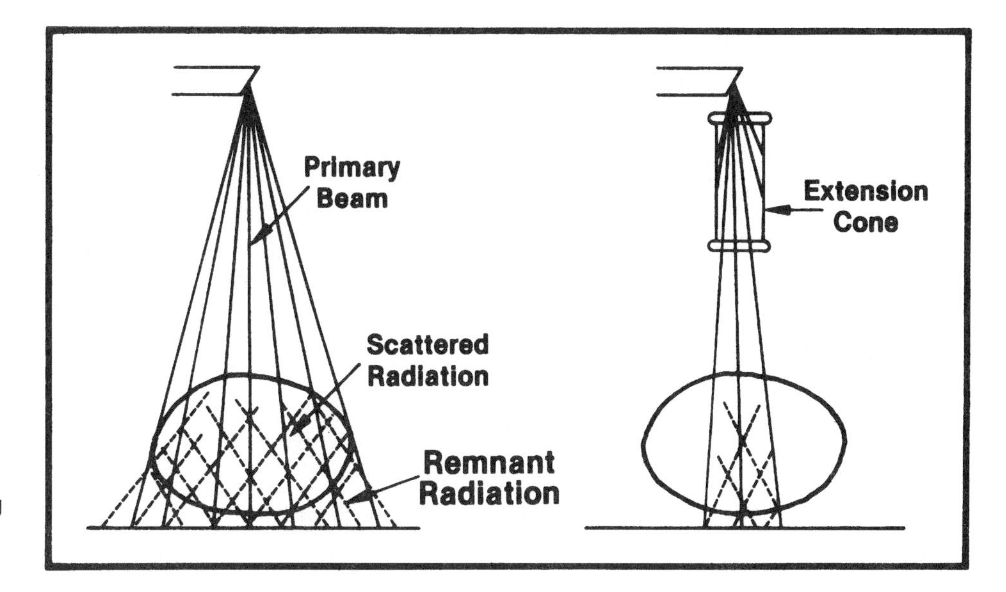

Figure 5-10
Use of Extension Cone To Reduce Radiographic Fog through Beam Restriction

One way to reduce the amount of secondary and scattered radiations produced by the body part is to limit the primary beam of radiation before it enters the patient's body. Ideally, the beam of radiation should be restricted to include only the area or structures of interest. In properly performed radiographic procedures, the beam of radiation is frequently smaller than the size of the film. Evidence of x-ray beam restriction demonstrated on the radiograph is a sign of good radiologic practice and the application of the principles of ALARA. A number of devices have been developed that restrict the beam of radiation, reduce the total exposure of the patient, and improve the radiographic contrast by reducing the levels of radiographic fog (unwanted radiographic density) upon the recorded image:

BEAM RESTRICTION DEVICES

1. **Aperture Diaphragms**
2. **Cones**
3. **Collimators**
4. **Lead Blockers**

Aperture Diaphragms

The use of aperture diaphragms is limited to specialized equipment. The aperture diaphragm is a device created from a flat metal sheet in which holes of different sizes and shapes have been cut. An advantage of this type of beam restriction is that the diaphragm can be specifically designed to suit a variety of film sizes and area shapes. However, it is somewhat inconvenient to use. With each different procedure, it is necessary to change manually the diaphragm to meet the requirements of the next examination. This means taking the diaphragm out of the diaphragm slot attached below the x-ray tube window and placing a new diaphragm into the slot. Additionally, since its attachment is located just below the x-ray tube window opening, the aperture diaphragm device cuts or restricts the beam at a considerable distance from the film. This allows for a penumbra (unsharpness) effect around the edges of the beam of radiation where it is recorded upon the film.

Cones

Another beam restriction device designed in the form of a cone is also located close to the x-ray tube window opening, but its similarity to the diaphragm ends there. The cone consists of a cylindrical tube of metal of a straight design, in which the area of the cone is the same at its proximal and distal ends; a flared design, in which the area of the cone is greater at its distal end than at its proximal end; or a straight design that allows the total length of the cylinder to be extended for even greater restriction of the beam. The cylindrical tubes of the cones are produced in different lengths and diameters and restrict the beam in a manner similar to the diaphragm proximally but provide an additional beam restriction distally, at the bottom of the cylinder's length. The longer the cylinder and the smaller the diameter of the opening, the more restrictive the beam limitation. As a result of this additional beam restriction at the distal end of the cone located several inches distant from the tube window, the unsharpness effect seen at the peripheral portions of the recorded image produced by a cone is less than that of the diaphragm.

It is also possible to purchase straight or flared design cones with a rectangular shape and openings that will limit the beam of radiation to a specific size of film at a certain SID, but the majority of cones are designed as cylinders. The x-ray beam emerging from the tube window produces an ever-widening beam of radiation directed toward the film. Unrestricted, the beam of radiation would form a circular image. The circular cone beam restriction devices also produce a circular beam of radiation. To cover the rectangular film sizes properly using a circular cone, it would be necessary to use a cone with a lower opening whose diameter would produce a circle larger than the actual film size. The cone coverage would have to be able to record the diagonal measurement of the film for total coverage of the radiograph. A review of the routine film sizes and their diagonal measurements are provided in Table 5-2.

For an abdomen examination on an adult using a circular-shaped cone, it would be necessary to use a cone device that would provide a 22" diagonal measurement in order to cover the entire 14" × 17" (36 cm × 43 cm) film surface. In this instance, large portions of the patient's body not included within the recorded image of the radiograph would be exposed to unnecessary radiation. The efficiency of the beam limitation, as well as the contrast improvement factor, would be diminished considerably with a cylindrical cone device since significant amounts of secondary and scattered radiations produced by portions of the body exposed to the primary beam outside of the structures of interest (size of the film) would be directed back toward

Table 5-2 Film Sizes and Their Diagonal Measurements

Film size (inches)	Diagonal measurement (inches)
5 × 7	8.5
8 × 10	12.75
10 × 12	15.5
11 × 14	17.75
14 × 17	22

the film and produce unwanted levels of radiographic density recorded upon the image as radiographic fog. Cones designed in the shape of a rectangle that would exactly cover the size of the film would be more advantageous to use in radiographic examinations of the abdomen. The application of the correct SID for the specific cone size used would be a major consideration for the proper application of the rectangular designed cone. In addition, like the diaphragm, cones must also be manually changed to adjust the desired beam restriction size to that required by different examinations, different-sized films, and different-sized patients.

A modification of the basic cylindrical cone, the extension cone is designed with two cylinders telescoping into one another. When collapsed, the beam restriction will be determined by the bottom diameter of the cone length. When the cone is extended, the longer cylinder length will further restrict the beam and limit the size of film that can be covered. The extension cone is more flexible in its application and can be used for more than a single film size, but its use is limited and it must still be manually changed to meet the requirements of different examinations. The extension cone also requires the application of a larger beam size to adequately expose any given film size.

Determination of Beam Size

It is important for the technologist to be able to determine the exact size of the beam of radiation that would result from the use of a selected cone or diaphragm. The size of the area covered by the beam restriction device is affected by the distance from the tube focus to the bottom of the beam-limiting device, the diameter or size of the opening of the beam-limiting device, and the SID. Formulas have been developed enabling you to calculate these factors and determine the actual size of the x-ray beam limitation. Let's apply these formulas in the following problems.

DIAPHRAGM FORMULA

$$\frac{\text{Anode-Diaphragm Distance}}{\text{Diameter of Aperture}} = \frac{\text{Source-Image Distance}}{\text{Size of Projected Image}}$$

1. A diaphragm is placed 3" (7.6 cm) from the tube focus. The SID is 60" (152 cm). The diaphragm has an opening cut into it 1" (2.54 cm) in diameter. How much coverage would you record?

   $3/1 = 60/x \qquad 3x = 60 \qquad x = 20''$

   A beam of radiation 20" (51 cm) would be recorded by this diaphragm. A 14" × 17" (36 cm × 43 cm) film has a diagonal length of 22" (56 cm) and would be adequately covered by the beam of radiation using this diaphragm, although some areas of the body outside of the recorded image would be exposed to unnecessary radiation. The use of any film smaller than 14" by 17" (36 cm × 43 cm) would be inappropriate with this diaphragm.

2. A diaphragm with a rectangular opening is to be placed 4" from the tube focus. The SID is 40" (102 cm). What must your diaphragm opening be in order to cover a 10" by 12" (25 cm × 30 cm) film?

   $4/x = 40/10 \qquad 40x = 40 \qquad x = 1''$

   $4/x = 40/12 \qquad 40x = 48 \qquad x = 1.2''$

A diaphragm with a rectangular opening of 1" by 1.2" (2.54 cm × 3.04 cm) would cover a 10" by 12" (25 cm × 30 cm) film size exactly at a 40" (102 cm) SID. If the SID changed or the size of the film changed, the diaphragm would also have to be changed. This diaphragm would be an improvement over the circular diaphragm described above, and the image would suffer from less radiographic fog.

CONE FORMULA

$$\frac{\text{Source-Image Distance} \times \text{Lower Diameter of Cone}}{\text{Distance from Tube Focus to Bottom of Cone}}$$

3. A new cylindrical cone has been obtained. The diameter at its lower rim is 5" (12.7 cm). The distance from the tube focus to its lower rim is 15" (38 cm). The procedure will use a 36" (91 cm) SID. What beam restriction can you expect using this cone?

$$36 \times 5 \div 15 = x \quad 180 \div 15 = 12''$$

A beam of radiation measuring 12" (30.4 cm) in diameter would be covered by the cylindrical cone device. An 8" by 10" (20 cm × 25 cm) film has a diagonal length of 12.75" (32.4 cm) and would be properly covered by the use of this cone at this SID. However, a 10" by 12" (25 cm × 30 cm) film with a diagonal length of 15.5" (39.4 cm) would not be properly covered with this cone. A film smaller than an 8" by 10" (20 cm × 25 cm) film would be inappropriate to use with this cone. (**Perform Experiment 58.**)

Collimators

The collimator incorporates the basic principles of both the diaphragm and the cone and refines and improves on them within a single, permanently attached apparatus that can be used for different film sizes, different SIDs, and different examinations. A collimator has a box-like appearance and is attached to the lower end of the x-ray tube housing adjacent to the tube window. Its precision design accurately defines the limits or edges of the x-ray beam with a minimal amount of image edge penumbra (unsharpness). It provides excellent beam limitation through the use of metal, lead-lined shutters placed within the device at different levels from the tube opening. These additional beam limitation devices limit the beam of radiation at several points distant from the tube focus before it is finally directed toward the patient. A unique advantage of the collimator is that the beam size can be adjusted with relative ease to the specific size required by the size of the structures of interest or the film size to be employed by the procedure. The beam limitation device of the collimator is continually adjustable and can restrict the beam to accommodate many different sizes of film and different structure sizes at different SIDs by the mere rotation of a dial or knob on the face of the collimator itself.

The most recent advance in beam restriction using collimators is the application of "positive beam" limitation. With positive beam restriction, the collimator is automatically adjusted to the size of the film placed in the Bucky tray. Sensors within the Bucky tray connected to the collimator automatically change the size of the openings of the beam restrictors to limit them to a maximum of the film size chosen for the procedure. The technologist cannot open the beam size larger than the film size used. If attempts are made to enlarge the beam restriction size manually, the collimator will automatically reduce the beam restriction to the size of the film placed in the Bucky tray. However, the technologist can still reduce the beam size to an area smaller than the film size by additional manual adjustments. The automated collimation control is limited to Bucky procedures. Manual collimator adjustments must still be performed for tabletop procedures. It is obvious that the collimator is a much more efficient device for the technologist to use and is more effective in performing the functions of beam restriction than is the diaphragm or the cone.

In addition, collimator devices are equipped with a light or laser beam that is projected toward the patient. This light or laser beam projects an image of the actual x-ray beam size and shape upon the body part being examined. The use of the projected light or laser beam upon the body part enables the technologist to center the part accurately for the examination and to adjust the beam limitation further to the structures of interest, which are often much smaller than the film size used for the examination. It is important that the projected light or laser beam be accurately aligned to the edges of the actual limits of the x-ray beam. With extended use and/or mishandling, the alignment of the light source with the x-ray beam may be compromised. A number of procedures have been developed to test the accuracy of the x-ray beam alignment. Such a test should be included in the quality assurance program of the radiology department. (**Perform Experiment 59.**)

Remember, the proper use of beam limitation devices is the primary technique to employ in your attempt to reduce radiographic fog and improve the radiographic contrast (visibility) of the recorded image. Beam limitation must be employed in all radiographic procedures, and, ideally, the devices should be adjusted to restrict the beam of radiation to expose only the area or structures of interest. (**Perform Experiment 60.**) Beam limitation devices actually restrict the size of the primary beam before it enters the patient. By restricting the primary beam of radiation, the technologist is thereby reducing the volume of tissue being exposed to radiation. This reduces the overall exposure of the patient and the amount of secondary and scatter radiations produced by the part. With severe beam limitation, there will be significant and noticeable decreases in the overall radiographic density of the recorded image. In order to maintain the radiographic density of the image when you "cone down" or restrict the beam, it may be necessary to increase the quantity of radiation. The necessary increase in exposure (mAs) depends on the thickness and opacity of the part being examined and on how severely you have restricted the beam.

Restricting the beam of radiation when changing from a 10" by 12" (25 cm × 30 cm) film to an 8" by 10" (20 cm × 25 cm) film requires little, if any, increase in exposure to maintain the radiographic density of the recorded image. However, changing from a 14" by 17" (36 cm × 43 cm) film to an 8" by 10" (20 cm × 25 cm) film will require an increase in mAs of approximately 50% in order to maintain the radiographic density of the recorded image. (**Perform Experiment 61.**)

Lead Blockers

An excellent way to improve the radiographic contrast by reducing the radiographic fog recorded upon the radiographic image is with the application of flexible sheets and lengths of rubberized lead cut into various rectangular sizes or different shapes to be used with specific examinations. These lead blocking devices are placed on the tabletop adjacent to the part being examined. The most effective application of lead blockers will place the sheets of lead as close as possible to the structures of interest or body part without their possible interference with the diagnostic requirements of the examination. Inappropriate use of lead blockers can potentially eliminate the recording of portions of the structures of interest or information required by the physician to make a proper diagnosis. Thus, although the use of lead blockers can enhance the radiographic contrast of the radiographic image, care should be exercised in their use. Lead blocking devices can and should be employed for both nongrid and grid tabletop examinations and with Bucky examinations when their use will not interfere with the recording of the structures of interest.

The contrast improvement factor associated with the utilization of lead blocking devices is dependent on the thickness of the body part and especially on the opacity of the tissues being examined. Low-density, high-fluid-level tissues, such as areas of the body possessing large deposits of fatty tissue, produce significant amounts of secondary and scattered radiations that must be prevented from reaching the film in order to produce a film of radiographic quality. The use of lead blockers for tabletop, nongrid examinations of even small body parts is recommended since there is nothing else to prevent the emerging secondary and scattered radiations produced by the body part from interacting with portions of the body beyond the structures of interest and outside of the beam limitation area. These emerging radiations will also interact with the surface of the examination tabletop and the cassette frame and will be directed toward the film. All of these multiple interactions can produce additional, unwanted radiographic density to be deposited within the recorded image.

With thicker body parts, whether employing a radiographic grid as a tabletop procedure or performing a Bucky examination, the procedure has the additional advantage of the contrast improvement factor associated with the use of a radiographic grid. Thus, the actual contrast improvement will differ between a nongrid procedure and a procedure employing a radiographic grid. However, there will be visible radiographic contrast enhancement whenever lead blocking devices are employed. Lead blocking devices are used in addition to appropriate beam limitation, not in the place of it.

Secondary and scattered radiations produced by the body part emerge out of the body in all directions, not just in the direction of the film. Areas of the body struck by the primary beam of radiation that are not within the area of the structures of interest will also produce secondary and scattered radiations that can travel in a direction toward your film and be recorded within your actual recorded image producing levels of radiographic fog.

Although beam restriction of the primary beam provides for significant contrast improvement, frequently the size and shape of the beam restriction does not coincide with the size or shape of the structures being examined. For example, the rectangular shape of the beam restriction of a collimator device does not coincide with, or resemble, the oval shape of the skull. In addition to the skull examination, there is a large number of other radiographic examinations in which many areas of the body and/or portions of the film lying outside the actual structures of interest will be exposed to both primary and/or exit radiation made up of remnant portions of the primary beam and secondary and scattered radiations produced by the body part and directed toward the radiographic film. These additional uncontrolled sources of radiation will produce additional radiographic density and reduce the radiographic contrast of the image as a result of radiographic fog.

In some examinations, the problem is not that the beam cannot be restricted to the shape of the part, but that the part to be examined is larger than the size of the beam restriction. As an example, the beam restriction for an abdomen examination on an adult will be restricted to cover a 14" by 17" (36 cm × 43 cm) film size. However, on a large patient, the size of the abdomen will often be wider than the 14" width of the film, and the length of the body will be considerably longer than the 17" film length. In these examinations, many areas outside the limited restrictions of the collimator beam will become a source of secondary and scattered radiations that could ultimately emerge from the body in the direction of the film to cause increased radiographic density (radiographic fog) over the recorded image. Therefore, the use of lead blocking devices help to enhance the action of the beam restrictors and further limit the amount of radiation reaching the film in areas of the film not necessarily identified within the structures of interest. All applications of lead blocking devices will reduce the levels of radiographic fog and improve the radiographic contrast of your recorded image.

Figure 5-11 illustrates the proper application of lead blocker placement in a lateral projection of the sacrum and coccyx. In this examination, significant amounts of secondary and scattered radiations would be produced within the thick body part as well as within areas of the body outside the structures of interest, such as the buttocks and posterior areas of the thighs. The buttocks, thighs, and areas beyond the posterior aspect of the spine in the areas covered by the examination would all add considerable secondary and scattered radiations that would be directed toward the radiographic film. Because of the shaping limitations of the collimator device to a rectangular shape, a strategically located number of flexible, rubberized lead blockers should be placed on the surface of the tabletop adjacent to the back of the patient and beneath the protruding tissue of the buttocks in order to absorb the secondary and scattered radiations produced by these areas. By effectively absorbing these radiations that do not contribute to the recording of the structures of interest, you will prevent them from reaching the film and causing levels of unwanted and uncontrolled radiographic density from being recorded upon the structures of interest. (**Perform Experiment 62.**)

Procedures Utilizing Lead Blockers

Remember that lead blocking devices that interfere with or cut off portions of the structures of interest due to improper application may cause the need for a repeat

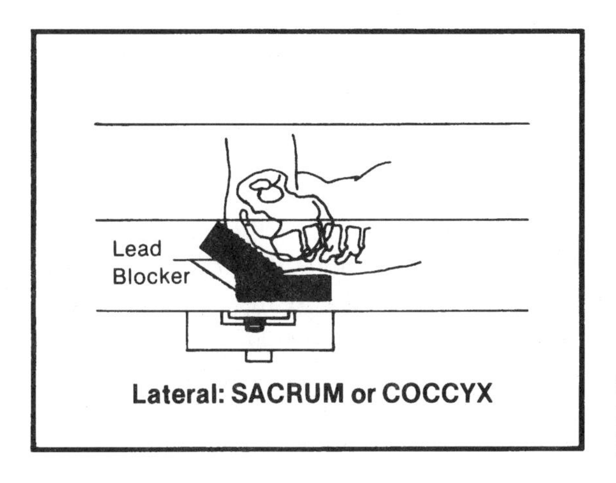

Figure 5-11
Use of Lead Blockers To Reduce Radiographic Fog

examination, so care should be employed in their application. The length of the rubberized lead sheets used as radiation blockers must be at least equal to the length of the film size or the structures of interest, whichever is longer. The width of the lead blockers must be sufficient so that they can absorb any potential radiation outside of the structures of interest that will contribute in a negative manner to the recording of the image.

Lead blocking materials can and should be used in most routine radiographic procedures. The use of lead blocking materials placed in a proper position close to the structures of interest or the body part being examined will not interfere with or affect the performance of the examination. However, their use will significantly improve the visibility of the structures of interest by reducing the amount of radiographic fog reaching the film and increasing the radiographic contrast within the recorded image. The following list identifies some of the examinations that would benefit from the use of lead rubber blocking sheets placed adjacent to the structures being examined:

- Anteroposterior and lateral forearm; lead blockers placed lengthwise along the medial and lateral sides of the forearm
- Anteroposterior and lateral humerus; lead blockers placed lengthwise along the lateral side of the humerus
- Anteroposterior shoulder; lead blockers placed above and along the lateral side of the shoulder
- Anteroposterior and lateral ankle; lead blockers placed along the medial and lateral sides of the ankle
- Anteroposterior and lateral leg; lead blockers placed along the medial and lateral sides of the leg
- Anteroposterior and lateral knee; lead blockers placed along the medial and lateral sides of the knee
- Anteroposterior and lateral femur; lead blockers placed along the medial and lateral sides of the thigh
- Anteroposterior and lateral hip; lead blockers placed along the lateral side of the hip
- Anteroposterior pelvis; a lead blocker can be placed on each side of the midline of the tabletop along the lateral sides of the pelvis, 8.5" from the center, for a total (width) distance between the lead blockers of 17". The blockers must be at least 14" long and should be taped in position. The patient is positioned over the blockers and centered to the midline of the table.
- Anteroposterior abdomen; a lead blocker can be placed on each side of the midline of the tabletop along the lateral sides of the abdomen, 7" from the center, for a total (width) distance between the lead blockers of 14". The blockers must be at least 17" long and should be taped in position. The patient is positioned over the blockers and centered to the midline of the table.
- Anteroposterior cervical spine; lead blockers placed along the two lateral sides of the neck
- Anteroposterior thoracic spine; a lead blocker can be placed on each side of the midline of the tabletop, 3" from the center, for a total (width) distance between

the lead blockers of 6". The lead blockers should be taped in position. The patient is positioned over these blockers and centered to the midline of the table.

- Lateral thoracic spine; lead blockers placed behind the patient's back at the level of the thoracic spine
- Anteroposterior lumbar spine; a lead blocker can be placed on each side of the midline of the tabletop, 4" from the center, for a total (width) distance between the lead blockers of 8". The lead blockers should be taped in position. The patient is positioned over these blockers and centered to the midline of the table.
- Lateral lumbar spine; lead blockers placed behind the patient's back at the level of the lumbar spine
- Lateral lumbosacral articulation; lead blockers placed behind the patient's back and beneath the protrusions of the buttocks at the level of the lumbosacral articulation
- Lateral sacrum and coccyx; lead blockers placed behind the patient's back and beneath the protrusions of the buttocks at the level of the sacrum and coccyx
- Lateral sternum; lead blockers placed in front of the sternum
- Skull series; lead blockers placed along the top and sides of the skull. Frequently a mask of rubberized lead with a cut-out design in the shape of the skull is used.

The use of flexible, rubberized lead blocking materials will significantly improve the visibility of the radiographic image. Failing to apply lead blockers in examinations where they can be used to advantage demonstrates a lack of concern on the part of the technologist for producing films of radiographic quality and a total disregard for the principles of ALARA by producing radiographs with excessive levels of radiographic fog. Poor quality radiographs may require repeat examinations and additional, unnecessary radiation to the patient.

Gonadal Shields

Flexible, rubberized lead blocking materials can also be utilized as gonadal shields. A number of commercially designed gonadal shields are available. However, using basic designs that are applicable to the male and the female patient, most departments construct their own gonadal shields from sheets of rubberized lead. Specially designed and shaped lead blockers employed for gonadal shielding must be of a minimal thickness of 2.5 mm lead equivalent. Gonadal shields are employed in radiography to improve the radiation protection of the patient. They are used in addition to the beam limitation devices, not in place of them. They provide additional protection to the highly radiosensitive gonad organs. Gonadal shields are required whenever the gonads lie within the primary beam and the use of the shields will not interfere with the recorded image needed for the diagnostic evaluation of the radiograph. When applied properly, the use of gonadal shields will reduce the genetically significant radiation dose to the patient. Gonadal shields, unlike the routine lead blocking devices, are specifically designed and utilized to improve the radiation protection of the patient and not to improve the radiographic quality of the image. As such, the use of gonadal shields is another example of good radiological practice and the application of the principles of ALARA.

In examinations of the pelvic region, including routine examinations of the femur, hip, pelvis, and abdomen, the radiation exposure of the gonads can be significantly reduced through the use of gonadal shielding. The male gonads lie within the scrotal sac below and external to the pelvic girdle, while the female gonads lie within the pelvic cavity. There are occasions when the information required by the examination may preclude the use of a gonadal shield. When it is necessary to visualize the symphysis pubis or pubic and ischial bones, it may be impossible to use gonadal shielding in the male, but it may still be possible to use a gonadal shield on a female for the same examination. There are many occasions when an examination of the abdomen will allow for the use of gonadal shielding in the case of a male, but the use of gonadal shielding for the female might be contraindicated because of its interference with the recorded image required for the diagnosis. Gonadal shields can and should be used for both male and female patients and should be employed on all patients (infants, children, and adults) whenever their use is not contraindicated by the information required to be recorded within the radiographic image. The design of the gonadal shield differs for the male and the female, and several different sizes of shields should be available to accommodate different patient sizes.

The previous discussion of gonadal shields identified those devices that are placed directly upon the surface of the patient's body and frequently secured in place with the use of adhesive or masking tape. Another type of gonadal shield that has been increasing in popularity is a device that can be attached to the collimator apparatus. This device is externally attached to the radiographic collimator usually with flexible metal tubing. The gonad shield can be moved into position by manipulating the flexible metal tubing when it can be used to advantage in the examination being performed. It can be moved out of the way when the shield would interfere with the requirements of the examination. Being permanently attached to the collimator apparatus, there is evidence that this type of gonadal shielding is more frequently employed during radiographic examinations due to its ease of operation and application than are devices that have to be physically placed and taped into position upon the patient.

Once the patient is positioned properly upon the examination table, the x-ray tube is brought over the patient, and the collimator light is turned on. The x-ray beam is centered to the part, and, if applicable, the gonadal shield attached to the collimator can be moved into position. The presence of the gonadal shield will cast a shadow within the beam of light projected from the collimator. By observing the shadow of the light projected upon the patient's body, the position of the gonadal shield can be adjusted in its proper position. The position of the shield will obliterate the shadow of the light in the area of the patient's body to be protected and accurately reflect the area where the primary radiation coming from the x-ray tube will be intercepted and eliminated from the radiographic image.

Radiographic Grids

As the primary beam of radiation traverses the patient's body, a portion of the radiation interacts with tissue and produces secondary radiation. These secondary radiations produced by the body tissue are usually of a shorter wavelength and lower penetrating ability than the primary radiation. However, some of the secondary radiation produced by the body may have sufficient energy to pass through the patient's body and reach the radiographic film. Secondary radiations produced by the body part will travel in many directions from its source of origin, just like the radiation produced in the x-ray tube. As such, those with sufficient energy to pass through the patient's body will be directed toward the film in many different directions, hindering the recording of a quality radiographic image.

Of greater concern to the production of radiographic fog is the scattered radiation produced by the body. In other interactions with body tissues, the primary beam upon striking a structure within the body is simply scattered (changed in direction) rather than being absorbed. Significant percentages of scattered radiations resulting from these interactions have energies nearly equal to the primary beam. They are able to pass through the body and reach the film from many different angles and directions (see Figure 5-12). This multidirectional path of the secondary and scattered radiations produced within the body frequently adds unwanted radiographic density and reduces the radiographic contrast (visibility) of the recorded image. (**Perform Experiment 63.**)

The radiographic density of every recorded image includes some secondary and scattered radiation. The radiographic quality of the recorded image is improved when the percentage of multidirectional scattered radiation reaching the film is kept to a minimum. Reducing the percentage of scattered radiation reaching the radiographic film is the primary function of a radiographic grid. In a nongrid chest examination, it has been determined that the radiographic density of the recorded

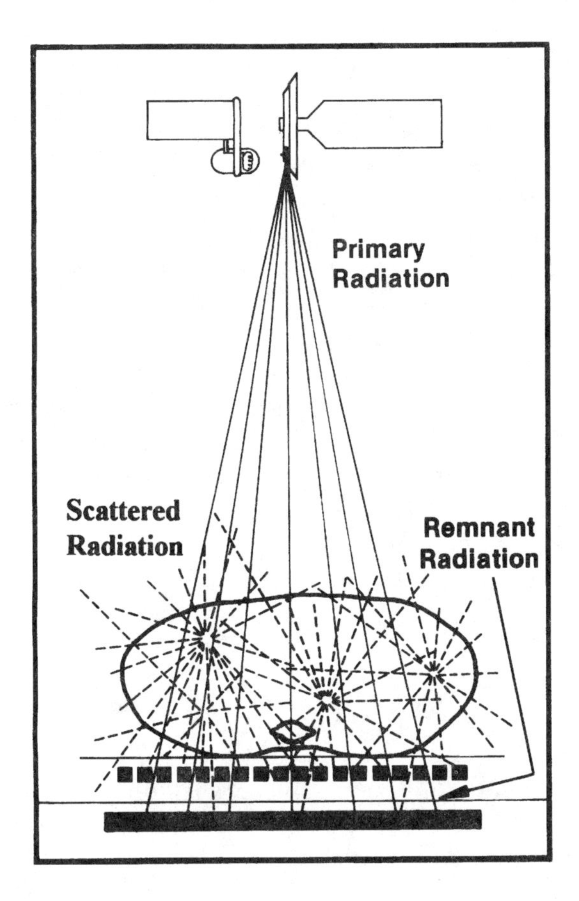

Figure 5-12
Use of a Grid To Reduce the Amount of Radiographic Fog Recorded by the Film

image represents a ratio of approximately 50% exit radiation from the primary x-ray beam of radiation and 50% scattered radiation produced by interactions with tissues contained within the body part. In a nongrid abdomen examination, the results are even worse; over 90% of the radiographic density of the recorded image is produced by scattered radiation produced by the body part. This higher percentage of scattered radiation results from the lower density, higher fluid content tissues that comprise the structures contained within the abdomen. Lower density tissues produce higher amounts of scattered radiation.

The radiographic grid is the most effective device for reducing the percentage of unwanted secondary and scattered radiation directed toward the film (Figure 5-12). The grid is a thin, flat, rectangular- or square-shaped device made of alternating strips of lead and a radiolucent substance. The grid is placed directly between the patient and the radiographic film. (**Perform Experiment 64.**)

The unique design of the grid allows for a much greater percentage of the exit radiation associated with the primary beam to reach the film, while absorbing most of the scattered radiation directed toward the film. Radiation that strikes and interacts with the lead strips of the grid is absorbed. Radiation striking the radiolucent material between the lead strips of the grid is permitted through the grid and allowed to record a radiographic density upon the film. The remnant portions of the primary beam exit the body in a predictable pattern, are permitted through the radiolucent materials between the lead strips of the grid, and are directed toward the film as a result of the design of the grid. The nature and direction of the secondary and scattered radiation produced by the body directed toward the grid and the film are unpredictable. The grid's design will prevent most of this unwanted, unpredictable radiation from passing through the grid so that it cannot reach the film. The effectiveness of a grid relates to how well it discriminates between the remnant portions of the primary beam and the secondary and scattered radiation emerging from the patient's body that attempts to pass through it.

Radiographic grids are expensive, precision devices that must be handled with care. They are thin, flat, rectangular- or square-shaped, wafer-like devices that can be damaged from mishandling, bending or striking of the edges or midsection of the grid against a hard surface, or improper storage. (They should be stored on their sides within a nonflexible, sturdy, protective case.) The precision design of the radiographic grid can be damaged by mishandling it or storing it flat and placing heavy materials on top of it. This will produce pressure upon the lead strips that can potentially cause them to change their shape, direction, or position, any of which can jeopardize the precise alignment of the lead strips. In most instances, a damaged grid cannot be repaired and can no longer be used in radiography as it will not function as intended and will produce numerous unacceptable patterns of grid lines and grid cutoff within your recorded image. Damaged grids must be replaced, and grids are expensive.

Physical Characteristics of Grids

A number of factors and influences must be examined to gain a complete understanding of the operation and effectiveness of a radiographic grid. Among these factors are the physical characteristics of a grid.

PHYSICAL CHARACTERISTICS OF RADIOGRAPHIC GRIDS
DESIGN: Linear or Crosshatched
GRID TYPE: Parallel or Focused
GRID RATIO
NUMBER OF GRID LINES PER INCH
GRID APPLICATION: Stationary or Moving

Grid Design

The overall, basic design of a grid affects its efficiency and the margin for error related to its application. Figure 5-13 shows three types of grid design.

Most radiographic grids are of the **linear design**. In the linear design grid, the lead strips are aligned adjacent to one another and placed lengthwise in the same direction within the structure of the grid. The linear design grid would be recorded as a pattern of thin lines lying adjacent to one another on the radiograph. The **crosshatch** or **criss-cross design** grid is a modification of the basic linear pattern. The crosshatch design grid uses two linear grids and aligns their linear patterns at right angles to each other. This significantly improves the efficiency and effectiveness of the grid function, but also considerably reduces the margin for error in its application. Using a regular linear design grid, the technologist can angle the central ray in the direction of the lead strip pattern of the grid's design. A crosshatch design grid does not permit you to angle your central ray in *any* direction. The central ray *must always be aligned and remain perpendicular* to the film for all radiographic examinations employing this type of grid design. This significantly limits the application of the crosshatch design grid and prevents its use with many routine radiographic procedures. The **rhombic crosshatch** grid design has properties and limitations in its application similar to those of the regular crosshatch grid.

Grid Type

There are two basic types of radiographic grids employed in radiography: (1) parallel and (2) focused.

Parallel Grids. **Parallel grids** are usually linear design grids constructed so that all of the lead strips lie parallel to each other within the design of the grid. If the lead strips were extended above the grid in straight lines, the strips of lead would never meet. Most grids employed in radiography as stationary grids are of the linear, parallel type design.

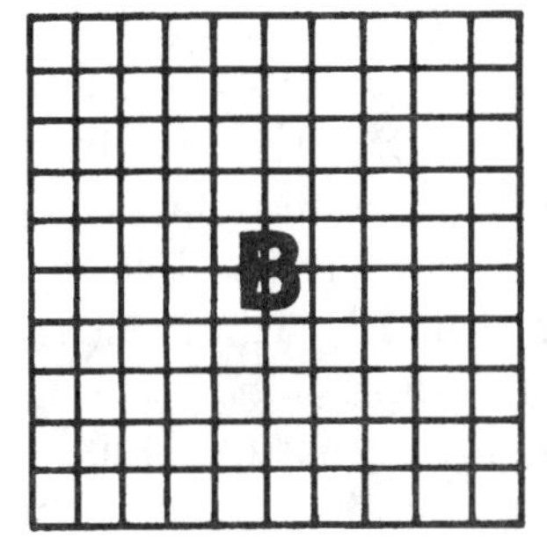

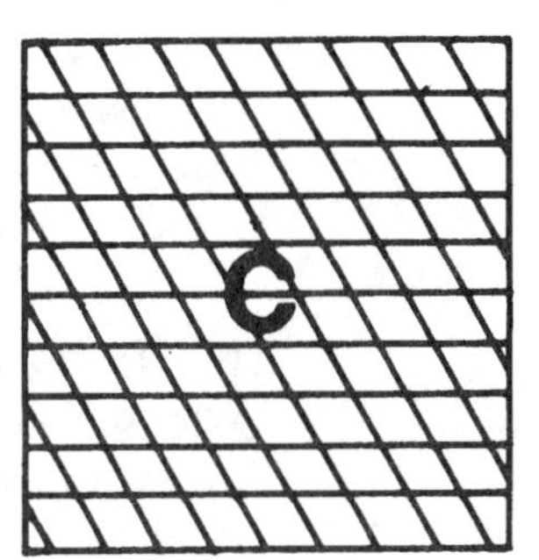

Figure 5-13
Three Types of Grid Design: (A), Linear; (B), Crosshatch; (C), Rhombic Crosshatch

Parallel grids used as stationary grid devices have numerous applications in radiography. Parallel grids are available in a variety of sizes that match the usual film (cassette) sizes used in the radiology department. They can be purchased separately, and, when required, they would be placed upon the front surface of a regular cassette and taped into position there. The cassette can then be used to serve a double-duty purpose, for both grid and nongrid procedures. Following their application, they can be removed from the cassette and stored until needed for another procedure, and the cassette can be placed back into service for nongrid procedures. Parallel grids purchased as separate devices are especially prone to damage from mishandling and pressure. The larger the size of the grid, the heavier it is because of the greater amount of lead contained therein and the more difficult to handle and protect from damage. Generally, the larger the size of the grid, the greater the potential for damage.

Parallel grids can also be incorporated into the frame of the cassette front during the manufacture of the cassette. Parallel grids permanently incorporated into the construction of the cassette have improved stability and less chance of damage from mishandling, which should extend its useful life. On the other hand, **grid cassettes** are extremely expensive, and they would be limited in their application in radiography to those procedures that require the use of a radiographic grid. They could not be used for the examination of small body parts or other radiographic examinations that do not normally require the application of a radiographic grid. When not in use, grid cassettes must be stored on their sides to prevent pressure damage to the lead strips of the grid. Figure 5-14 is a diagram of the cross-sectional design of a parallel grid.

Parallel Grids and SID. One of the factors that influences and produces limitations to the effective application of a parallel grid is the SID. Although the application of a parallel grid does allow for considerable latitude in the use of different SIDs, it is also a factor that limits its effectiveness. A parallel grid can perform its function effectively at a 40" (102 cm) SID from the smallest of film sizes up to and including a 14" by 17" (36 cm × 43 cm) film. With a parallel grid, as the SID is increased, the portions of the beam of radiation directed toward the film would include less of the more divergent, peripheral portions of the beam and more of the perpendicular, central portion of the beam. Therefore, as the SID is increased, the efficiency of the parallel grid is increased. A parallel grid would function more effectively at eliminating secondary and scattered radiations from penetrating through it at a 72" (183 cm) SID than at 40" (102 cm) since the major portions of the x-ray beam intercepted by the grid would represent the more perpendicular, central portion of the x-ray beam, which is aligned to the parallel pattern of the alignment of the lead strips contained within the parallel grid. Therefore, parallel grids are ideal choices for procedures that may be performed at increased SIDs.

However, at reduced SIDs, a parallel grid would not perform as effectively, and, depending on the film size required by the examination, **grid cutoff** at the lateral edges of the radiographic image may occur. Using a 14" by 17" (36 cm × 43 cm) film with a parallel grid at an SID less than 36" (91 cm) will result in evidence of remnant radiation associated with the primary beam being absorbed in the areas that coincide with the lateral edges of the grid (Figure 5-14). If the SID were reduced further, the grid cutoff would increase and influence greater portions of the radiographic image. At reduced SIDs, grid cutoff occurs because the lead strips aligned parallel with one another no longer coincide with the divergent pattern of radiation directed toward

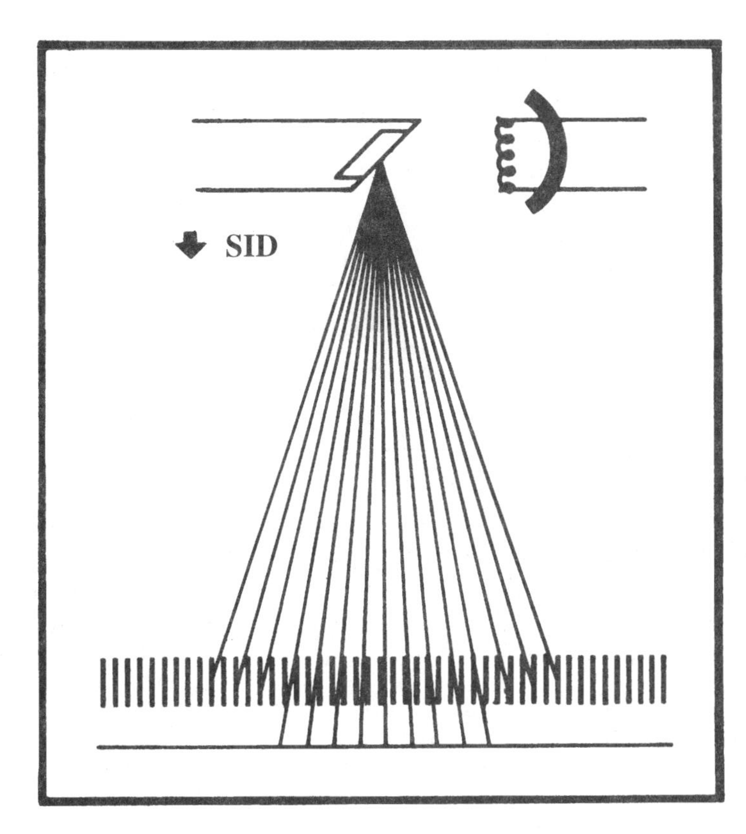

Figure 5-14
Primary Beam Absorption (Grid Cutoff) by the Lateral Edges of a Parallel Grid Caused by a Reduction of the SID

them. This lack of alignment between the divergent pattern of radiation directed toward the grid and the actual lead strip pattern of the grid's parallel design would be greater in the peripheral portions of the beam, which are directed toward the extreme lateral edges of the grid. With grid cutoff, the portions of the radiographic image recorded further from the midline of the body will experience greater and greater amounts of radiographic density loss until no amount of radiographic density will be recorded at all. Grid cutoff can prevent all radiation from passing through it if the angle of the radiation beam is opposed to the direction of the lead strips. Grid cutoff is also identified by the recording of a strange pattern of variegated grid lines within the peripheral portions of the recorded image. (**Perform Experiment 65.**)

Although a reduced SID would produce primary beam cutoff in the periphery of a 14" by 17" (36 cm × 43 cm) film, smaller film sizes, such as an 8" by 10" (20 cm × 25 cm) or a 10" by 12" (25 cm × 30 cm) film, would not be influenced. Even at a reduced SID, smaller film sizes would still be utilizing only the more perpendicular, central portion of the x-ray beam that is properly aligned with the parallel lead strips of the grid. The divergent, greater angled rays in the periphery of the x-ray beam that would be cut off at the edges of the grid would not be included within the film size at all.

Parallel Grids and Central Ray. When using a parallel grid, it is vital for the central ray to be arranged so that it is directed perpendicular (at 90°) with the surface of the grid. With parallel grids, it is possible to angle the central ray in the same direction that the linear pattern of the grid lines are arranged. On top of the grid itself will be imprinted information including a symbol that indicates the direction of the grid lines. With rectangular-shaped grids, it is easy to determine the direction of the grid lines as they will be aligned in the direction of the length of the grid. Thus, it is possible to angle your central ray in either direction toward the top or bottom of the length of the grid. However, it is not possible to angle your central ray in a direction that opposes the alignment of the parallel lead strips. Angling in a direction opposing the alignment of the lead strips will produce severe grid cutoff of a very distinctive pattern and design (Figure 5-15).

In Figure 5-15, angling of the central ray will permit the normally divergent beam of radiation at the peripheral portion of the total beam to be aligned more properly with the direction of the grid's lead strips. The one side of the divergent portion of the beam of radiation that would normally be cut off is now aligned with the lead strips and will produce a positive radiographic density on the film corresponding to that side of the film. However, the normally perpendicular central ray is now more divergent and angled slightly against the direction of the grid lines and will begin to be affected by grid cutoff. In addition, the peripheral portion of the x-ray beam at the opposite side of the film from the central ray will produce complete grid cutoff of the radiographic image. The radiographic image would possess proper radiographic density on one side of the film, which would gradually decrease as it goes toward the middle of the film and be almost devoid of any radiographic density at the opposite side of the film. The margin for error related to the angulation of the central ray against the direction of the grid lines is examined in the section on grid ratio. (**Perform Experiment 66.**)

Focused Grids. The **focused grid** is designed differently from the parallel grid design. The divergent pattern of the x-ray beam emerging from the x-ray tube can be a cause of grid cutoff in the peripheral portions of the recorded image when the parallel grid lines and the angle of the divergent beam no longer will permit radiation

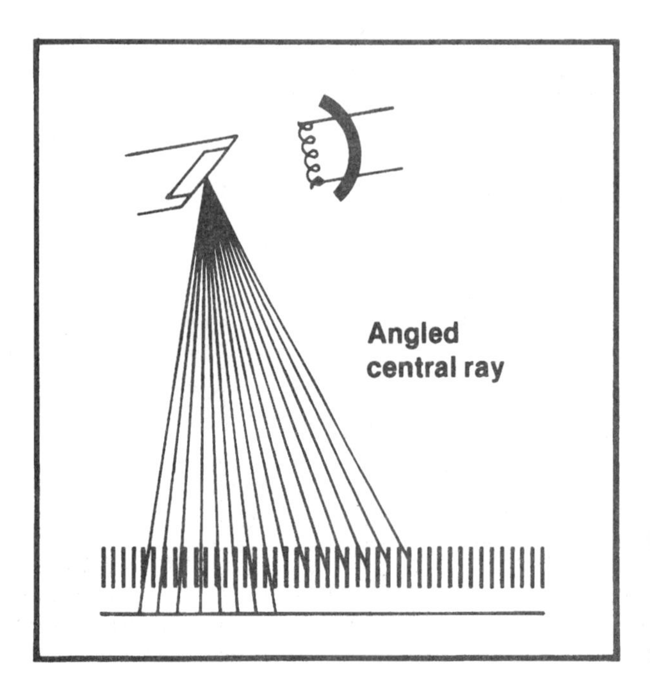

Figure 5-15
Parallel Grid Cutoff Produced by Angling the Central Ray in a Direction Opposing the Lead Strips

to pass through the grid. The focused grid is designed to take advantage of the divergence of the x-ray beam as it spreads after leaving the x-ray tube. Similar to parallel grids, the focused grid's lead strips are arranged in a parallel pattern in the midsection of the grid. However, as the lead strips are placed in the grid further from the middle of the grid, they are designed to incline slightly toward the center of the grid. In fact, if the lead strips of the focused grid were extended above the grid, they would eventually meet at a point above the grid. The location where the pattern of grid lines would meet above the grid is known as the **point of convergence** (Figure 5-16).

A measurement taken from the point of convergence above the focused grid to the middle of the surface of the grid is known as the **focusing distance** of the grid. It is obvious that the focused grid is most effective at reducing radiographic fog and improving the radiographic contrast of the recorded image when the SID is aligned with the focusing distance of the grid. At this distance, the ratio of primary beam permitted to pass through the grid is at its maximum. The maximum grid clean-up associated with the grid cutoff of secondary and scattered radiation passing through the grid would also be aligned with the focusing distance. As a result of the unique design of the grid, which takes advantage of the actual divergent pattern of the x-ray beam, the focused grid is more effective in its contrast improvement factor than the parallel grid.

Focused Grids and SID. Using a focused grid at an SID other than the focusing distance recommended by the manufacturer will result in primary beam grid cutoff in the peripheral portions of the recorded image similar to the cutoff pattern identified with a parallel grid in Figure 5-14. However, depending on the grid ratio of the focused grid, which we will examine later in this chapter, there is a margin for

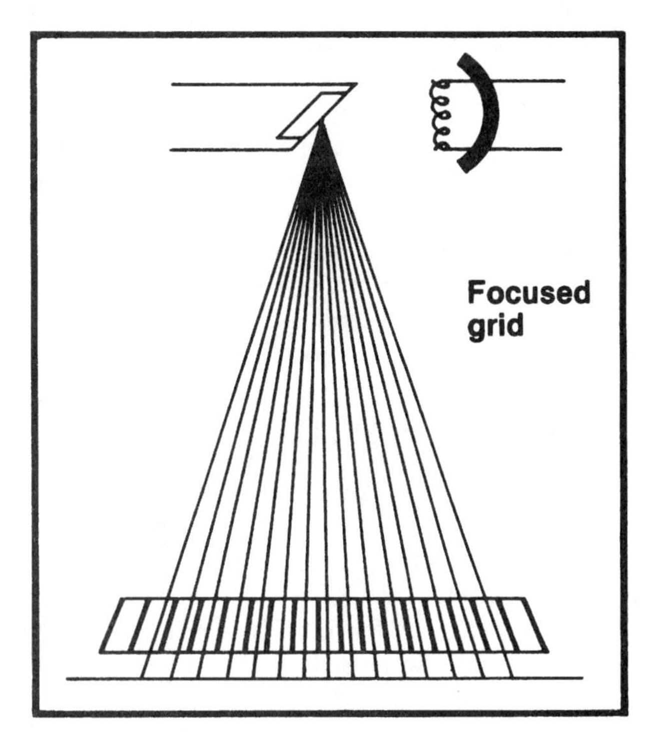

Figure 5-16
Use of a Focused Grid To Utilize the Divergence of the X-ray Beam

error above and below the focusing distance that can be used without producing appreciable amounts of grid cutoff. The margin for error when using a focused grid at other than the focusing distance decreases as the grid ratio increases.

For a focused grid with a grid ratio of 8:1, the margin for error applied to the focusing distance is approximately 25%. That is, if the focusing distance of the grid is 40" (102 cm), it would be safe to increase or decrease the SID by 25% without appreciable amounts of grid cutoff. This enables the 8:1 focused grid with a focusing distance of 40" (102 cm) to be used in an SID range from 30" to 50" (76 cm to 127 cm). With higher ratio grids, the margin for error would be significantly less. The grid cutoff associated with distance changes, like the parallel grid, usually is associated with the use of large film sizes. However, unlike the parallel grid, which can improve in efficiency with an increased SID but be adversely influenced by decreasing the SID, the focused grid can be critically affected by SID changes in either direction. Therefore, the SID factor is a more critical consideration in the application of a focused grid than with a parallel grid. (**Perform Experiment 67.**) The focusing distance of a focused grid is clearly imprinted upon the surface of the grid.

Focused Grids and Central Ray. Not unlike the central ray precautions described with the application of a parallel grid (Figure 5-15), when using a focused grid, the central ray should be arranged so that it is directed perpendicular with the surface of the grid. A linear-designed focused grid will also permit the central ray to be angled in the same direction as the linear pattern of the grid lines are arranged. Most focused grids are used with Bucky devices and are therefore permanently installed beneath the tabletop of the x-ray examination table. The linear pattern of the lead strips of the focused grid are arranged so that they lie in a parallel arrangement with the longitudinal axis of the tabletop. Therefore, it is possible to angle the central ray toward the head or foot of the patient lying upon the length of the tabletop without experiencing grid cutoff. It is not possible to angle your central ray from side to side along the tabletop without experiencing severe grid cutoff. (**Perform Experiment 68.**)

When you use a focused grid, it is also important that the central ray be aligned to the center (midpoint) of the grid and not deviate to one side or the other. Although you can move your central ray along the length of the tabletop and adjust your Bucky tray and film to be centered to the beam of radiation, you cannot off-center your central ray to either side of the midline of the tabletop. Off-centering the central ray to either side of the midline of the tabletop will produce a pattern of grid cutoff (Figure 5-17). The cutoff resulting from an off-centering of the central ray to the midpoint of the grid will be more severe and visible when using large film sizes, such as a 14" by 17" (36 cm × 43 cm) film. (**Perform Experiment 69.**)

Focused Grid Tube Side. Because of the special alignment of the lead strips of a focused grid with the divergent pattern of the x-ray beam, it is designed to be used with a designated tube side facing the x-ray tube. The tube side of the focused grid is imprinted upon the surface of the grid. The designated tube side of the focused grid when placed into the Bucky apparatus must be facing the x-ray tube. If the grid surface were reversed, and the tube side of the grid were facing the film, there would be severe grid cutoff in the areas away from the center of the grid on both sides (Figure 5-18). With the focused grid inverted or reversed from its proper position within the Bucky apparatus, the ever-widening divergent radiations in the periph-

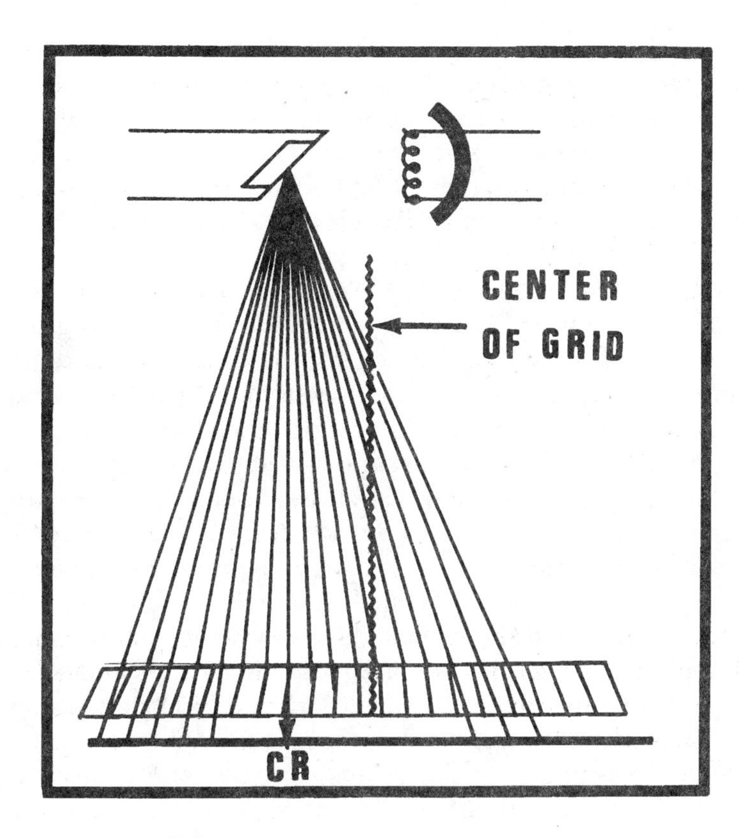

Figure 5-17
Grid Cutoff Resulting from the Off-centering of the Central Ray to a Focused Grid (CR, Central Ray)

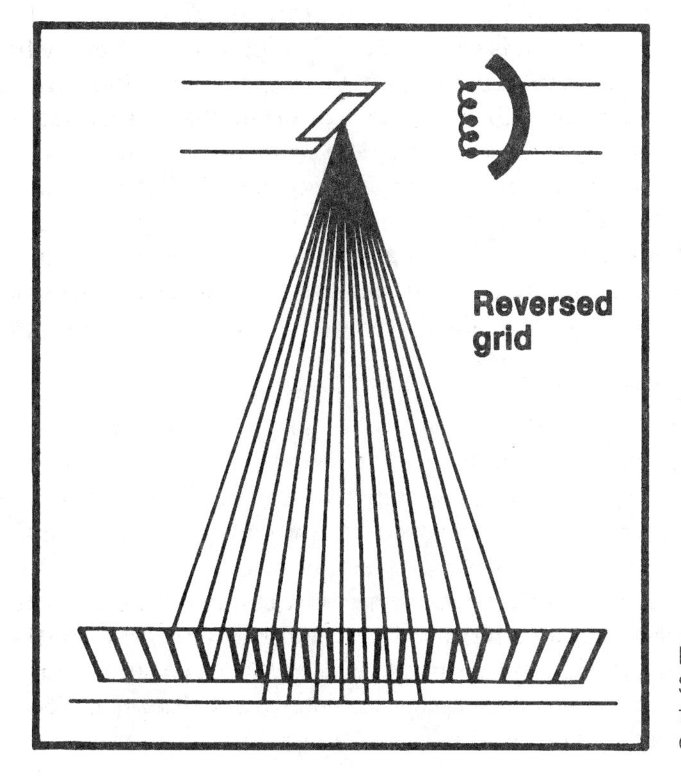

Figure 5-18
Severe Cutoff Resulting from the Reversal of a Focused Grid

eral portions of the x-ray beam are no longer aligned with the focused pattern of the grid, but oppose it. The x-ray beam in these areas is no longer able to pass through the grid, and this would be evidenced by a loss of radiographic density in the periphery of the recorded image due to grid cutoff. This influence would be more evident with examinations requiring large films such as a 14" by 17" (36 cm × 43 cm) than with those requiring smaller film sizes. (**Perform Experiment 70.**)

Grid Ratio

Grid ratio has a major influence on the physical attributes of grid efficiency. Grid ratio refers to certain physical characteristics built into the grid during manufacture and ultimately reflects the total lead content of the particular grid. Basically, the greater the lead content contained within a grid, the greater the contrast improvement factor associated with the use of that grid. The ratio of a grid refers to a comparison between the actual height of the lead strips used to absorb radiation employed in the manufacture of the grid and the distance between each strip filled in with the radiolucent material that will allow radiation to pass through (Figure 5-19). Generally, as the grid ratio increases, the total amount of lead utilized in the construction of the grid increases, and the functional efficiency of the grid also increases. However, as the grid ratio increases, the margin for error in its application decreases.

Most grids are manufactured with 8:1, 12:1, and 16:1 grid ratios. An 8:1 grid ratio specifies a grid where the measurement of the height of the lead strips is eight times greater than a measurement of the distance between the lead strips. As the grid ratio increases, the margin for error in its application decreases. In our introduction to the differences between parallel and focused grids, we learned that angling against the direction of the lead strips of the grid produces grid cutoff. As the grid ratio increases, the margin for error associated with angling against the direction of the lead strips decreases. The margin for error when using an 8:1 ratio grid is 7°. This means that severe grid cutoff will begin to occur as you approach an angle of the central ray against the direction of the lead strips of the grid of 7°. For a 12:1 grid ratio, the margin for error is reduced to 5.25°, and with a 16:1 grid ratio, the margin for error is only 3.25°. As you can see, as the grid ratio increases, the technologist must be more careful to apply the principles that influence its efficiency. The majority of grids sold as separate grids employ an 8:1 grid ratio. The 12:1 and 16:1 ratio grids, because of their reduced margin for error in their application, are mainly used with stationary and moving grid applications in which the grids are permanently affixed to the apparatus. The majority of Bucky devices use either a 12:1 or 16:1 ratio grid.

Number of Grid Lines/Inch

Another major influence on the physical attributes of grid efficiency is the number of lead strips contained within an inch or centimeter. The number of lead strips per inch not only influences the efficiency of the grid, but also affects the physical appearance of the recorded image. When radiographic grids are used as stationary devices, there will be the appearance of a pattern of lines representing the actual lead strips of the grid recorded within the radiographic image. In areas of the image represented by the lead strips, the radiation will have been absorbed by the lead strips, and the radiographic image will be recorded as white lines. The actual radiographic density of the image is made up of the radiation that has passed through the radiolucent material between the lead strips of the grid. The pattern of

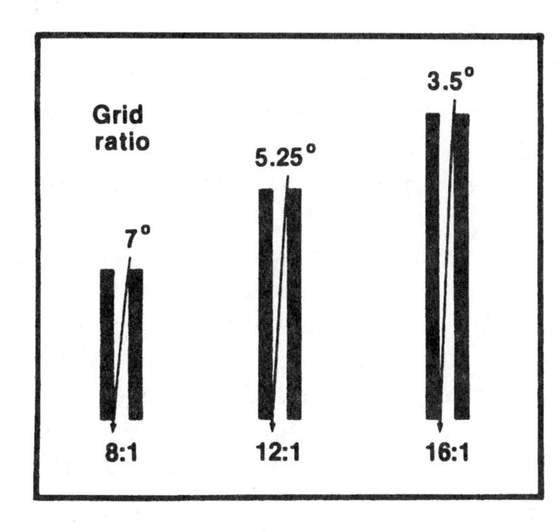

Figure 5-19
Representation of Grid Ratios and the Maximum Degree of Angling before Grid Cutoff Occurs

the lead strips recorded within the image can produce an objectionable appearance in the image depending on the number of lines per inch and the thickness of the appearance of the lines recorded.

Radiographic grids are specified as 60-line grids, for example, which means that there are 60 lead strips contained within an inch. Most radiographic grids are manufactured to contain 80, 100, or up to 120 or 150 lines per inch. In general, the greater the number of lead strips per inch, the greater the efficiency of the grid. As the number of lead strips increases within the design of the grid, the total lead content of the grid increases. However, as the number of lead strips per inch increases during the construction of the grid, the actual width of each lead strip decreases. Eventually, the relationship between the thickness of the lead strips and the number of lead strips contained within the space of an inch reaches a point of diminishing returns. As the lead strips become thinner in order to accommodate greater numbers of lines per inch, their ability to absorb high-energy, greater penetrating radiation associated with higher kilovoltage diminishes. The number of lead strips per inch is specified on the surface of the grid.

Physical Efficiency of the Radiographic Grid

The overall physical efficiency of a radiographic grid is directly related to the total amount of lead contained within the grid. It is a combination of the grid ratio and the number of lines per inch that specifies the total quantity of lead contained within the grid. There are many ways to manipulate the relationship between the ratio of the grid and the number of lead strips per inch during the manufacture of the grid that will influence the physical efficiency of the grid.

The manufacturer can manipulate the relationship of the grid ratio and numbers of lead strips per inch and, thus, the overall lead content of the grid in a number of ways depending on the requirements of the radiographic applications in which they will be employed. A radiographic grid can be manufactured with different grid ratios and the same number of lead strips per inch by changing the height of the lead strips. In this instance, the radiographic grid with the higher grid ratio would measure a greater overall thickness. It is also possible to maintain the ratio of the grid by using the same height of the lead strips but changing the number of lines per inch by adjusting the thickness of the lead strips. Figure 5-20 demonstrates two 8:1 ratio grids. In the upper diagram, the lead strips are quite thick, and there is a total of only 5 lines per inch in the grid. In the lower diagram, the height of the lead strips is the same, the space between the lead strips is the same (therefore maintaining the grid ratio), but there are 16 lines per inch. This adjustment in the number of lead strips per inch was accomplished by changing the thickness of the lead strip itself.

In the comparison in Figure 5-20, the 8:1 ratio grid with 5 lines per inch would have a greater efficiency and contrast improvement factor than the 8:1 ratio grid with the 16 lines per inch due to the greater thickness of the lead strips and concentration of the lead. This would be especially true when using higher energy, greater penetrating radiation associated with higher kilovoltage. However, if both grids were used as stationary grid devices, the recorded image utilizing the 8:1 ratio with the 5 lines per inch would produce a far more objectionable appearance of grid lines within the recorded image than the image recorded by the grid with 16 lines per inch.

The kilovoltage applied to the x-ray tube will also influence grid efficiency. As the kilovoltage applied to the tube increases, interactions within the body between the radiation and body tissues increase, and the resultant secondary and scattered radiations produced possess higher energy. The radiation resulting from higher kilovoltage penetrates through the body in a more forward direction that is similar to the direction of the emerging remnant portions of the primary radiation. As a result, more scattered radiation is capable of passing through the grid and producing radiographic fog, which reduces the visibility of the recorded image. One way of dealing with increased radiographic fog caused by the higher kilovoltage is to select grids with higher efficiency. Usually, the choice would be to select a grid with a higher grid ratio. An 8:1 ratio grid is generally effective for radiographic procedures employing up to 85 kVp. For procedures using kilovoltage above this level, a higher ratio grid, such as a 12:1 or a 16:1 grid, should be used.

Application of Grids

Shortly after the discovery of x-rays, around the turn of the century, radiographic grids were developed to improve the visibility of the recorded image. The first radiographic grids were made of fairly thick lead strips with thin pieces of wood slats placed between the strips. The earliest grids were simply placed on top of the film

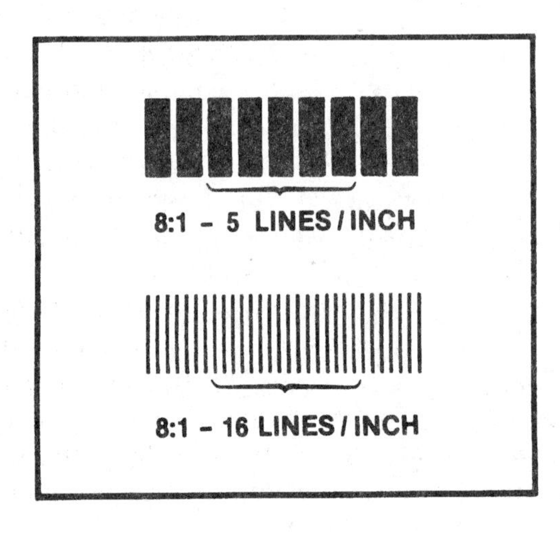

Figure 5-20
Adjusting the Thickness of the Lead Strips in Order To Maintain the Same Grid Ratio with Different Numbers of Lines per Inch

holder between the patient and the film and maintained in that position during the exposure, not unlike the process used today of taping a separate grid onto the front of a cassette. We would refer to this application of radiographic grids as stationary grids since the grid is not in motion during the exposure. The use of stationary grids still has considerable application in today's technology. Unfortunately, the early grid devices recorded significant patterns of thick grid lines on the film. In the earlier grids, this was quite objectionable and detracted from the appearance of the image. Modern stationary grids are of such precision design and use so many lines per inch (up to and above 100 lines per inch) that at normal viewing distances, their presence is hardly noticeable.

As technology advanced, it was discovered that if the grid was made to move during the exposure, the appearance of the objectionable grid lines would not be recorded upon the film. A number of mechanisms have been developed to produce a movement of the grid during the exposure that eliminates the recording of their appearance within the radiographic image. With moving grids, it is possible to increase the efficiency of the clean-up of scattered radiation by using grids with thicker lead strips and/or higher grid ratios and still not have the objectionable grid lines recorded within the recorded image. (**Perform Experiment 71.**)

High-efficiency grids are utilized in grid moving mechanisms that are located beneath the surface of the x-ray examination tabletop. The first really effective devices for moving the grid during the exposure were developed by a person whose last name was Bucky. Since then, the many different mechanisms developed to move a grid during the exposure have been referred to as **Bucky devices.** In fact, the tray used to hold the film within the grid moving mechanism during the exposure is referred to as the Bucky tray, regardless of the mechanism to which it is attached. Earlier movements operated against a spring-loaded device and had to be manually set. Current devices are integrated with the circuitry of the control panel and operate automatically. The initial problems included how to get the motion started, how to keep it going during the exposure, and how to regulate its speed of movement. It was important that the motion of the grid be smooth and continue during the entire exposure. If the exposure began before the grid movement started or was brought up to speed, or if the exposure continued after the grid movement stopped, a very objectionable image of grid lines would be recorded upon the film. The early grid movement devices required the technologist to set the time of the grid movement manually. The technologist would set the total movement of the grid to a time longer than the selected time of exposure. In theory, this appeared sound, but in actual operation, there were many additional factors that made a more accurate evaluation between the manually set grid movement time and the time of exposure necessary. Current grid movement devices utilize integrated circuitry and reciprocating motor mechanisms that keep the grid moving at a constant speed back and forth across the surface of the radiographic film at all times before, during, and after the actual time of exposure.

The appearance of grid lines is no longer a problem associated with moving grid mechanisms. The appearance of grid lines on a recorded image when using a Bucky device would indicate that something is wrong with the operating mechanism or that the Bucky movement is switched off. Many pieces of equipment have an on/off switch related to the movement of the Bucky. Since the operation of a modern Bucky device requires the movement of the grid before, during, and after the radiographic exposure, with some Bucky mechanisms there may be a slight delay in the actual exposure when the technologist pushes the exposure button on the control panel in order to allow the movement of the Bucky mechanism to start and come up to speed prior to the actual exposure. However, for tabletop procedures not employing the Bucky, it may be important to be able to rely on an instantaneous exposure. This may be advantageous in examinations of patients having difficulty in holding their breath and comatose patients who cannot hold their breath, and with problems of potential motion when examining uncooperative infants, children, or adult patients. Switching off the Bucky will eliminate the slight delay associated with the movement of some grid mechanisms. When using pieces of x-ray equipment that possess an on/off switch for the Bucky mechanism, the technologist must be sure that the mechanism is turned on when utilizing the Bucky, or his or her radiograph will have the objectionable appearance of grid lines within the recorded image.

Functional Characteristics of Grids

FUNCTIONAL CHARACTERISTICS OF RADIOGRAPHIC GRIDS

- **Grid Selectivity**
- **Contrast Improvement Factor**

Grid Selectivity

Since the radiographic grid lies in the path of *all* radiation directed toward the film, a percentage of the remnant portions of the primary radiation will be absorbed by the grid in addition to secondary and scattered radiations. Ideally, a radiographic grid should be able to transmit a high percentage of the emerging remnant portions of the primary radiation to the film in order to minimize the exposure to the patient. At the same time, it should be able to absorb the maximum amount of unwanted secondary and scattered radiations emerging from the body in order to record an image possessing satisfactory radiographic contrast and visibility of the recorded details.

Grid selectivity is identified as the ratio of the exiting primary radiation transmitted to the film as compared with the exiting secondary and scattered radiations transmitted to the film. The greater the ratio of primary to secondary and scattered radiation transmitted to the film, the greater the selectivity of the grid and the better the visibility of the image recorded. Grid selectivity is influenced by the grid ratio and the number of lines per inch and how the grid is applied during the procedure.

Contrast Improvement Factor

The ultimate criterion of grid efficiency is the improvement in radiographic contrast between a recorded image produced without the application of a grid and a similar image produced with the application of a grid. Since it is the function of a grid to allow the primary radiation to pass through it and selectively prevent the scattered radiation from passing through it, in order to increase the radiographic contrast and improve the visibility of the recorded details, the ultimate consideration of the efficiency of a grid must be specified as the **contrast improvement factor.** The contrast improvement factor is the ratio of the radiographic contrast visible on the film with the application of a grid divided by the radiographic contrast visible on the film produced without the grid. A measurement of the radiographic densities between several adjacent structures on the recorded image using a radiographic grid compared with the radiographic densities of the same adjacent structures on the recorded image from a nongrid procedure will enable you to visualize the contrast improvement.

However, an actual determination of the contrast improvement factor will be difficult for the technologist to make. Manufacturers of radiographic grids provide a graphic illustration of the contrast improvement factor for the various combinations of grid ratios and numbers of lines per inch that they produce. The contrast improvement factor specified by the manufacturer is a good guideline to employ in the purchasing of radiographic grids. However, the contrast improvement factor is influenced and dependent on the selection of proper radiographic exposure factors. It is influenced by the exposure (mAs), the kilovoltage, the volume of tissue irradiated (field size), and the opacity and thickness of the tissues and structures contained in the part to be examined. Therefore, although the contrast improvement factor may be specified by the manufacturer, it is only within the practical application of the grid by the technologist during routine radiographic procedures that the ultimate improvement associated with the use of the radiographic grid can be determined.

Use of Radiographic Grids

The use of radiographic grids significantly improves the visibility of the image, and they are useful in a wide range of radiographic procedures. However, it is important to recognize that the use of grids also significantly reduces your radiographic density as compared to the nongrid procedure. Therefore, the use of grids is not advocated in examinations of low density tissue or small body parts. It would be inappropriate to employ a grid in an examination of a wrist, for example, since you would have to increase your exposure factors in order to produce the necessary radiographic density with the grid, and it would unnecessarily increase the patient's exposure. Using radiographic grids in procedures where they are not required would be contrary to the principles of ALARA. When should a radiographic grid be employed?

When To Use a Grid

- **Over 10 cm Thickness of Body Part**
- **Above 60 kVp**

A general rule of thumb would be to employ a grid when examining areas of the body measuring over 10 cm. At and above this body thickness, significant amounts of scattered radiation will be produced, indicating the need for a grid. Furthermore, the quantity of radiation needed with thick body parts to produce the desired density only adds to the production of secondary and scattered radiations produced by the body tissues. Additionally, when examining tissues that require more than 60 kVp for penetration, a grid should be used. At kilovoltage higher than 60, the radiation interaction patterns with the body tissues increase significantly, and the amount of secondary and scattered radiations also increases. This increase is especially apparent in the areas of the body containing high fluid content, lower density tissues.

As with the use of any radiographic device, you must learn how to alter your exposure factors in order to control and maintain the radiographic density of your recorded image when going from a nongrid technique to one that employs the use of a radiographic grid.

One method of compensation that has been suggested is to increase your kilovoltage from the nongrid technique. This method increases the exposure to your patient less than other methods of compensation. However, this method does not take into consideration grid efficiency factors related to different grid ratios or the limitations imposed on grids with increased numbers of lead strips per inch (thinner lead strips) that cannot operate as effectively at higher kilovoltage. Additionally, by increasing the kilovoltage, you are altering the entire absorption characteristics of the subject contrast and may seriously affect the radiographic scale of contrast in the radiographic image.

If an increase of 20 kVp between a nongrid and an 8:1 ratio grid procedure enables you to maintain the radiographic density, would a 20-kVp increase produce the same effect if a grid with a 12:1 ratio was used? If your nongrid procedure was performed using 60 kVp, then the same procedure performed using a grid would increase your kilovoltage to 80 kVp using this method of compensation. However, if your nongrid procedure originally employed 80 kVp, would your increase to 100 kVp produce the same influence as the change from 60 to 80 kVp? Kilovoltage influences the exposure rate of the x-ray beam, the penetrating ability of the x-ray beam, and the intensification factor of intensifying screens. Is it logical to assume that a 20-kVp increase will produce the same influence in the 60-kVp range as it would in the 80-kVp range? (**Perform Experiment 72.**) Compensation to maintain radiographic density between a nongrid and a grid procedure by increasing the kilovoltage may not be applicable or appropriate in all radiographic situations. A 10-kVp increase may be appropriate in one instance, whereas a 20-kVp increase may be required in another.

Another method of compensation to maintain radiographic density when changing from a nongrid to a grid procedure is to increase the mAs according to the physical characteristics (grid ratio) of the grid. Unfortunately, this method, although more controlled and predictable, increases your patient's total exposure. However, with the availability and application of higher speed film-screen recording systems, the amount of exposure increase is not as significant as it was in the past. An increase in the mAs to compensate for the application of different grid ratios appears to be a method that has greater application over a wider range of radiographic procedures than an increase in the kilovoltage. Table 5-3 illustrates the adjustment of the mAs in order to maintain radiographic density between nongrid and grid procedures. (**Perform Experiment 73.**) If a nongrid procedure required 5 mAs, the same procedure performed employing an 8:1 ratio grid would require 20 mAs; a 12:1 ratio grid would require a 25-mAs exposure.

Using adjustments in the mAs to compensate for changes from nongrid to grid procedures (Table 5-3) can also compensate for changes from the application of one grid ratio to another. There are occasions when the technologist may have to consider these relationships. A procedure performed in the radiology department using a Bucky with a 12:1 ratio grid may have to be performed the next day as a portable examination using an 8:1 ratio grid. Although there are other factors that must be taken into consideration, such as the different influence of a focused grid versus a parallel grid and the influence of the moving grid versus a stationary grid between the procedure performed in the radiology department and the procedure to be performed as a portable examination, the mAs adjustment related to the different grid ratios employed is a good place to begin your exposure factor adjustments.

Table 5-3 Grid/mAs Relationship

Grid ratio	Adjustment
6:1	Use 3 × the nongrid mAs
8:1	Use 4 × the nongrid mAs
12:1	Use 5 × the nongrid mAs
16:1	Use 6 × the nongrid mAs

Intensifying Screens

Although few, if any, nonscreen (direct exposure) procedures are performed in radiography today, due to the excessive amount of exposure required, a comparison of the influences between nonscreen and film-screen recording system radiography related to radiographic fog is still appropriate to consider. The use of different speed film-screen recording systems influences the amount of mAs required to produce a desired radiographic density. The quantity of radiation used has a major influence on the amount of secondary and scattered radiations produced by the body part, and, therefore, there is a relationship between the use of intensifying screens, radiographic fog, and radiographic contrast. Intensifying screens also affect the radiographic contrast scale. There are inherent properties of the screen phosphors that produce a significantly higher, shorter scale contrast than the same procedure performed as a nonscreen procedure. In addition, radiographic density differences attributed to kilovoltage change when using intensifying screens are far more significant, because of the intensification factor of screens, than the same kilovoltage change when performing the examination as a nonscreen procedure. (**Perform Experiment 74.**)

Radiographic Contrast Summary

We have now examined the importance of radiographic contrast to radiographic quality. We have identified those factors that affect contrast in the image and have discovered that the radiographic contrast scale that is quite appropriate in one examination might be improper for another. We have examined methods to produce and control the radiographic contrast scale. In addition, we have discovered the multiple factors and influences that can detract from the visibility of the recorded detail by producing radiographic fog. We have also reviewed and examined various methods employed to reduce or eliminate the influences of radiographic fog. Basically, radiographic contrast is controlled and maintained by reducing the amount of secondary and scattered radiations reaching your film. Radiographic contrast and, therefore, the visibility of the recorded image are always improved by a reduction in radiographic fog. Let's summarize what we have learned.

Subject Contrast. Because of their differences in thickness and opacity, the multiple structures of the body absorb different amounts and types of radiation passing through them. It is this attenuation of the beam produced by the subject contrast contained within the body part that ultimately contributes to the radiographic contrast of the radiographic image. The patient factors affecting subject contrast are influenced by the quality of the beam (kilovoltage). Changes in kilovoltage alter the absorption characteristics of the tissues.

Film Contrast. Film contrast relates to the inherent properties of film speed and latitude and the ways the different quantities and qualities of radiation affect them. Radiographic contrast can be considered a quantity factor and relates to the total number of useful densities recorded within the radiographic image, or it can refer to the quality of the differences between radiographic densities recorded between adjacent structures of interest.

Radiographic Contrast Scale. The number of useful densities recorded on the film and the ratio of difference between them are referred to as the **radiographic scale of contrast.** High contrast indicates wide density differences and fewer numbers of recorded densities within the useful range. Low contrast indicates more narrow density differences and greater numbers of recorded densities within the useful range. The greatest exposure latitude exists when the kilovoltage necessary to penetrate the structures of interest has been chosen. At that kilovoltage, the maximum allowable adjustment in the mAs between a minimum and maximum that will still produce a satisfactory radiographic density will be available. Minor adjustments of the kVp/mAs relationship can allow the technologist to change the radiographic contrast scale to meet the specific needs of an examination, physician, or patient condition. A radiographic image with a proper radiographic contrast scale will permit the visibility of all of the structures of interest.

Radiographic Fog. Radiographic fog is the most significant factor contributing to a loss of image visibility. It can be caused by a number of factors. Many of these factors were examined in Chapter 4 introducing radiographic density. In this chapter, we limited our investigation to the factors of secondary and scattered radiations and their influence on the production of radiographic fog. After learning of their influences, we examined methods to control and eliminate this unwanted factor. We learned that subject contrast can contribute to radiographic fog, especially in lower density, fluid-filled tissues. The SID can also contribute to fog, and, in limited application, the air-gap technique can be employed to reduce this effect. We discovered that the proper relationship of kilovoltage to mAs is the major controlling factor in the production of radiographic contrast. Radiographic fog levels will be minimal when the kilovoltage necessary to penetrate the part and the appropriate mAs have been selected to provide for proper radiographic density.

Beam Restriction. Beam restriction provides a method to reduce significantly the exposure to your patients, reduce the production of secondary and scattered radiations, and reduce radiographic fog. Beam restriction increases radiographic contrast and the visibility of the recorded image. Of the beam-limiting devices available, the collimator is the most effective and efficient. In addition to beam restriction, the use of lead blocking devices can also improve the radiographic contrast of your image by further reducing the radiographic fog levels recorded upon your radiographic image. The use of gonadal shielding was introduced to remind technologists of their responsibility to their patients in applying the principles of ALARA.

Radiographic Grids. Radiographic grids are the most effective devices for reducing the amount of unwanted radiation directed toward the radiographic film. The proper application and use of radiographic grids will significantly improve the visibility of the image and increase the radiographic contrast. Any restrictions imposed on the application of a radiographic grid would depend on the type and design of the grid. Generally, as the ratio of the grid and the number of lines per inch increase, the efficiency of the grid improves. Radiographic grids significantly reduce the overall radiographic density of your recorded image compared with a nongrid procedure, and your technical factors will have to be adjusted (increased) in order to maintain the radiographic density of your image.

Intensifying Screens. The use of intensifying screens enables you to reduce your overall exposure, thereby reducing the amounts of secondary and scattered radiations produced. Intensifying screens also affect the radiographic contrast scale. Intensifying screen phosphors inherently produce an image with a higher scale of contrast compared to a similar procedure performed without screens. Using intensifying screens requires a more accurate assessment of the required exposure because there is far less margin for error (exposure latitude) than there is with a nonscreen procedure.

Radiographic Contrast Review Questions

Name ____ Date __________________

1. Describe the function of radiographic contrast to the radiographic quality of the recorded image.

2. Radiographic contrast can be described as a quantity and a quality factor of the radiographic image. Explain.

3. The differences in the intensity of x-ray transmitted by a particular body part because of its differing absorption characteristics is known as __________ ______________.

4. Describe an area/body part/structure that would be identified as possessing good subject contrast.

5. Describe an area/body part/structure that would be identified as possessing poor subject contrast.

6. Differentiate between the factors and qualities that influence the subject contrast of the areas/body parts/structures described in Questions 4 and 5.

7. Does a single structure of the body, such as a femur, possess subject contrast? Explain.

8. Describe how the following factors would affect tissue thickness and therefore influence subject contrast:

   a. patient age:

   b. patient sex:

   c. patient development:

   d. patient condition:

9. Describe how the following factors could influence subject contrast:

   a. Cellular composition:

   b. Compactness of tissue:

   c. Materials between tissue:

   d. Status of hollow organs:

10. Body tissues vary considerably in radiopacity. Provide examples of four naturally occurring densities within the body, arranging them in order from least to most dense.

11. Define the term **Angstrom unit.**

12. Using the following formula, 12.4/kVp, determine the minimum wavelength of a 70 kVp beam and a 100 kVp beam.

13. Of what significance is the minimum wavelength contained within a given beam of x-radiation?

14. The ability of a film to respond to a known quantity of radiation is referred to as the ______________ of the film.

15. A consideration of the differences in radiographic density that a specific film is able to record is known as the ______________ of the film.

16. The subject contrast of body tissues is demonstrated on the x-ray film within certain defined density ranges. The useful diagnostic range of radiographic densities that will demonstrate the tissues of the body is:

    a. 0.2 to 2.50
    b. 0.4 to 2.55
    c. 0.4 to 2.75
    d. 0.5 to 3.20

17. Using the diagrams provided, answer the following questions:

0.2	0.5	0.8	1.1	1.4	1.7	2.0	2.3	2.6	2.9	3.2	**Film (A)**

0.2	0.7	1.2	1.7	2.2	2.7	3.2					**Film (B)**

A. Which film provides for greater latitude? _____
B. Which film has the higher ratio between densities? _____
C. Which film is best used in an area of the body exhibiting poor subject contrast? _____
D. How many useful body densities are demonstrated by Film A? (number) _____
E. How many useful body densities are demonstrated by Film B? (number) _____
F. Which film would be considered a film that possesses high contrast? _____

18. The visible difference between adjacent densities demonstrated on the film that results from subject and film characteristics is known as _____.

    a. density scale
    b. radiographic contrast
    c. subject contrast
    d. film contrast

19. The difference between the minimum and maximum useful density produced on the film is known as the _____.

    a. film speed
    b. fog
    c. contrast scale
    d. density ratio

20. A radiographic image that demonstrates short, abrupt changes from the minimum to the maximum useful density is said to possess _____.

    a. considerable latitude
    b. long-scale contrast
    c. low contrast
    d. high contrast

21. In areas of poor subject contrast, it could be advantageous to select an exposure technique that will produce _____.

    a. higher contrast
    b. lower contrast
    c. longer scale of contrast
    d. extended density tones

22. In areas of good subject contrast, it could be advantageous to extend the scale of contrast in order to demonstrate more structural differences within the recorded image. Which of the following exposure techniques would best accomplish this goal? _____.

    a. high mAs, low kVp
    b. high mAs, high kVp
    c. low mAs, high kVp
    d. low mAs, low kVp

23. The scale of contrast produced on the film is controlled by the _____.

    a. mA/time relationship
    b. mAs/kVp relationship
    c. mAs/distance relationship
    d. kVp/distance relationship

24. As the kilovoltage is increased and the density on the film increases, a compensation in the mAs is required to maintain film density. How will the scale of contrast exhibited on the film be influenced? _____.

    a. lengthened
    b. shortened
    c. fogged
    d. unchanged

25. Fog does not destroy the recorded detail of a radiograph, since it is not a factor that influences or controls this consideration of radiographic quality. However, fog does influence the _____ of the recorded image.

    a. density
    b. contrast
    c. visibility
    d. all of the above

26. Describe how the production of secondary and scattered radiations is affected by the

    a. quantity of tissue

    b. density of tissue

27. In areas of the body measuring less than 10 cm or requiring 60 kVp or less, _____ provide adequate protection from the factor of fog.

    a. beam limitation devices
    b. radiographic grids
    c. lead blockers
    d. gonadal shields

28. When examining structures of the body that require more than 60 kVp or that are thicker than 10 cm, _____ are necessary to provide adequate protection from fog.

    a. beam limitation devices
    b. radiographic grids
    c. lead blockers
    d. gonadal shields

29. You will always require the necessary _________ to penetrate a part. No increase in _________ will be able to correct for inadequate penetration.

30. A much longer scale of contrast could be produced with the use of (*circle one*) nonscreen direct exposure/intensifying screen cassettes. Explain your answer.

31. In areas of poor subject contrast, (*circle one*) nonscreen direct exposure/intensifying screen cassettes would be beneficial. Explain your answer.

32. For an obese patient, how would the radiographic contrast be affected by a posteroanterior projection of the abdomen versus an anteroposterior projection? Why?

33. Describe the application of the air-gap principle and how it would affect the radiographic contrast of the recorded image.

34. As x-rays traverse the body, some of the energy is absorbed and other portions of the x-ray beam interact with the tissues of the body, producing _____ radiation.

    a. primary beam
    b. scattered
    c. secondary
    d. all of the above

35. As x-rays traverse the body, some of the energy is deviated from its normal path and continues to travel through the body with minimal loss of energy in a new direction. These radiations are known as _____.

    a. primary beam radiation
    b. scattered radiation
    c. secondary radiation
    d. all of the above

36. Radiation produced by the body in this manner (Question 35) would chiefly affect the _____ of the radiographic image.

    a. distortion (size)
    b. distortion (shape)
    c. sharpness of details
    d. visibility of details

37. The primary method employed to reduce the production of secondary and scattered radiations within the body is the use of _____.

    a. beam limitation devices
    b. lead blockers
    c. radiographic grids
    d. intensifying screens

38. In what manner would the following factors affect the percentage of radiation interaction with body tissues:

    a. a coned-down 8" by 10" (20 cm × 25 cm) procedure of a part versus a 14" x 17" (36 cm × 43 cm) procedure

    b. an increase in the thickness of the part

    c. an increase in the kilovoltage applied to the x-ray tube

39. The *chief* influence on the interactions of body tissue with radiation is _____.

    a. milliamperage
    b. intensifying screens
    c. kilovoltage
    d. source-image distance

40. Finish this statement: Ideally, the radiographic beam should be limited

    ____

    ____.

41. In what manner is the application of beam limitation devices an example of the principles of ALARA?

42. Describe the appearance and construction of an aperture diaphragm.

43. Identify an advantage and a disadvantage of the aperture diaphragm as a beam limitation device.

44. A diaphragm is placed 2" (5 cm) from the tube target as in the diagram provided. The SID is 36" (91 cm), and you want to cover an image width diameter of 8" (20 cm). What would your diaphragm opening be? (Show all work.)

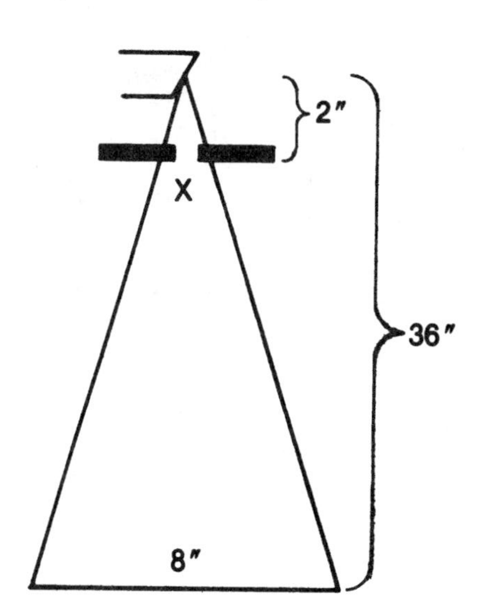

45. You have a diaphragm with a rectangular opening of 1/2" × 3/4". It is placed 3" from the anode, and you are using a 30" SID. What size film will this diaphragm adequately cover? Solve for length and width. (Show all work.)

46. In a comparison between the aperture diaphragm and the cone, which device would be considered more effective/efficient? Explain your choice.

47. Differentiate between the following types of cones: a straight design, cylindrical-shaped cone; a flared design, cylindrical-shaped cone; a straight design, extension cone; a rectangular design, flared cone.

48. Describe a disadvantage of a cone used as a beam limitation device.

49. A new cone has been purchased that has a lower opening of 5" (12.7 cm). The length of the cylinder is 10" (25 cm), and the distance from the top of the cylinder to the anode is 4" (10 cm). Indicate the radiation coverage on the film when the SID is 36" (91 cm). (Show all work.)

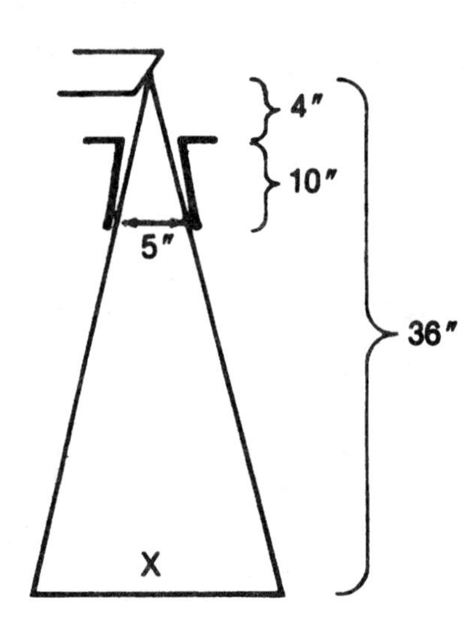

50. Considering the image size coverage indicated in Question 49, what size film could you properly use with that cone? (Consider the film diagonal.)

51. A new cone has been purchased (extension cone) that has a lower opening of 4" (10 cm). The length of the collapsed cylinder is 12" (30 cm), and the distance from the cylinder top to the anode is 3" (7.6 cm). When extended, the length of the cylinder is 20" (51 cm). Determine the image coverage of this cone when collapsed and when extended using an SID of 40" (102 cm). (Show all work.)

52. A box-like apparatus that incorporates the best features of the aperture diaphragm and the cone is called a ____________________.

53. Describe at least two advantages of the type of beam limitation device identified in Question 52.

54. Describe the function and advantages of positive beam collimation.

55. What is the purpose for using lead blocking devices in the performance of some examinations? In what manner does the use of lead blocking devices influence radiographic fog and the visibility of the recorded image?

56. Describe the positive application of lead blocking devices for at least two different radiographic examinations.

57. Gonadal shields must be a minimum thickness of _____ equivalent.

    a. 1.5-mm aluminum   c. 1.5-mm lead

    b. 2.5-mm aluminum   d. 2.5-mm lead

58. What is the main purpose of a grid device? Describe how the radiographic grid accomplishes this purpose.

59. Describe the basic construction of a grid.

60. With the aid of the diagram provided, describe and illustrate the principle of the grid's operation.

    A. Draw in the cone of radiation from the anode to include the entire patient.

    B. Indicate what occurs in the body, and label the primary, secondary, scattered, and image-producing radiations on the diagram.

    C. Describe the principle of a grid's operation as demonstrated by your completed diagram.

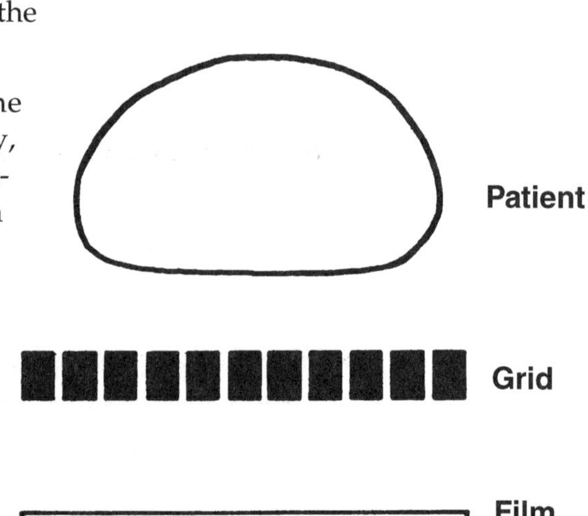

61. Grid efficiency is determined by the overall lead content and how the lead is distributed. Name the two major physical factors related to grid efficiency.

62. Grids whose lead strips are aligned so they travel in one direction along the length of the grid are called ______________ grids.

63. Grids whose lead strips are aligned so that two patterns of strips are at right angles to each other are called ______________ grids.

64. Grids designed with the lead strips arranged so that, if they were extended above the grid, they would never meet are called ______________ grids.

65. Grids designed so that the lead strips are angled toward the center of the grid as you go toward the periphery are called ______________ grids.

66. Why has the specialized grid design described in Question 65 been developed?

67. In comparing the types of grids described in Questions 64 and 65, which grid would you expect

    a. to be functionally more efficient?

    b. to allow for a greater margin of error in use?

68. What do you mean when you say that you're using a stationary grid? In what manner does this affect your radiographic image?

69. Most stationary grids are (*circle one*) focused/parallel.

70. The relationship of the heights of lead strips to the distance between the strips is known as ____.

71. Review the grid pattern diagrams provided below, and answer the questions that follow. (Use the *letters* in your answers.)

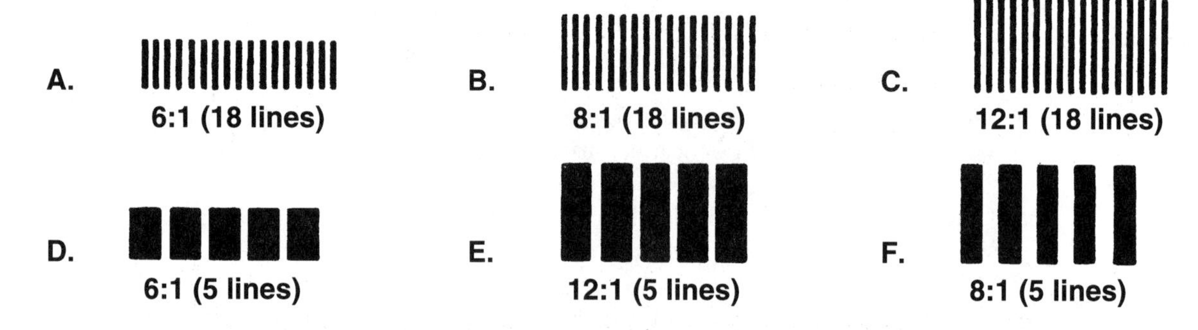

A. Grid Pattern Diagrams A, B, and C all represent the same number of lines per inch, and the thickness of the lead strips and the spaces between them are the same. However, each represents a different grid ratio. Explain how this increase was achieved for each grid.

B. Grid Pattern Diagrams E and F represent the same number of lines per inch and a similar thickness of the grid itself. Again, each represents a different grid ratio. Explain how this increase was achieved.

C. Grid Pattern Diagrams A and D represent the same grid ratio but different numbers of lines per inch. How was this adjustment in grid design made?

D. Comparing the 12:1 grid in Diagram C with the 12:1 grid in Diagram E, which, if either,

would provide greater grid efficiency? _____

would be more radiographically objectionable? _____

would allow a greater margin for error? _____

Explain your answer.

E. Comparing the 8:1 grid in Diagram B with the 8:1 grid in Diagram F, which, if either,

would provide greater grid efficiency? _____

would be more radiographically objectionable? _____

would allow a greater margin for error? _____

Explain your answer.

72. In a comparison of an 8:1 grid to a 12:1 grid of similar design,

which grid contains more lead equivalent? _____

which grid is considered more efficient? _____

which grid allows for less margin for error? _____

73. Can a parallel grid be used effectively at a 72" (183 cm) SID? Explain your answer.

74. Can a parallel grid be used effectively at a 25" (63 cm) SID when performing an anteroposterior open-mouth, odontoid process projection of the cervical spine? Explain your answer.

75. Can a parallel grid be used effectively at a 25" (63 cm) SID for an abdomen examination on an adult if the collimator can be opened wide enough to cover the film size? Explain your answer.

76. Can the grid device used in a routine table Bucky apparatus be used effectively at 72" (183 cm) for a chest examination? Explain your answer.

77. An 8:1 parallel grid generally allows for a margin for error in angulation of the tube against the grid lines of __________ degrees.

78. An 8:1 focused grid generally allows for a __________% adjustment in the SID above or below the recommended distance.

79. The recommended distance to use with a focused grid is referred to as the ______ ________________ of the grid.

80. If the grid lines on a focused grid were extended above the grid, they would meet at a point above the grid. This point would be referred to as ___________ _________ of the grid.

81. The majority of focused grids are used as (*circle one*) stationary/moving grids.

82. For an abdomen radiograph, describe the appearance of the film if you off-center the tube to a parallel grid. Be *specific* in your answer.

83. A moving grid is usually designed with (*circle one*) thicker/thinner lead strips. Explain the reason(s) for your choice.

84. The designation of a "tube side" would be seen on a (*circle one*) focused/parallel grid.

85. Describe the appearance of an abdomen radiograph if you improperly employ a grid with reference to the "tube side" factor.

86. Describe and compare the density and radiographic quality of a radiograph of the abdomen when the same exposure factors are used for a nongrid procedure that have been employed on a satisfactory film of the abdomen using an 8:1 ratio grid.

87. When adjusting from a nongrid procedure to a grid procedure, the compensation method that would produce the least increase in patient exposure would be:

    a. to increase your mAs
    b. to decrease your distance
    c. to increase your kilovoltage
    d. to decrease your mAs

88. Centering would be (*circle one*) more/less critical when using a crosshatch grid as compared to a linear grid with only one pattern.

89. A convenient method of changing from nongrid to grid procedures is to increase the __________ according to the grid ratio employed.

90. A tabletop procedure of a lateral knee was performed using 1.66 mAs and 60 kVp at a 40" (102 cm) SID. You are requested to repeat the examination using the Bucky apparatus of the radiographic table. A 12:1 ratio grid is contained within the Bucky apparatus. Adjusting the mAs to maintain radiographic density, the new mAs employed would be __________ mAs.

91. Comparing a parallel grid incorporated in a cassette with a parallel grid purchased and used separately, complete the following.

    Grid in cassette

    a. an advantage:

    b. a disadvantage:

    Loose grid

    a. an advantage:

    b. a disadvantage:

92. Angling in opposition to the direction of the grid lines while employing a 12:1 ratio grid will cause primary beam absorption when the improper angle approaches __________ degrees.

93. Describe the functional characteristic of grid selectivity.

94. Describe the functional characteristic of the grid contrast improvement factor.

95. Compare the radiographic contrast of an image produced as a nonscreen procedure with that of an image produced when a film-screen recording system is employed.

Radiographic Contrast Analysis Worksheet

Let's review our analysis of radiographic contrast with the following exercise. A satisfactory diagnostic radiograph of the pelvis was produced using the following technical factors:

100 mA
1 sec (time of exposure)
70 kVp
40" (102 cm) SID
1-mm focus size
Minimum OID
Regular screen-type film

100 speed film-screen system
8:1 ratio, 100-line moving grid
Proper collimation
2.5-mm aluminum filtration
92°F development at 90 sec
Normal thickness and opacity of the part

Without compensation, the changes listed in the table below are made one by one. Indicate the effect, if any, each change has on the radiographic contrast of the radiographic image.

1. If the radiographic contrast and visibility of the image are improved, mark a plus (+) in the space provided.
2. If the radiographic contrast and visibility of the image are decreased, mark a minus (–) in the space provided.
3. If the radiographic contrast and visibility of the image are unchanged, mark a zero (0) in the space provided.

Radiographic Contrast

Proposed change	Effect	Proposed change	Effect
Use a 2.0-mm focus	_____	Change to a 5:1 ratio, 100-line moving grid	_____
Use 10" (25 cm) OID	_____	Change to a 12:1 ratio, 100-line moving grid	_____
Reduce the SID to 25" (63 cm)	_____	Remove all collimation	_____
Increase the SID to 48" (122 cm)	_____	Remove all filtration	_____
Use screen-type film in a cardboard holder	_____	Develop at 100°F for 90 sec	_____
Use a 200 speed film-screen recording system	_____	Body thickness reduced by an atrophic condition	_____
Use 200 mAs	_____	Pathological condition reduces opacity of body tissues	_____
Increase the kVp to 90	_____		
Omit the use of a grid	_____		

6 Quality Control

Quality control and the implementation of quality assurance programs in radiology departments have been readily accepted in recent years as a method of standardizing the output of the various examinations, reducing repeat examinations, and providing a measure of accountability. Quality control is not just for the large radiology department employing dozens of technologists and ancillary personnel using many different pieces of equipment and containing multiple x-ray examination rooms. The design and implementation of a quality assurance program does not have to be a time-consuming, arduous process. Nor does it represent an expensive cost factor within the overall budget of the department. On the contrary, quality assurance programs have proved their effectiveness at providing better overall operating standards for equipment and less equipment downtime and have proved cost effective by reducing the numbers of repeat exposures and examinations and permitting the correction of small problems before they progress to large, expensive repairs.

Quality Assurance

Most technologists react to quality assurance by identifying with one aspect of the total program: film processing control. Although playing an important role in the accountability of the department, film processing control is only one facet of a multifunctional process. Even the smallest of departments can implement a program of quality control enabling them to manage better the radiological performance of their equipment, materials, and staff. In actuality, it is easier for the smaller department to set up a program of quality control because the number of personnel using the equipment and possibly abusing the materials employed for the radiographic examinations is less. In smaller, less busy departments where the equipment is used less often, the various procedures and tests associated with the quality assurance program may not have to be performed as often. The larger, busier departments use equipment and materials more frequently and by larger numbers of personnel, producing a greater chance of error, abuse, damage, and wearing out of the equipment and materials used for the radiographic procedures.

In large radiology departments, it is not uncommon to find a quality control technologist specifically hired to attend to the quality assurance program within the department. In addition, they may have a health physicist and a maintenance engineer to participate in these activities. Still, quality assurance is the function of everyone working within the radiology department, not just the quality assurance personnel. Therefore, it is necessary for all personnel working directly with the equipment or materials involved in producing radiographic examinations to be provided an in-service education program to familiarize them with the operation of the quality assurance program within your department. They must be familiar with the personnel, the duties and responsibilities of each, and the specific accountability expected of them as members of the health care team. Many technologists will be assigned a specific role and duties associated with the quality assurance program. By involving others and providing responsibility for functional attributes of the overall program, each member of the team begins to identify with his or her role and performs his or her functions with a contributory attitude toward the good of the whole. In smaller departments, delegating responsibilities for different facets of the overall quality assurance program helps to prevent the activities from becoming a task rather than a contribution to good practice. We will divide our investigation of quality control by identifying the routine activities, responsibilities, and advantages of the implementation of the program. We will examine the areas of equipment factors, material factors, and personnel factors, all of which are involved in the overall process. The process begins with proper planning, the identification of goals, and the assignment of personnel duties and responsibilities.

Reject-Film Analysis

A good place to begin is to determine where your department is related to the overall quality of the radiographic procedures performed there. One yardstick used to evaluate this factor is to identify the numbers of repeat and rejected films your department is producing. Someone should be assigned to document and analyze the reason for every repeat or rejected radiograph performed in the department. The room number where the procedure was performed must be identified, and the radiographer who performed the procedure must also be identified. Ideally, all technologist personnel should use lead markers to put their initials on every radiograph they produce. This will aid in the reject-film analysis.

The rejected films must be recorded and classified according to errors associated with positioning, exposure factors, motion, film processing, fog, static, or artifacts. A basic form can be developed to include all of these factors plus the identification of the room and radiographer. The reject-film analysis should be carried on for a period of at least one month. At the end of the month, a critical review of the reject-film log sheets will enable you to determine any immediate problems that need to be addressed and to identify trends and areas that require improvement.

Reject-film analysis provides an excellent subject for an in-service education program. Reviewing samples of the rejected films as well as reviewing the statistics of the department performance over the period of a month is often a very revealing and informative program. Technologists are seldom aware of their repeat rate or the reasons for their repeats. Individual meetings can be held with personnel who appear to have specific problems related to positioning errors, exposure factor errors, and so forth in an effort to provide them with positive feedback and enable them to become aware of the areas that they should be addressing in their efforts to reduce their rejection rate. Film handling and film processing errors can be addressed with the darkroom personnel in an effort to reduce these problems. Eventually, after the implementation of a quality assurance program, a reject-film rate of more than 5% would be considered unacceptable.

Material Factors

Many of these factors have been introduced within the various chapters of the book, as have the tests necessary to evaluate them. We will review these factors toward a better understanding of their contribution to the entire quality assurance process.

Cassettes

X-ray cassettes should be routinely checked for damage to their frames, warping, loose hinges, and evidence of age, all of which could potentially produce light leaks into the cassette. Cassettes that show evidence of light leaks and radiographic fog upon the film should be removed from service and submitted to the quality control technologist for repair. The cassette front and back should be routinely cleaned with a noncorrosive cleaning solution. It is not uncommon for splashes of barium sulfate or iodinated contrast media to dry upon the surface of the cassette and create an artifact that will be present within all radiographs taken with that cassette.

Intensifying Screens

The surface of the intensifying screens should be routinely inspected for scratches, wear, cracks, age, discoloration, embedded materials, dirt, and other artifacts that will affect its image-producing properties. Cassettes with intensifying screens exhibiting any of the problems identified above should be removed from service and checked by the quality control technologist to determine whether they can be repaired or have to be replaced. Ideally, all cassettes used within the department will have small identification marks within the cassette that register on all radiographs produced with that cassette. A number of commercial identification systems are available to number the cassettes internally that serve this purpose. Problems with a particular cassette will be far easier to locate if there has been a cassette identification system established. The outside back of the cassette must have a similar identification to coincide with the internal marking. Intensifying screens should be routinely cleaned. Although a capful of any mild liquid soap product diluted in a quart of water can be used, it is more effective to employ one of the nonabrasive commercial screen cleaning preparations available. These solutions also add a thin antistatic layer to the surface of the screens. The screens should be allowed ample time to air-dry before placing unexposed film in them. They should not be rubbed dry, as this can cause damage to the protective layer of the screen.

Film-Screen Contact

As discovered in Chapter 2, "Recorded Detail," poor film-screen contact is a major contributor to image unsharpness. All cassettes within the department should routinely be checked and tested for film-screen contact. Cassettes determined to have problems should be removed from service and checked by the quality assurance technologist to determine whether they can be repaired.

Radiographic Grids

Stationary, wafer-type grids are prone to potential damage through abuse, misuse, improper storage, or simply excessive or uneven pressure or weight placed upon them during radiographic examinations. Radiographic grids should be routinely checked for evidence of warping or misalignment of the lead strips. A simple test can be performed by placing the grid upon the surface of a cassette and exposing it to a small exposure. The radiographic image can be evaluated for evidence of density changes that are often associated with grid cutoff. Similar tests should be performed for all grid cassettes within the department. Grids that demonstrate evidence of problems should be removed from service and evaluated further by the quality control technologist.

Radiographic Fog

One of the major sources of radiographic fog is light leaks into the darkroom, usually from around the entrance. The single-door entrance is the major contributor to light leaks into the darkroom. Tests for light leaks into the darkroom should be routinely performed. Weather stripping and other materials used for light insulation around the passbox doors and around the darkroom entrance can deteriorate with age or wear out through use and allow for small, unnoticeable light leaks that can fog your film. Another source of radiographic fog within the darkroom is unsafe safelights. Filters can wear out, safelight housings can become damaged, improper wattage bulbs can be placed into the safelight housing, all of which can produce an unsafe working environment around exposed film. The safety of safelights should routinely be tested as part of the overall quality assurance program.

Film Artifacts

Fingerprints or smudges on the radiographic image can be evidence of dirty, oily hands by the film processing personnel. Crinkle marks or crescent marks are evidence of mishandling of the film. Static electricity marks should make the quality assurance technologist aware that there may be areas of the work surface within the darkroom that require grounding. Additional radiographic fog can be caused by overactive processing chemical solutions. Check the accuracy of the temperature of the various solutions. Various chemical stains demonstrate evidence of possible processing errors, including exhausted chemicals, poor chemical replenishment rates, and dirty crossover roller systems. Linear scratches upon the surface of the film can be evidence of film processor transport problems. A routine maintenance program for the operation of the automatic processor was introduced in Chapter 4, "Radiographic Density." It would be advantageous to review that section to understand its significance to the overall quality assurance program.

Equipment Factors

The quality control program will include a series of tests and controls associated with the operation of the x-ray units and other equipment used in the production of the radiographic image.

Milliampere-Seconds (mAs) Output

The technologist must have an assurance that the x-ray equipment he or she is utilizing will accurately provide the output of exposure consistent with the technical exposure factors selected. The output of x-ray equipment must be routinely tested and evaluated. The various pieces of equipment and different radiographic rooms should be calibrated and their output integrated in such a manner that technologists can be assured of their exposure factor selection regardless of the room they are using or the equipment they have chosen. The calibration between two-phase and three-phase x-ray-generating equipment must be conducted by a proper service technician.

Milliampere Output

The technologist must be able to select a specific focus size, and the exposure rate output must be accurate for the milliampere setting selected. The selection of a different size focus should not alter or significantly affect the exposure rate output for the same milliampere settings. A selection of a 200 mA setting must provide for 200 mA regardless of whether the technologist is choosing the small or the large size focus. A simple test of the milliampere output was introduced in Chapter 4, "Radiographic Density," although a more accurate assessment of the exposure rate may be required within an operational quality assurance program. Of equal importance is that the milliampere output is constant between different pieces of x-ray equipment and for different x-ray rooms. The routine evaluation of the output of milliamperes is an important facet of the testing program for equipment.

Milliampere output must be routinely adjusted by the service technician or engineer as the x-ray tube gets older and the number of exposures applied to the tube increases. With age and usage, the x-ray tube filament becomes thinner as the tungsten evaporates from its surface. This produces a slow but steady increase in the heat produced and the electrons released for the same filament current; therefore, there will be a slow but steady increase in the milliampere output associated with the tube operation even though the actual milliampere setting selected by the technologist remains the same. A small decrease in the filament current by the service technologist can help to maintain a constant tube output.

Timer Accuracy

In Chapter 4, "Radiographic Density," we considered the problems associated with a timer mechanism not operating properly. Tests for the accuracy of the x-ray timing mechanism should be routinely included in your quality assurance program. Simple tests with a spinning top or a synchronous spinning top can be performed, but for the assessment of the accuracy of extremely short exposure times of less than 1/60 (.0166) sec, other devices provided by the service technician must be employed.

Automated Exposure Control

The use of automated exposure control devices can help to reduce the need for repeat examinations if they are used properly. The accuracy of the positioning of the patient, the centering of the central ray over the automated exposure device, and the centering of the image receptor are all factors that contribute to the accuracy of the automated exposure control device. In all instances, the kilovoltage necessary to penetrate must be employed. Automated exposure devices all have a timer override. That is, a maximum time is set to prevent patient overexposure in case the positioning is not proper and the automated device is not reading an accurate section of the body thickness and opacity.

Collimator Control

The collimator is the most frequently employed beam limitation device in the performance of radiologic procedures. It is important that the manual or automatic controls accurately identify the actual beam coverage selected. If the technologist chooses a 6" by 8" (15 cm × 20 cm) image size, the collimator should provide for this exact beam limitation at the chosen SID. In addition, the light detection device of the collimator must be accurately aligned with the lead shutters of the collimator so that the projected light image upon the patient's body accurately reflects where the beam of radiation is directed and the exact size of the exposure coverage. The accuracy of the beam limitation and the alignment with the projected light beam of the collimator device should be routinely checked and included in a quality assurance program.

Film Processing

The basic control of film processing includes the standardization of the time and temperature of the development of the recorded image. With manual processing, this was frequently a problem due to many variables, the human element, and the potential for uncontrolled sight development.

Automated control of film processing has eliminated many of the variables, including most of the problems associated with the human element. Automated film processing maintains an accurate control of the temperature of the chemical solutions, the activity and replenishment rates of those solutions, and the time of processing by means of a roller transport system. Guidelines associated with the maintenance and operation of the automatic film processor are provided in Chapter 4, "Radiographic Density."

To integrate film processing control into the quality assurance program, the processor must be routinely checked and tested. Sensitometry is the accepted method recommended for the routine testing of the processor. Film strips produced by a sensitometer are used to produce controlled radiographic density measurements. Pre-exposed strips are also available for departments who do not have access to a sensitometer. Film strips are processed daily and their radiographic density levels plotted against standardized performance levels. Minor changes and problems can be frequently observed and corrected before any major problems become evident. The routine maintenance of automatic film processors is frequently assigned to service contracts. It is important to be sure the contract includes adherence to the quality assurance standards that you have established for your department and that the service technicians are aware of your program and perform their functions within the parameters you have established.

Preventive Maintenance

Accurate recordkeeping forms the basis for all of the material and equipment factors testing and evaluation processes we have thus far identified. Each radiographic room and piece of equipment should have its own record log that includes dates of maintenance procedures and performance standards. All tests performed as part of the quality assurance program, the date they were performed, and the evaluation of the test should be recorded. The film processing portion of the quality assurance program should also be well documented with daily records kept of the film sensitometry as well as routine daily cleaning and maintenance schedules and weekly or monthly maintenance programs for the automatic film processor.

The planning, designing, assigning of personnel, and implementation of a quality assurance program will help to reduce equipment downtime and repeat examinations and provide for a greater accountability of the responsibility to provide radiological services of high quality. Periodic reject-film analyses should be conducted on a routine basis in order to provide information as to how well the overall quality assurance program is performing. A quality assurance program also represents an example of concern for the patient and adherence to the principles of using radiation levels "as low as reasonably achievable" (ALARA) by utilizing your equipment within its proper operating parameters and thereby minimizing potential repeat examinations and reducing the total radiation exposure to the patient.

Standardized Exposure Technique Charts

An important facet of any quality control program is the development and implementation of **standardized exposure technique charts.** The use of standardized exposure technique charts is an essential step toward the accountability of the quality assurance process necessary to avoid the problems associated with the technical guessing game. The development of a chart based on recommended exposure factors and accessories can help to standardize the performance of the radiological examination, significantly reduce repeat films, and improve the overall quality of the radiographic images produced.

Basically, a standardized exposure technique chart provides the technologist with exposure, equipment, and accessory guidelines to be used in the routine examination and is based on the average thickness and opacity of the body part for each of the multiple examinations performed within the radiology department. Standardized exposure technique charts are especially useful to orient new personnel to the equipment and expected levels of quality performance within a radiology department. When established, the technique chart will determine the accepted levels of the scale of contrast, the amount of patient exposure, and the margin for error (exposure latitude) available to the technologist when choosing specific exposure settings.

The use of exposure charts is, however, not absolute, but simply one additional tool available to technologists in the performance of their responsibilities. Earlier in our analysis of radiographic quality, we introduced the big X in the radiographic examination: the patient. The patient factors of composition of the tissue, opacity of the tissue, thickness of the part, general characteristics of physique and development, and the presence of pathology must all be considered when applying the principles of a standardized exposure guide. It is essential for the performance of the examination that proper information be provided on the examination request form. Information as to the reasons for the examination and the condition of the patient, including his or her diagnosis, is essential to the proper performance of the radiological examination. It is the responsibility of the technologist to elicit this information from the patient if it is not available on the examination request. To fail to do so would be considered negligent.

Is the reason for the examination of the chest to investigate the presence of potential lung pathology, or is it to investigate the possible presence of rib fractures due to a recent trauma? Is that increased thickness of the part associated with your measurement of the abdomen a result of the presence of increased air associated with an intestinal obstruction, or the condition of ascites associated with the presence of increased fluid within the abdominal cavity, or perhaps the presence of a tumor? All of these factors must be considered by the technologist when applying the principles of exposure factor selection and utilizing the standards established by the technique guide. Thus, the guidelines established by the standardized exposure technique chart must be adjusted for the considerations of the patient factors.

Optimum Kilovoltage

Throughout our analysis of radiographic quality, we indicated that the kilovoltage necessary to penetrate the part should always be used, as this allows for the greatest exposure latitude in the selection of the minimum to maximum mAs that will produce a diagnostic radiographic image. The relationship of the kilovoltage and mAs was established so that the kilovoltage is used to provide penetration of the part and the mAs selected to provide for the required radiographic density. Optimum kilovoltage and technique charts using this concept are slightly different from the basic concept of choosing the kilovoltage necessary to penetrate. The optimum kilovoltage provided in standardized technique charts may be different from the kilovoltage necessary to penetrate the part as identified in Chapter 5, "Radiographic Contrast." This is because in establishing the basic guidelines for the development of the standardized technique chart, the physicians have determined the radiographic scale of contrast desired for the examinations performed within the department. Therefore, the kilovoltage selected for the exposure guide provided to the personnel performing the examinations will often indicate a kilovoltage higher than that actually needed to penetrate in order to produce the desired radiographic scale of contrast agreed upon by the physicians. This kilovoltage is referred to as **optimum kilovoltage.** Optimum kilovoltage technique charts generally produce radiographic images with a longer scale of contrast than those using a variable kilovoltage. The radiographic scale of contrast will be extended; therefore, more total information will be recorded and visible within the radiographic image. The actual diagnostic value of the radiographic image will be increased.

Preparation of the Optimum Kilovoltage Technique Chart

The goal in every radiology department is to produce consistent and repeatable quality radiographic images with a minimum of exposure to the patient and a maximum margin for error (exposure latitude), which will reduce the numbers of reject-films and repeat examinations. In order to meet this goal, the department must develop, test, and adopt the principles of exposure guidelines and establish standardized exposure technique charts to be used in every radiographic room and with each piece of radiographic equipment. A program of in-service education must be provided so that all technological personnel are familiar with the standards established and know how to utilize the exposure guidelines to advantage.

The following considerations are part of the process of developing the guidelines necessary to establish standardized technique charts. First of all, there must be agreement by the physicians related to the overall quality (radiographic scale of contrast) of the radiographic image. This will enable the quality control technologist to establish the optimum kilovoltage range for the various examinations performed within the department.

The various pieces of equipment and accessories must be tested and calibrated according to established standards as outlined in the section on quality assurance of this chapter. The output of the x-ray units and the condition of the cassettes, intensifying screens, grids, and collimators should be checked and adjusted to meet the required standards. Standards for the film processing system must be established, tested, and maintained as part of the quality assurance program. Only when these standards have been established can the development and eventual implementation of a standardized technique chart be successful.

Calipers for the measurement of all body parts must be available within every radiographic room, and the technological personnel must be trained in using these devices properly and measuring the thickness of the part for every radiographic procedure. The use of standardized exposure technique charts can be successful only when the technologists assume their responsibility in helping to eliminate the technical guessing game and adopt the standards established by the exposure guidelines. Essential to these guidelines is the measurement of every body part to be examined.

Every technique chart developed must be based on specific parameters of operation. Each technique chart will be unique related to the equipment and accessories utilized for the performance of the radiological examination. The speed of the film-screen recording system must be identified. A basic standard should be established related to a specific speed system. For example, the exposure technique chart can be based on a 200 speed system. Knowing this, the technologist can adjust his or her selected exposure factors when employing a faster or slower speed system for any given examination. The grid ratio for the recommended exposures must be identified. Again, when employing a grid ratio different from the grid ratio established within the guidelines of the technique chart, the technologist can adjust the exposure factors in order to produce a consistent radiographic image.

After all of the basic development and preparation have been completed, suitable testing of the recommended exposure standards should be performed using phantom parts to establish basic parameters of exposure for a given measured thickness. Optimum kilovoltage levels for every examination and body part should be established; the mAs necessary to produce the desired radiographic density should be identified for the normal thickness of a given body part. After testing has been completed, a draft of the technique chart for each radiographic room should be provided to each of the technologists and posted within the radiographic room in a prominent place within the control panel.

Implementation of the Optimum Kilovoltage Technique Chart

Once the basic guidelines for a standardized exposure technique chart have been established, the chart should be implemented for use within the department. During the initial phase of implementation, a careful reject-film analysis must be undertaken. Every rejected film must be identified and the reasons for its rejection established. Particular attention must be given to those films rejected on the basis of exposure factors. Therefore, during the initial phase of the implementation of standardized exposure techniques, specific basic information for every examination performed must be recorded by the technologist. This includes the actual measurement of the thickness of the part using the caliper in centimeters; the mAs selected for the part; and any change or deviation from the established parameters of the technique chart, such as a different grid ratio or film-screen recording system speed, SID, and so forth. This information must be recorded for every examination performed during the initial phase of the introduction of the standardized technique chart. The information can be included on the original x-ray examination request or on separate forms specifically designed for that purpose.

A detailed analysis of the rejected films compared with the actual technical exposure factors used will help the quality control technologist to refine and improve the basic recommended exposure techniques. As additional examinations are performed on the patient with reduced thickness of part and increased thickness of part, additional mAs values can be established for utilization by the department. Remember, a minimum of 25% to 30% change in the mAs is necessary to produce a visible radiographic density change within the recorded image. Within a month of implementing a standardized exposure technique chart, the accuracy level of the factors covered by the chart will have been established, and the repeat film rate will have been reduced significantly. It is especially important that all new technologists be introduced into the quality control procedures of the department and instructed how to use the technique charts properly. Emphasis must be placed on their responsibility and accountability related to the performance of their radiological examinations and the use of the exposure parameters established by the department. A reject-film analysis should be performed on all new personnel for at least a month. The quality control technologist can determine the areas, if any, that must be reviewed with the new technologist in order to improve the quality of his or her performance and to maintain the overall radiographic quality of the department.

Table 6-1 shows a sample chart for an optimum kilovoltage, variable mAs technique chart for a single-phase, fully rectified x-ray unit using a 200 speed film-screen recording system and a 12:1 ratio grid for the upper extremities. Table 6-2 shows a similar chart for the lower extremities. Table 6-3 shows a similar chart for the spine, abdomen, and chest. Table 6-4 shows a similar chart for the skull. Milliampere-seconds for the average and the –2 cm and +2 cm thicknesses of the various body parts and examinations have been provided. The charts also provide space for the insertion of mAs selections for reduced and increased tissue thickness. Remember the technique chart is simply a guideline, and its implementation within any radiographic department must be performed after an appropriate process of planning and testing.

Frequently within any radiology department there will be a mixture of single-phase and three-phase generating x-ray units. The standardized exposure technique chart established for a single-phase generating unit cannot be transferred to and utilized within a room using a three-phase generating unit. The x-ray produced with a three-phase generating unit differs from that produced with a single-phase unit. Three-phase generating units produce a source of radiation that is more constant and has less fluctuations. The kilovoltage is maintained at a nearly constant peak;

Table 6-1 Upper Extremity and Shoulder Girdle Optimum kVp–Variable mAs Technique Chart

Single-Phase Generator, 200 Film-Screen Recording System, 12:1 Ratio Grid

X-ray exam	Projection	kVp	Grid	Inch SID	Remarks	Avg cm thick-ness	mAs	−2 cm thick-ness mAs	−4 cm thick-ness mAs	+2 cm thick-ness mAs	+4 cm thick-ness mAs	+6 cm thick-ness mAs
FINGERS	ALL	60	No	40		2	1.25	.90		1.80		
HAND	AP & OBL	60	No	40		3–4	1.60	1.20		2.40		
	LAT	60	No	40		8–10	3.00	2.25		4.50		
WRIST	AP & OBL	60	No	40		3–4	1.60	1.20		2.40		
	LAT	60	No	40		8–10	3.00	2.25		4.50		
FOREARM	AP & OBL	60	No	40		7–8	3.50	2.60		5.25		
	LAT	60	No	40		8–9	4.50	3.40		6.75		
ELBOW	AP & OBL	60	No	40		7–8	4.00	3.00		6.00		
	LAT	60	No	40		8–9	5.00	3.75		7.50		
HUMERUS	AP & LAT	60	No	40		12–14	7.00	5.25		10.50		
SHOULDER	AP	80	Yes	40	Lead blockers	12–14	20	15		30		
	TRANS-LAT	80	Yes	40		30	80	60		120		
CLAVICLE	PA & AX-PA	80	Yes	40		12–14	20	15		30		
SCAPULA	AP	80	Yes	40		12–14	20	15		30		
	LAT	80	Yes	40	Lead blockers	12–14	30	22.5		45		

Note: AP = anteroposterior, OBL = oblique, LAT = lateral, PA = posteroanterior, AX-PA = axial-posteroanterior, TRANS-LAT = transthoracic lateral.

Table 6-2 Lower Extremity and Pelvic Girdle Optimum kVp–Variable mAs Technique Chart

Single-Phase Generator, 200 Film-Screen Recording System, 12:1 Ratio Grid

X-ray exam	Projection	kVp	Grid	Inch SID	Remarks	Avg cm thick-ness	mAs	−2 cm thick-ness mAs	−4 cm thick-ness mAs	+2 cm thick-ness mAs	+4 cm thick-ness mAs	+6 cm thick-ness mAs
TOES	ALL	60	No	40		2–3	1.25	.90		1.80		
FOOT	AP & OBL	60	No	40		6–7	3.00	2.25		4.50		
	LAT	60	No	40		8–9	4.00	3.00		6.00		
OS CALCIS	AXIAL	60	No	40	Lead blockers	9	5.00	3.75		7.50		
	LAT	60	No	40		7	4.00	3.00		6.00		
ANKLE	AP & OBL	60	No	40	Lead blockers	9	5.00	3.75		7.50		
	LAT	60	No	40		7	4.00	3.00		6.00		
LEG	AP & LAT	60	No	40	Lead blockers	11–13	7.00	5.25		10.50		
KNEE	PA	70	Yes	40	Lead blockers	11–13	16	12		24		
	LAT	70	Yes	40	Lead blockers	10–12	12					
PATELLA	PA	70	Yes	40	Coned down	11–13	16	12		24		
	AXIAL & LAT	70	Yes	40	Lead blockers	9–10	12	9		18		
FEMUR	AP & LAT	80	Yes	40	Gonadal shield	18–20	30	22.50		45		
HIP	AP & LAT	80	Yes	40	Gonadal shield	20	60	45		90		
	Crotch LAT	90	Yes	40	Gonadal shield	24–26	80	60		120		
PELVIS	AP	80	Yes	40	Gonadal shield	20–22	60	45		90		

Note: AP = anteroposterior, OBL = oblique, LAT = lateral, PA = posteroanterior.

Table 6-3 Spine, Abdomen, and Chest Optimum kVp–Variable mAs Technique Chart

Single-Phase Generator, 200 Film-Screen Recording System, 12:1 Ratio Grid												
X-ray exam	Projection	kVp	Grid	Inch SID	Remarks	Avg cm thick-ness	mAs	−2 cm thick-ness mAs	−4 cm thick-ness mAs	+2 cm thick-ness mAs	+4 cm thick-ness mAs	+6 cm thick-ness mAs
CERV-SPINE	AP & ODON	80	Yes	40		12	20	15		30		
	OBL & LAT	80		72	Upright	12	20	15		30		
THOR-SPINE	AP*	80	Yes	40	*Anode heel	20	80	60		120		
	LAT*	80	Yes	40	and shallow	30	100	75		150		
	SWIMMERS	80	Yes	40	breath	30	100	75		150		
LUMB-SPINE	AP	80	Yes	40	Gonadal shield	20	80	60		120		
	OBL	80	Yes	40	Gonadal shield	24	120	90		180		
	LAT	90	Yes	40	Gonadal shield	28	160	120		240		
	L5-S1 JUNC	90	Yes	40	Gonadal shield	30	180	135		270		
LUMBO SACR ARTIC	AP	80	Yes	40	Gonadal shield	22	100	75		150		
	LAT	90	Yes	40	Gonadal shield	30	180	135		270		
SACROILIAC JOINTS	AP	80	Yes	40	Gonadal shield	20	80	60		120		
	RPO & LPO	80	Yes	40	Gonadal shield	24	120	90		180		
SACRUM	AP	80	Yes	40	Gonadal shield	20	80	60		120		
	LAT	90	Yes	40	Lead blockers	30	160	120		240		
COCCYX	AP	80	Yes	40	Coned down	20	60	45		90		
	LAT	80	Yes	40	Lead blockers	30	100	75		150		
ABDOMEN	AP	80	Yes	40	Gonadal shield	20	80	60		120		
	OBL	80	Yes	40	Gonadal shield	24	90	67.50		135		
	LAT	80	Yes	40	Gonadal shield	28	120	90		180		
	LAT DECUB	80	Yes	40	Gonadal shield	24	80	60		120		
CHEST FOR LUNGS	PA	120	Yes	72	Inspiration	20	6	4.5		9		
	OBL	120	Yes	72	Inspiration	26	8	6		12		
	LAT	120	Yes	72	Inspiration	30	12	9		18		
RIBS	AP/PA	80	Yes	40	Suspended	20	20	15		30		
	OBL	80	Yes	40	respirations	24–26	30	22.50		45		
STERNUM	PA-OBL	60	Yes	40	Shallow breath	20	60	45		90		
	LAT	70	Yes	40	Suspend respir	30	80	60		120		

Note: AP = anteroposterior, ODON = odontoid, OBL = oblique, LAT = lateral, LAT DECUB = lateral decubitus, PA = posteroanterior, RPO = right posterior oblique, LPO = left posterior oblique.

* = applicable only to AP and LAT projections.

Table 6-4 Skull Optimum kVp–Variable mAs Technique Chart

Single-Phase Generator, 200 Film-Screen Recording System, 12:1 Ratio Grid												
X-ray exam	Projection	kVp	Grid	Inch SID	Remarks	Avg cm thick-ness	mAs	−2 cm thick-ness mAs	−4 cm thick-ness mAs	+2 cm thick-ness mAs	+4 cm thick-ness mAs	+6 cm thick-ness mAs
SKULL	TOWNE	80	Yes	40	Upright except	20	60	45		90		
	CALDWELL	80	Yes	40	trauma	20	40	30		60		
	LAT	80	Yes	40	patient	15–16	20	15		30		
	BASAL	90	Yes	40		20	80	60		120		
SINUSES	CALDWELL	80	Yes	40	Upright	20	40	30		60		
	WATERS	80	Yes	40		20	60	45		90		
	LAT	70	Yes	40		15	20	15		30		
	BASAL	90	Yes	40		20	80	60		120		
FACIAL	CALDWELL	80	Yes	40	Upright except	20	40	30		60		
BONES	WATERS	80	Yes	40	trauma	20	60	45		90		
	LAT	70	Yes	40	patient	15	20	15		30		
ZYGOMA	BASAL	60	No	40		20	20	15		30		
MANDIBLE	PA	80	Yes	40	Upright except	14	20	15		30		
	LAT	60	No	40	trauma	14	10	7.50		15		
	AP CONDYLE	70	No	30	patient	14	10	7.50		15		
NASAL	WATERS	80	Yes	40	Upright except	20	60	45		90		
BONES	LAT	60	No	40	trauma	15	1.5	1.12		2.25		
	INTRAORAL	60	No	40	patient	2	100	75		150		
OPTIC FORAMEN	RHESE	80	Yes	40	Upright except trauma patient	20	60	45		90		
MASTOIDS	TOWNE	80	Yes	40	Upright except	20	60	45		90		
	STENVERS	80	Yes	40	trauma	24	70	50		100		
	MAYERS	80	Yes	40	patient	20	30	22.50		45		
	LAWS	60	No	30		15	20	15		30		

Note: LAT = lateral, PA = posteroanterior, AP = anteroposterior.

therefore, the technical exposure factors (mAs) required for a three-phase generating unit will be less than those required for a single-phase generating unit. A separate technique chart must be developed for the three-phase generating equipment utilized within the department. A proper testing program similar to that produced in establishing the exposure guidelines for the single-phase unit technique chart must be performed. You will find that the exposure values for the three-phase generating units will be able to be reduced by approximately 50% over the exposure factors established by the single-phase generating units. Table 6-5 provides a basic, blank technique chart that can be reproduced and used to establish exposure technique guidelines within your radiology department. The use of standardized exposure technique charts is a positive example of the principles of ALARA by establishing guidelines for the utilization of the x-ray equipment and standards for the selection of exposure factors that will provide for consistent, repeatable quality of the recorded images, thereby reducing the numbers of reject films and repeated examinations.

Radiation Safety

Principles of ALARA

The performance of all radiological procedures is directed to minimize the radiation exposure of the patient and personnel. If radiation control guidelines have been established, and the personnel have been instructed in their application, the occupational exposure of personnel and the radiation exposure of patients can be kept to a minimum. This concept is embodied within the principles of ALARA. We have introduced these principles throughout *An Analysis of Radiographic Quality* and will re-examine the principles related to minimal dose radiography here.

Protective Equipment and Apparel

During radiographic procedures, the technologist should remain outside of the examination room behind the protective barriers of the control panel. During fluoroscopic procedures when required to be in the room, the technologist is required to wear protective apparel. Protective garments include lead-impregnated aprons and gloves. These garments come in lead equivalents of 0.25, 0.50, and 1.0 mm. Within the diagnostic range of radiography, most departments use lead-equivalent protective garments of 0.50 mm.

During procedures in which patients will be seated adjacent to the examination table and scattered radiation from the exposure may potentially reach the areas of the patient's gonads, the patient should also be provided with protective garments to eliminate this unnecessary exposure. Patients should be seated or arranged upon the table with a consideration to the direction of scattered radiation.

Holding of Patients during Examinations

Technologist personnel should never be used to hold patients during x-ray examinations even when protective apparel is utilized. For infants, children, and adults who are unable to cooperate, position achievers and maintainers such as

Table 6-5 Optimum kVp–Variable mAs Technique Chart

_____-Phase Generator, _____ Film-Screen Recording System, _____ Ratio Grid

X-ray exam	Projection	kVp	Grid	Inch SID	Remarks	Avg cm thick-ness	mAs	−2 cm thick-ness mAs	−4 cm thick-ness mAs	+2 cm thick-ness mAs	+4 cm thick-ness mAs	+6 cm thick-ness mAs

positioning sponges and tape as well as part- and whole-body immobilization devices should be employed. Relatives or friends of the patients who accompany them for the examination can be used to help with the examination. They should be provided with protective apparel. As a last resort, other hospital employees not directly associated with the radiology department can be asked to assist with the examination or to hold patients. Protective apparel should always be provided these personnel.

Unnecessary Examinations

Most radiology departments have established guidelines for the proper practice of radiological services that include the principles of ALARA. Included in these guidelines is the elimination of unnecessary examinations. There should be no "routine" x-ray examinations. No x-ray examination should be performed without a specific and precise medical indication.

Pregnancy Considerations

The technologist who becomes pregnant should inform her supervisor. The supervisor will check the technologist's past radiation exposure monitoring reports to determine whether any additional precautions are necessary. In most cases, no additional precautions are required. However, it may be the decision of the supervisor to remove the pregnant technologist from fluoroscopic procedures and portable radiographic examinations during the term of her pregnancy since these two activities have been shown to result in the greatest amount of occupational exposure to the technologist.

The female patient of childbearing years should be a consideration in the guidelines for the practice of radiation safety in all radiology departments. The major concern is the unknown or unsuspected pregnancy during the first month of pregnancy. Many departments have adopted the principle of the ten-day rule. The ten-day rule refers to the ten-day interval following the onset of menstruation. During this period, the chances of pregnancy are least likely. All examinations that include radiation to the abdomen and pelvis should be limited to this ten-day period. Of course, if the examination is required for the health of the patient, it should be conducted. However, care should be taken to minimize patient exposure and to shield the fetus from the primary beam whenever possible. It is important that female patients of childbearing ages be questioned as to whether they are or could potentially be pregnant. Frequently, this is done in the form of a questionnaire rather than a direct inquiry of the patient. Evidence of pregnancy or suspected pregnancy must be reported to the radiologist or physician in charge in order to determine whether the examination should be performed, canceled, or delayed until a later time during the pregnancy.

Application of the Principles of ALARA

There are numerous factors, techniques, and choices that the technologist employs in the conduct of the radiological examination that relate to the proper application of the principles of ALARA.

A proper analysis of radiographic quality is a basic consideration in the application of the principles of ALARA, demonstrating the technologist's knowledge and understanding of the principles of the production of the recorded image and the proper use of equipment, materials, and exposure factors. A properly educated and experienced technologist will avoid the problems inherent in the technical guessing game. Choosing the equipment to be used for a given examination and selecting the accessories and exposure factors are frequently trade-offs that the knowledgeable technologist understands will produce some loss of image quality. The loss of image quality is often associated with the recorded detail. Changing to a higher speed film-screen recording system, immobilizing the part, and reducing the time of exposure to eliminate patient motion are examples of the positive application of the principles of ALARA in an effort to avoid the need for a repeat examination.

Direct exposure radiography is contraindicated to the principles of ALARA and is no longer advocated in general radiography procedures.

The use of beam restriction devices, which help to eliminate unnecessary exposure to portions of the patient's body not included in the recorded image, is an example of the principles of ALARA. Beam restriction should be employed in all radiographic examinations. The size of the beam of radiation should be restricted to record the structures of interest and not the actual film size. Proper beam restriction improves the overall visibility of the recorded image by reducing the radiographic fog and thus the potential for a needed repeat examination as well as reducing the overall radiation within the primary beam directed toward the patient. The use of lead blockers placed adjacent to the part to be examined will further eliminate radiographic fog, improve the visibility of the image, and potentially reduce the need for repeat examinations.

The rules of good radiologic practice require the use of gonadal shielding in all applicable examinations. In general, gonadal shields should be used in all patients who are potentially reproductive, that is, all patients under the age of 45 and perhaps even older males. Gonadal shields should be employed whenever the gonads lie in or near the primary beam. However, gonadal shielding should be used only when its presence will not interfere with the required diagnostic information for the examination.

In the area of exposure factors and accessories, the kilovoltage necessary to penetrate the part should always be used. Insufficient kilovoltage requires the use of increased mAs to produce the required radiographic density of the recorded image, and increased mAs is associated with increased patient exposure. The use of radiographic grids significantly improves the visibility of the recorded image by preventing scattered radiation from reaching the film. However, the use of radiographic grids also requires an increase in the exposure to the patient due to the absorption characteristics of the grid. Therefore, radiographic grids should not be employed in radiographic procedures of body parts that measure less than 10 cm that use less than 60 kVp. Their use in these procedures would be contraindicated and contrary to the principles of ALARA.

Finally, beam filtration provides an excellent example of the positive application of the principles of ALARA. The primary beam of radiation directed toward the patient consists of a wide range of wavelengths and penetrating abilities. A significant portion of the beam is comprised of low-level, less penetrating radiation of long wavelengths. This radiation does not have the penetrating ability to pass through the patient's body and participate in the recording of the radiographic image. The use of beam filters of at least 2.5-mm aluminum equivalent will harden the beam of radiation and eliminate most of the low-level, less penetrating radiation from the primary beam. Beam filtration reduces the patient's overall radiation exposure and significantly reduces his or her radiation skin dose.

Also included in the application of the principles of ALARA is the implementation of quality controls identified in this chapter in the sections on quality assurance and standardized exposure technique charts.

It is important that technologists recognize their responsibility in the application of x-rays on humans during their practice of radiologic technology. All means available to minimize the patients' exposure consistent with the requirements of the procedure being performed and the diagnostic requirements of the radiographic quality of the recorded image should be used.

Quality Control Review Questions

Name ______________________________________ Date ________________

1. Describe the purpose of quality controls and the implementation of a quality assurance program within the radiology department.

2. How will a reject-film analysis contribute to the establishment of quality controls within the radiology department?

3. Provide a brief statement indicating how each of the following material factors would be involved in a quality assurance program: cassettes, intensifying screens, film-screen contact, radiographic grids, radiographic fog, and film artifacts.

4. Provide a brief statement indicating how each of the following equipment factors would be involved in a quality assurance program: milliampere output, timer accuracy, automated exposure control, collimator control, and film processing.

5. Discuss the purpose of a preventative maintenance program and the need for accurate recordkeeping in the implementation of a quality assurance program within the radiology department.

6. Discuss the importance of the development of a standardized exposure technique chart for use in the radiology department. Describe its purpose within the quality assurance program.

7. Can one standardized exposure technique chart be utilized for all of the equipment and units within the department? Explain.

8. If your technique chart is based on a 100 speed film-screen imaging system, what must you do when you are employing a 400 speed film-screen imaging system?

9. If your technique chart is based on a 12:1 grid ratio Bucky grid mechanism, and you are performing the procedure using a stationary 8:1 grid ratio parallel-grid cassette, what adjustments, if any, must you make in order to maintain the radiographic density of the recorded image?

10. Of what significance is the patient's diagnosis, condition, and the purpose of the examination written on the examination request when employing standardized exposure technique charts? How would these factors influence your choice of exposure factors?

11. Describe the purpose and importance of calipers and the measuring of the patient's body part thickness to the quality of the radiographic image when using standardized exposure technique charts.

12. Differentiate between the kilovoltage necessary to penetrate the part and the use of optimum kilovoltage in a standardized exposure technique chart.

13. Discuss the principles of ALARA related to establishing guidelines for the conduct of good radiologic practice procedures within the radiology department.

14. Describe the positive use of protective apparel for the technologist, for the patient, and for assisting personnel related to the principles of ALARA.

15. Elaborate on the basic statement that technologists should not hold patients during examinations. Why has this principle been adopted?

16. What do we mean by the statement that there should be no "routine" x-ray examinations?

17. Describe the purpose and application of the 10-day rule related to the performance of x-ray examinations on females of childbearing years.

18. Why is direct exposure radiography contrary to the principles of ALARA?

19. Discuss the functions and contributions of the radiologic technologist toward the application of the principles of ALARA.

7 Radiographic Quality Summary

A recorded image that is of radiographic quality must possess excellent structural detail with a minimum of distortion. Additionally, the image must possess proper radiographic density and contrast. With these two short sentences, we have identified and qualified all of the principles, factors, and relationships required of a film of radiographic quality.

In the preceding chapters of this book, we have analyzed radiographic quality with the above definition in mind. We have identified two major concepts that represent the proper balance contained within this definition: (1) the image details and (2) the image visibility. We have also referred to these concepts as the **geometric properties** and **photographic properties** of the radiographic image. You have discovered through your analysis of radiographic quality that the proper balance between these two concepts is necessary to produce a film with optimal interpretative information. The basic factors associated with the geometric properties of the image are recorded detail and distortion. The basic factors associated with the photographic properties of the image are radiographic density and radiographic contrast.

You have investigated and examined each of these basic factors, both as individual influences on radiographic quality and as interrelated, interdependent factors. You have examined a number of principles and formulas and have performed experiments related to the multiple factors that control, influence, and otherwise affect each of these four basic factors associated with the production of the radiographic image. Performing an analysis of radiographic quality in this manner has enabled you to develop a thorough understanding of these principles and a confident working relationship with the technical exposure factors and formulas that influence the production of the radiographic image. This analysis of radiographic quality has been only a brief introduction to the major factors that influence the production of a quality radiograph. There are other factors and influences to consider, but they, too, can be analyzed in a similar way.

Take a good look at each radiographic image you produce. Take another look—a more critical one! If the recorded image is nondiagnostic, analyze it carefully to discover the reasons for its inferior quality before you attempt to repeat the examination. Is the lack of recorded detail the result of motion due to breathing, or does the blur associated with your image result from the quantum mottle due to the selection of an ultrafast imaging system? Is the poor visibility of the details the result of an improper kVp/mAs relationship, or does it result from the excessive radiographic fog from inadequate beam limitation? By critically examining each radiograph you produce in this manner, you will be able to make the appropriate choices in the adjustments necessary to produce a proper diagnostic result—choices based on your knowledge and understanding of the multiple factors that influence the recording of the radiographic image.

If the film you have produced is diagnostic, review it again with a critical eye toward improving the radiographic quality. Consider the image's fine points; analyze the factors and relationships in the recorded image that could be improved. A careful film analysis and critique should be a basic process for every radiographic image you produce. Only in this way can you expect to eliminate the errors inherent in the technical guessing game. Only in this way can you expect to achieve success in your efforts to improve your technical skills and competence in order to produce films that are of radiographic quality.

A review of the "Analysis of Radiographic Quality Diagram" provided in this chapter can help you to coordinate your efforts toward improving your technical skills. The diagram provides a quick review of all of the factors that control and influence the production of a film of radiographic quality.

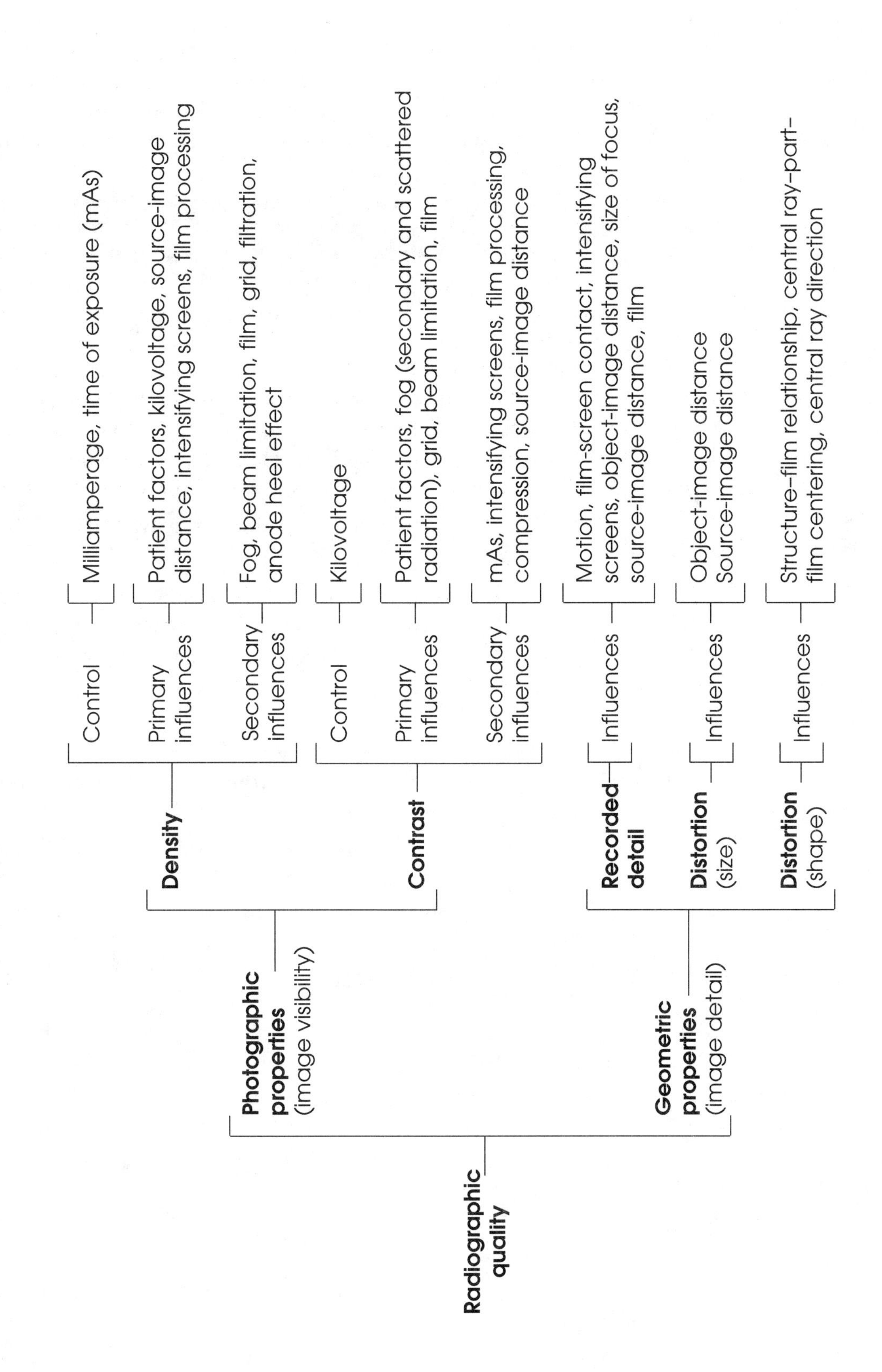
Analysis of Radiographic Quality Diagram
Radiographic quality
Photographic properties (image visibility)
Geometric properties (image detail)
Density
Contrast
Recorded detail
Distortion (size)
Distortion (shape)
Control
Primary influences
Secondary influences
Control
Primary influences
Secondary influences
Influences
Influences
Influences
Milliamperage, time of exposure (mAs)
Patient factors, kilovoltage, source-image distance, intensifying screens, film processing
Fog, beam limitation, film, grid, filtration, anode heel effect
Kilovoltage
Patient factors, fog (secondary and scattered radiation), grid, beam limitation, film
mAs, intensifying screens, film processing, compression, source-image distance
Motion, film-screen contact, intensifying screens, object-image distance, size of focus, source-image distance, film
Object-image distance
Source-image distance
Structure–film relationship, central ray-part-film centering, central ray direction

Bibliography

Burns, E. F. 1992. Radiographic Imaging: A Guide for Producing Quality Radiographs. W.B. Saunders, Philadelphia, Pa.

Campeau, F., and Phelps, J. 1993. Limited Radiography, Delmar, Albany, N.Y.

Carroll, Q. B. 1993. Fuch's Principles of Radiographic Exposure, Processing and Quality Control, 5th Ed. Charles C Thomas, Springfield, Ill.

Christensen, E., Curry, T., and Dowdey, J. 1978. An Introduction to the Physics of Diagnostic Radiology, 2nd Ed. Lea & Febiger, Philadelphia, Pa.

Cullinan, A. M. 1987. Producing Quality Radiographs. J.B. Lippincott, Philadelphia, Pa.

Haus, A. G., and Dickerson, R. E. 1993. Characteristics of Kodak Screen-Film Combinations for Conventional Medical Radiography, Technical and Scientific Monograph, No. 3, Eastman Kodak, Rochester, N.Y.

Hiss, S. S. 1993. Understanding Radiography, 3rd Ed. Charles C Thomas, Springfield, Ill.

International Commission on Radiological Units. Methods of Evaluating Radiological Equipment and Materials (ICRU Report 10f). Handbook 89. U.S. Department of Commerce, National Bureau of Standards, Superintendent of Documents, Washington, D.C.

Malott, J. A. 1993. The Art and Science of Medical Radiography, 7th Ed. Mosby-Year Book, St. Louis, Mo.

McKinney, W. E. J. 1988. Radiographic Processing and Quality Control. J.B. Lippincott, Philadelphia, Pa.

Selman, J. 1994. Fundamentals of X-Ray and Radium Physics, 8th Ed. Charles C Thomas, Springfield, Ill.

Sprawls, P. 1993. Physical Principles of Medical Imaging, 2nd Ed. Aspen Publishers, Gaithersburg, Md.

Thompson, T. 1979. Cahoon's Formulating X-ray Technique, 9th Ed. Duke University Press, Durham, N.C.

Tortorici, M. 1992. Concepts in Medical Radiographic Imaging: Circuitry, Exposure & Quality Control. W.B. Saunders, Philadelphia, Pa.

Analysis Experiments

Performing the following experiments when they are introduced within the text of the book will enable the student technologist to achieve a practical, working knowledge of the equipment, materials, and exposure factors vital to an understanding of radiographic quality. The information and skills gained by the performance of each experiment are frequently progressive in nature, and the knowledge and experience learned through performing one series of experiments will often assist the student in the understanding and performance of additional experiments.

The following materials are recommended for use with the various experiments in this analysis of radiographic quality. Depending on the type of radiographic equipment and the materials available to perform the experiments, it may be necessary to apply other techniques or to substitute alternative apparatus, material, or equipment for those suggested for use within the specifications of the experiments provided in this book. In addition, although suggested procedural parameters are provided for each experiment, they, too, may have to be adapted to meet the specific requirements or limitations of your equipment and available materials.

However, all of the suggested experiments can be performed using ordinary radiographic equipment normally found in any routine radiology department or well-equipped school x-ray laboratory. As a result of the need to adapt the experiments to the requirements of the equipment and materials available to the student, specific suggestions for technical exposure factors have been avoided. Instead, in most instances, a requirement for the production of a specific radiographic bone density of 1.0 to 1.2 has been recommended. Producing and maintaining an appropriate radiographic bone density of 1.0 to 1.2 will become easier after the performance of but a few experiments. This concept will enable the student to provide the baseline exposure and produce the initial radiographic image on which all the other exposures and radiographic images within each experiment are based.

An important aspect to the proper performance of the experiments is to complete the conclusions section for each experiment and have them reviewed by the instructor. Completing these experiments will provide the student with a real, practical knowledge and understanding of the principles related to the production of recorded images of radiographic quality. It will enable him or her to develop skills that can be transferred to the performance of radiographic examinations within the radiology department. It will also assist in the understanding of the principles of using levels of radiation that are "as low as reasonably achievable" (ALARA) and the importance of applying these principles in the performance of every radiographic examination. Finally, the performance of radiological experiments within the controlled environment of the radiological laboratory will enable the student technologist to gain confidence and develop the skills necessary to avoid the errors inherent in the technical guessing game.

Suggested Materials for Experiments

1. Manual and automated film processing facilities
2. Densitometer
3. Resolution test pattern (resolution grid capable of resolving up to at least 10 line pairs [LP]/mm)
4. Penetrometer (aluminum step wedge)
5. Screen contact test grid, wire mesh, or several boxes of unused paper clips
6. Spinning top
7. Synchronous spinning top
8. Filters, four 1-mm-thick aluminum filters
9. Paraffin, one 3" block, 5" by 8" (13 cm × 20 cm)
10. Orthopedic casting material
11. Tape measure, capable of measuring up to 80" (203 cm)
12. Dry bones (different loose bones from a skeleton; scapula, femur, humerus, metacarpals, os calcis, talus, clavicle, lumbar vertebra)
13. Radiographic phantoms (thorax, pelvis, ankle, knee, elbow, hand and wrist)
14. X-ray film, nonscreen and screen-type film (different speed and contrast)
15. Intensifying screens, up to four different film-screen speed combinations
16. Cassettes, different sizes
17. Nonscreen film holders (cardboard holders or disposable packs)
18. Radiolucent positioning sponges, different sizes and shapes
19. Lead rubber blockers, strips of different sizes and shapes for film division
20. Parallel grids, different ratios (8:1, 12:1) and sizes (10" × 12" [25 cm × 30 cm] and 14" × 17" [36 cm × 43 cm]), separate and/or within-grid cassettes
21. Bucky, moving grid mechanism (a 12:1 ratio grid is preferable)
22. Beam limitation devices, aperture diaphragms, cones, and collimators

Name ______________________________ Date ______________

EXPERIMENT 1
RADIOGRAPHIC QUALITY—Image Visibility versus Image Detail

Objective

To determine the relationship between the image visibility and the image detail of the recorded image.

Procedure

Set up your equipment to perform a tabletop procedure using a 10" by 12" (25 cm × 30 cm) direct-exposure (cardboard) film holder with screen-type film. Select the small focus of the tube, and adjust the SID to 40" (102 cm). Place a small, thick bone (i.e., os calcis, talus, sacrum) at the midpoint of your film holder, collimate your beam to include a margin 1" (2.54 cm) beyond the bone's edges, and direct your central ray to enter the midpoint of the centered bone using a perpendicular angle. Two exposures will be made on a single film. **Note:** The procedure can also be performed using a slow (detail) film-screen recording system.

Exposure 1

1. Selecting the small focus, use a low (50) kVp technique, and select the necessary mAs to achieve a radiographic bone density of 1.0 to 1.2.
2. Expose the film.

Exposure 2

1. Keeping everything else stationary, remove the bone from the film holder.
2. Cover one half of the exposed film (one half of the recorded bone image) with lead rubber blockers.
3. Reduce your technical factors (mAs) by 50%, and take an additional exposure of the unprotected film.
4. Process your film.

Analysis

Your film will demonstrate the recorded image of the bone divided into two different radiographic densities, each one possessing distinct and different image visibility. Examine the recorded details of the bone in the portion of the film representing your initial exposure. Now observe the recorded details within the film portion that received the additional exposure. Pay particular attention to the recorded details adjacent to the areas where the two different radiographic densities meet.

Conclusions

1. Which of the recorded images could be considered an image of radiographic quality? Explain the reasons for your selection.

2. Does it appear that the observed radiographic quality difference is attributed to a loss of the actual recorded details of the image or to the loss of the visibility of the recorded details? Explain the reasons for your choice.

3. Examine the recorded details of both image densities closely. Are the actual recorded details of the bone represented by the image that received the additional exposure less sharp or defined compared to the recorded details of the adjacent image that received the proper exposure? Provide reasons for your choice.

4. Will an image recorded with well-defined, sharp details automatically ensure that you will produce a film that is of radiographic quality? Explain your answer.

5. What can you conclude about the relationship between the image detail and the image visibility of the recorded image from this experiment?

Name ____ Date ____________________

EXPERIMENT 2
RADIOGRAPHIC QUALITY—Image Detail versus Image Visibility

Objective

To determine the relationship between the image detail and the image visibility of the recorded image.

Procedure

Set up your equipment to perform a tabletop procedure using a 10" by 12" (25 cm × 30 cm) direct-exposure (cardboard) film holder with screen-type film and a dry bone (e.g., a femur). Divide your cassette in half lengthwise using a lead rubber blocker. Select the small focus of the tube and adjust the SID to 40" (102 cm). Center the proximal end of the femur (femoral head, neck, and trochanter) on the unprotected half of the cassette, and collimate your beam to include the entire film section. Direct your central ray to enter the midpoint of the divided film section using a perpendicular angle. **Note:** The procedure can also be performed using a slow (detail) film-screen recording system.

Exposure 1

1. Use a low (50) kVp technique, and select the mAs necessary to achieve a radiographic bone density of 1.0 to 1.2. See Experiment 1.
2. Expose the film section.

Exposure 2

1. Remove the bone, and adjust the lead blocker to cover the exposed section of the film.
2. To the distal end of the femur, tie a piece of string of sufficient length so that it will reach back to the control booth.
3. Place the bone back on the film, centering the proximal end of the bone to the uncovered, unexposed film section.
4. Adjust the centering of the central ray appropriately.
5. Using the same exposure factors, expose the bone while pulling on the string, creating motion of the bone. (Only a minimum of motion is necessary for this procedure.)
6. Process the film.

Analysis

Your film will demonstrate two different recorded images of the femur. This will enable you to compare their radiographic quality. Examine the radiographic density and the contrast scale visible in each recorded image. Now compare the recorded details in each image.

Conclusions

1. Which of the recorded images represents an image of radiographic quality? Explain the reasons for your selection.

2. Do you believe the radiographic quality difference between the two images is attributed to poor photographic properties and, thus, a loss of image visibility or to a loss in the actual sharpness of the details of the recorded image? Explain the reasons for your selection.

3. Review the image visibility of both recorded femurs. Is the overall visibility of the image represented by the stationary bone's image better than that attributed to the moving bone's image, or does it just appear that way because of the differences in the actual sharpness of the details of the recorded bone images? Explain the reasons for your answer.

4. Will an image possessing a proper balance of the photographic properties of density and contrast automatically ensure that you will produce a film that is of radiographic quality? Explain your answer.

5. What can you conclude about the relationship between the image visibility and the image detail of the recorded image from this experiment?

Name ______________________________ Date ______________

EXPERIMENT 3
RECORDED DETAIL—Motion Unsharpness versus Time of Exposure

Objective

To demonstrate that motion unsharpness can be controlled by a reduction in the time of exposure with little appreciable loss of radiographic quality.

Procedure

A total of five radiographs using a small body part phantom (e.g., ankle, foot, elbow, or wrist) will be taken with a number of suggested adjustments in the technical factors in order to reduce the required time of exposure.

Film 1

1. Using a 40" (102 cm) SID, select the 100 mA station on the control panel using the small focus. Employ a low 50 kVp, and select the correct time of exposure to provide a radiographic bone density of 1.0 to 1.2.
2. Employ a slow speed (detail) film-screen recording system with an 8" by 10" (20 cm × 25 cm) cassette.
3. Center your central ray using a perpendicular angle, and collimate your beam to the part size.
4. Expose and process your film.

Films 2, 3, 4, and 5

For the remaining films, set up your procedure as for Film 1. Then make the following progressive adjustments one by one, exposing and processing your film after each.

Film 2: mA/Time Relationship

Double your milliamperage and reduce your time of exposure to maintain the same mAs.

Film 3: Imaging System Speed

In addition to the mA/time adjustment indicated for Film 2, change to a more rapid imaging system speed and adjust your mAs by reducing your time of exposure by the appropriate amount for the new imaging system speed.

Film 4: Source-Image Distance

In addition to the mA/time adjustment and the increased imaging system speed indicated for Film 3, reduce your SID to 30" (76 cm), and adjust your mAs by reducing your time of exposure according to the inverse square law.

Film 5: kVp/mAs Relationship

In addition to the mA/time adjustment, the increased imaging system speed, and the reduced SID indicated for Film 4, increase your kilovoltage by 15%, and reduce your mAs by 50% of the value used with Film 4 by a further reduction in your time of exposure.

Analysis

You will have five recorded images for comparison of overall radiographic quality and the progressive effect of employing different materials, technical factors, and times of exposure toward the maintenance of radiographic density in order to eliminate problems associated with possible motion unsharpness.

Conclusions

1. For each film, select the same specific area within one of the bones, and draw a circle around it. Examine its density with a densitometer, and record the density and exposure factors for each film as indicated below:

Film 1

____Milliamperage
____Time of exposure
____mAs selected
____Film-screen speed
____SID
____Kilovoltage
____Image density

Film 2

_____Milliamperage
_____Time of exposure
_____mAs selected
_____Film-screen speed
_____SID
_____Kilovoltage
_____Image density

Film 3

_____Milliamperage
_____Time of exposure
_____mAs selected
_____Film-screen speed
_____SID
_____Kilovoltage
_____Image density

Film 4

_____Milliamperage
_____Time of exposure
_____mAs selected
_____Film-screen speed
_____SID
_____Kilovoltage
_____Image density

Film 5

_____Milliamperage
_____Time of exposure
_____mAs selected
_____Film-screen speed
_____SID
_____Kilovoltage
_____Image density

2. Is it possible to reduce motion unsharpness by the manipulation of exposure factors? Explain your answer.

3. Compare the overall radiographic quality of the images. Does the radiographic quality differ from Film 1 to Film 5? Describe the difference in image detail and image visibility, if any, that has occurred.

4. Compare the radiographic density of the images. Does the density differ from Film 1 to Film 5? Describe the difference, if any, that has occurred. What factors may have contributed to the difference noted?

5. Considering the detrimental effects of motion unsharpness, are exposure factor changes to reduce the time of exposure justified and purposeful? Explain your answer.

6. Considering the problems of patient motion, describe how this experiment demonstrates the positive application of the principles of ALARA.

Name ______________________________ Date ______________

EXPERIMENT 4
RECORDED DETAIL—Material Unsharpness: Radiographic Film Speed

Objective

To demonstrate the speed ratio and different photographic properties between nonscreen and screen-type film.

Procedure

All exposures are performed as tabletop procedures using 8" by 10" (20 cm × 25 cm) films, 40" (102 cm) SID, at 50 kVp with a small size focus.

Film 1: Nonscreen Film

If available, use nonscreen film in a disposable holder, or place a nonscreen film in a direct exposure (cardboard) film holder. Using a perpendicular angle, center your central ray to the film, and collimate your beam to a 4" (10 cm) square. Select the mAs in order to produce a radiographic density of 2.0; expose and process your film.

Film 2: Screen-type Film, Same Exposure

Place a screen-type film in a direct exposure (cardboard) film holder. Set up as for Film 1. Using the same technical factors, expose and process your film.

Film 3: Screen-type Film, Adjusted Exposure

Place a screen-type film in a direct exposure (cardboard) film holder. Set up as for Film 2; however, adjust your mAs in order to produce a radiographic density of 2.0 (same as Film l). Expose and process your film. **Note:** Several exposures may have to be taken to produce the desired radiographic density of 2.0 for Films 1 and 3.

Analysis

Place the radiographs side by side on film illuminators in order to compare their appearance. You can now compare the film speed differences, if any, between nonscreen and screen-type radiographic film.

Conclusions

1. Compare the radiographic density of Film 1 and Film 2. Measure with a densitometer, and record this value in the spaces provided below:

   Film 1: _____ radiographic density
   Film 2: _____ radiographic density

2. Are the radiographic densities of Film 1 and Film 2 similar? Describe the difference, if any, and discuss the reasons for this difference.

3. The (*circle one*) nonscreen/screen-type film is faster.

4. Film 1 and Film 3 should have similar radiographic densities. Indicate the mAs required to produce the radiographic density in Films 1 and 3. Record these values in the spaces provided below:

   Film 1: ______mAs
   Film 3: ______mAs

5. Determine the difference in film speed between the nonscreen and screen-type film emulsions used for this experiment.

6. Of what importance is the difference in film speed between nonscreen and screen-type film to the radiologic technologist performing direct exposure radiography versus intensifying screen radiography?

Name ______________________________ Date ______________

EXPERIMENT 5
RECORDED DETAIL—Material Unsharpness: Radiographic Film Resolution

Objective

To investigate the loss of recorded detail between two types of radiographic films: (1) direct exposure (nonscreen) film and (2) screen-type film.

Procedure

Using a dry bone and a resolution grid capable of resolving up to 10 LP/mm, two radiographs will be produced. Set up your procedure to perform a tabletop exposure. All exposures should be made with the same size focus, a 40" (102 cm) SID, and the same 50 kVp.

Film 1: Direct Exposure Screen-type Film

1. Place a sheet of screen-type film in an 8" by 10" (20 cm × 25 cm) direct exposure (cardboard) film holder, and place it on the tabletop.
2. Place a small bone (e.g., a talus, os calcis, or metacarpal) and the resolution grid right next to one another on the film surface, centering the grid to the midpoint of the film holder.
3. Direct your central ray to the midpoint of the grid using a perpendicular angle, and collimate the beam to include the grid and the bone.
4. Select appropriate exposure factors (mAs) to produce a 1.0 to 1.2 bone density.
5. Expose and process your film.

Film 2: Direct Exposure Nonscreen Film

Except for the following, repeat the procedure outlined for Film 1.

1. Change to nonscreen film placed in a direct exposure (cardboard) film holder or used in its own disposable film holder.
2. Adjust your mAs as needed because of film speed differences in order to produce a radiographic density similar to that of Film 1. (See Experiment 4 for speed differences between nonscreen and screen-type film.)

Analysis

You can now compare the recorded details of the image produced with nonscreen film to those produced with screen-type film. Review the recorded details of the bone and the resolution grid in each radiograph. Determine the greatest number of line pairs per millimeter recorded in each radiographic image. Record this resolution information and the exposure factors employed for each radiograph in the appropriate spaces provided below.

Conclusions

1. Using a densitometer, record the radiographic density of each film, selecting an area adjacent to both the bone and the resolution grid for this purpose.

   Film 1: _____ radiographic density
   Film 2: _____ radiographic density

2. Compare the recorded details of the resolution grid, and indicate the image resolution of each in the space provided below:

   Film 1: _____ LP/mm screen-type film
   Film 2: _____ LP/mm nonscreen film

3. Compare the recorded details of the bone's image in Film 1, taken with screen-type film, to the recorded details of the bone's image taken with the nonscreen film (Film 2). Is there an obvious difference in the appearance of the recorded details between the two radiographic images? Explain.

4. Does there appear to be any difference in the resolution of the image details as indicated in Question 2 between the two different types of radiographic film?

5. What would you conclude about different types of radiographic film and their influence on radiographic recorded detail? On their contribution to image unsharpness?

Name ______________________________ Date ______________

EXPERIMENT 6
RECORDED DETAIL—Material Unsharpness: Intensifying Screens

Objective

To investigate the loss of recorded detail attributed to the use of intensifying screens.

Procedure

Using a dry bone and a resolution grid capable of resolving up to 10 LP/mm, three radiographs will be produced. Set up your equipment to perform a tabletop procedure. All exposures should be made using the small focus, the same 40" (102 cm) SID, and the same 50 kVp.

Film 1: Direct Exposure

1. Place a sheet of screen-type film in an 8" by 10" (20 cm × 25 cm) direct exposure (cardboard) film holder.
2. Place the resolution grid on the film holder, being sure it is in absolute contact with the surface. Place a small dry bone (e.g., an os calcis, talus, or metacarpal) right next to the grid on the film holder.
3. Direct your central ray to the midpoint of the grid using a perpendicular angle, and collimate the beam to include the grid and the bone.
4. Select the appropriate mAs to produce a 1.0 to 1.2 bone density.
5. Expose and process your film.

Film 2: Slow (Detail) Film-Screen Recording System

Choose a 10" by 12" (25 cm × 30 cm) cassette with the slowest (detail) available recording system, and arrange your materials exactly as above in Film 1. Using the screen intensification factor for the kilovoltage selected (Chapter 4), adjust your exposure factors (mAs) in order to maintain the same radiographic density as in Film 1. Expose and process the film.

Film 3: Rapid Film-Screen Recording System

Choose a 10" by 12" (25 cm × 30 cm) cassette with the fastest available recording system, and arrange your materials exactly as in Film 1. Using the intensifying screen speed factor in Table 4-3 (Chapter 4), adjust your exposure factors (mAs) in order to maintain the same radiographic density as in Film 1. Expose and process the film.

Analysis

You can now compare the recorded details of three different imaging systems. Carefully examine the recorded details of the dry bone and resolution grid images in Film 1. Determine your highest level of visual acuity (the greatest number of line pairs per millimeter that you can see clearly recorded). Determine the highest level of resolution for Films 2 and 3. Circle the same area on the radiographic image of the bone on each radiograph, and determine the radiographic density with a densitometer. Record the image resolution, the exposure factors (mAs), and the bone density achieved in each radiograph in the appropriate space provided below:

Levels of Image Resolution, Exposure, and Density

Film 1:	_____ LP/mm	_____ mAs	_____ bone density
Film 2:	_____ LP/mm	_____ mAs	_____ bone density
Film 3:	_____ LP/mm	_____ mAs	_____ bone density

Conclusions

1. Compare the recorded details of the bone's images in the nonscreen (Film 1) and the slow speed recording system (Film 2) procedures. Is there a loss of recorded detail? Is it an obvious loss? Explain your answer.

2. Compare the recorded details of the bone's images in the nonscreen (Film 1) and the fast speed recording system (Film 3) procedures. Is there a loss of recorded detail? Is it an obvious loss? Explain your answer.

3. Compare the recorded image detail losses between Film 1 and Film 2 with the image losses produced between Film 1 and Film 3. Describe the differences, if any.

4. Compare the recorded details of the bone's image in the slow recording system (Film 2) with the recorded details of the bone's image in the fast recording system (Film 3). Is there a loss of recorded detail? Is it as obvious a loss as that which occurred between the nonscreen and slow speed system procedure? Between the nonscreen and fast speed system procedure? Explain your answer.

5. Determine the actual material unsharpness attributed to the slow speed recording system and to the fast speed recording system.

   Example: If the nonscreen procedure detail level was 4.3 LP/mm, and the slow speed recording system procedure detail level was 2.8 LP/mm, you can determine the material unsharpness of the screens as follows:

   4.3 LP/mm nonscreen

   –2.8 LP/mm slow speed recording system

   1.5 LP/mm loss attributed to the intensifying screen

   1.5/4.3 = 0.348-mm material unsharpness loss

   Material Unsharpness: Slow speed recording system _______ mm

   Fast speed recording system _______ mm

6. The least amount of unsharpness would be recorded with a (*choose one*) fast screen/slow screen/nonscreen procedure.

7. Discuss the influence of screen speed as it relates to image detail. Of what practical value is this knowledge to the technologist?

8. Considering the exposure factors (mAs) employed for each radiograph, describe the advantages of screen radiography as they relate to the patient and to the overall technical procedure.

9. How does the application of different film-screen recording system speeds relate to the principles of ALARA?

Name ______________________________ Date ______________

EXPERIMENT 7
RECORDED DETAIL—Material Unsharpness: Film-Screen Contact Test

Objective

To demonstrate a test for film-screen contact.

Procedure

Three radiographs will be produced tabletop using an 8" by 10" (20 cm × 25 cm), a 10" by 12" (25 cm × 30 cm), and a 14" by 17" (36 cm × 43 cm) slow speed (detail) recording system cassette. All exposures should be made using the small focus, 40" (102 cm) SID, 50 kVp, and a perpendicular angle of the central ray centered to the middle of the cassette. You will need a film contact test screen (a wire mesh embedded in a firm plastic frame) for this procedure. **Note:** If a contact test screen is not available, you can use large, unused paper clips spread out evenly over the surface of the cassette front. For the larger film size, you may require more than one box of paper clips.

Film 1: 8" by 10" (20 cm × 25 cm)

1. Place the cassette on the tabletop, center the central ray over the cassette, and collimate to the size of the film.
2. Place the contact test screen upon the cassette front so that the center of the screen is aligned with the center of the cassette, being sure that it is level, even, and touching the cassette at all four corners. If using paper clips, spread them out evenly all over the surface of the cassette front, being sure they are flat and in contact with the cassette.
3. Select the appropriate mAs to produce an overall radiographic density of 1.5 to 1.8.
4. Expose and process your film.

Film 2: 10" by 12" (25 cm × 30 cm)

Set up your procedure exactly as for Film 1; open the collimator to cover the film size. Using the same exposure (mAs), expose and process your film.

Film 3: 14" by 17" (36 cm × 43 cm)

Set up your procedure exactly as for Film 1; open the collimator to cover the film size. Using the same exposure (mAs), expose and process your film.

Analysis

Examine the radiographic image of the contact test screen on each radiograph separately. Look for any differences in the recorded details of the image or for minor differences in the radiographic density from one area or section of the radiograph to another. Observe the radiograph directly, then move to the side of the illuminator and look back at the radiographic image from a slightly different angle. Are there any areas that appear blurred within the recorded details or within the radiographic image of the film contact test screen in Film 1? If so, encircle these areas. Are there any differences in the radiographic density of the radiographic image of the film contact test screen in Film 1? If so, encircle these areas.

Perform the same visual analysis for Film 2 and Film 3, and encircle any areas that appear blurred or that have changes in the pattern of radiographic density. Remove any cassette demonstrating a possible film-screen contact problem from service, and report your findings to your instructor, supervisor, or the quality control technologist assigned to the department. Check the outside of the cassette of any radiographic image that demonstrates a possible film-screen contact problem for frame damage, warping, or extraneous materials on the front or back of the cassette that may have caused the problem. In the darkroom, remove the unexposed film, and replace it into the film bin. Now check the inside of the cassette for any internal damage, intensifying screen damage, or the presence of extraneous matter. If you discover the possible problem, write your findings on a sheet of paper, attach it to the cassette, and give it to the appropriate personnel for possible repair.

Conclusions

1. Since motion unsharpness also produces a blurred radiographic image, how does the technologist differentiate between motion and poor film-screen contact?

2. Discuss the possible significance of a small area of poor film-screen contact within the radiographic appearance of the chest located in the lower lobe of the right lung to the diagnostic interpretation of the recorded image.

3. Would you normally expect greater problems associated with poor film-screen contact with a small or large size film? Why?

4. Discuss the importance of a quality assurance program that includes routine film contact testing to the performance of high-quality radiographic procedures.

5. In what manner is the application of a quality assurance program that includes the testing of cassettes for film contact an example of the principles of ALARA?

Name ______________________________ Date ______________

EXPERIMENT 8
RECORDED DETAIL—Material Unsharpness: Film-Screen Contact

Objective

To demonstrate the loss of recorded detail attributed to improper film-screen contact.

Procedure

Using a dry bone and a resolution grid capable of resolving up to 10 LP/mm, two radiographs will be produced. Set up your equipment to perform a tabletop procedure using a 10" by 12" (25 cm × 30 cm) cassette with a rapid film-screen recording system (same as used in Experiment 6, Film 3). All exposures should be made using the small focus, 40" (102 cm) SID, and 50 kVp.

Film 1: Good Film-Screen Contact

1. Center the resolution grid to the cassette, being sure it is in absolute contact with the surface. Place the proximal end of a dry bone (e.g., a humeral head, a femoral head, neck and trochanter) right next to the resolution grid on the film holder.
2. Direct your central ray to the midpoint of the resolution grid using a perpendicular angle, and collimate the beam to include the grid and the part of the bone adjacent to the grid.
3. Select the appropriate mAs to produce a 1.0 to 1.2 bone density. (Film 3, Experiment 6.)
4. Expose and process your film.

Film 2: Poor Film-Screen Contact

Prepare your second film holder in the darkroom to create a poor screen contact of 2 mm in the area of the film on which the radiographic image will be recorded:

1. Open the cassette, remove the film, and place four small objects of 2 mm thickness (e.g., a nickel) on the front screen of the cassette 3" in from each of the four corners.
2. Replace the film in the cassette on top of the objects, and close the cassette securely.
3. Following the same steps outlined in Film 1, expose and process your film.

Analysis

Review the radiographic images of both films on an illuminator. You can now compare the recorded details of the image produced with good film-screen contact with those produced with poor film-screen contact of 2 mm. Carefully examine the recorded details of the dry bone's images. Now review the recorded details of the resolution grid's images. Determine the resolution pattern that represents the greatest number of line pairs per millimeter that you can see clearly defined in each radiographic image.

Conclusions

1. Compare the overall appearance of the radiographic image of the dry bones produced with a poor contact level of 2 mm with the image produced with good film-screen contact. Describe any differences noted related to image detail and image visibility.

2. Using the contact unsharpness formula, determine the loss of recorded detail that would result from a poor contact level of 2 mm when a fast film-screen recording system (as in Experiment 6, Film 3)is used. The factor of geometric unsharpness is not included in this problem.

   New Unsharpness = Screen Unsharpness × Poor Contact²

   New Unsharpness: _______ mm

3. Examine the resolution grid image details and record the greatest number of line pairs per millimeter demonstrated in the radiograph taken with good film-screen contact, Film 1. Compare this with the resolution grid image with poor film-screen contact, Film 2. Record the resolution pattern of the poor film-screen contact image. Would the greatest loss of recorded detail be attributed to loss of resolution (number of line pairs per millimeter) or to loss of the sharpness of the details resolved? Explain your answer.

   Film 1: _______ LP/mm

   Film 2: _______ LP/mm

4. Consider the influence of poor film-screen contact, and compare its effect on image quality to the influence of screen unsharpness described in Experiment 6. Which influence appears more detrimental to the production of a film of radiographic quality? In what manner are these two factors interrelated? Explain your answers.

Name ____ Date ________________

EXPERIMENT 9
RECORDED DETAIL—Geometric Unsharpness: Size of Focus

Objective

To demonstrate the effects of different sizes of focus on the sharpness of the recorded details.

Procedure

Using a dry bone and a resolution grid capable of resolving up to 10 LP/mm, two radiographs will be produced. Set up your equipment to perform a tabletop procedure using a slow (detail) film-screen recording system cassette, 8" by 10" (20 cm × 25 cm), 40" (102 cm) SID, and 50 kVp. All technical factors for both radiographs should be the same except for the size of focus selected.

Film 1: Small Focus

1. Place the resolution grid on the cassette, centered to its midpoint. Place a small bone (e.g., a talus, os calcis, or metacarpal) on a 2" thick radiolucent sponge, and arrange the sponge on the cassette adjacent to the resolution grid.
2. Using a perpendicular angle, center your central ray to the midpoint of the grid, and collimate your beam to include the bone and the grid.
3. Select the small focus of the tube, and using the 100 mA station on the control panel, determine the appropriate time of exposure to produce a 1.0 to 1.2 bone density.
4. Expose and process your film.

Film 2: Large Focus

1. Repeat Steps 1 and 2 as for Film 1.
2. Select the large focus of the tube while maintaining the 100 mA station on your control panel.
3. Using the same exposure factors, expose and process your film.

Analysis

This experiment is not intended to measure accurately the size of either focus. However, it will enable you to observe and determine the geometric unsharpness and loss of recorded detail attributed to a small and large focus of a given x-ray tube in a practical manner.

Compare the recorded details of the resolution grid and bone in each radiograph. Determine the greatest number of lines pairs per millimeter recorded in each radiographic image. Using the nominal focus-size specifications provided by the x-ray tube manufacturer located on your tube rating chart, determine the geometric unsharpness attributed to each recorded image by using the geometric unsharpness formula.

Conclusions

1. Review the radiographic images of the bone on Film 1 and Film 2. Is there an obvious loss of recorded details from one image to the next? Which image demonstrates the better radiographic quality? Why?

2. Review the radiographic images of the resolution grid on Film 1 and Film 2. Is there an obvious loss of recorded details from the image of one resolution grid to the next? Is this loss as appreciable as that demonstrated in the images of the bone? Explain the reasons for your conclusion.

3. Identify the nominal focus-size specifications provided by the manufacturer and the other geometric unsharpness factors employed in the production of the two radiographs. Record this information in the appropriate spaces provided.

   _______ mm, small focus _______ OID
   _______ mm, large focus _______ SOD
   _______ SID

   Using the geometric unsharpness formula, determine the unsharpness associated with the use of the small and large focus using the factors identified above: Focus Size × OID/SOD

   Film 1: _______ geometric unsharpness, small focus
   Film 2: _______ geometric unsharpness, large focus

4. Record the greatest number of line pairs per millimeter that are resolved in each of the recorded images.

   Film 1: _______ LP/mm, small focus

   Film 2: _______ LP/mm, large focus

5. Considering the results of this experiment, describe the role of the size of focus in producing geometric unsharpness, and indicate how you would use this information to advantage to produce an image with minimal geometric unsharpness.

Name ______________________________ Date ____________

EXPERIMENT 10

RECORDED DETAIL—Geometric Unsharpness: Object-Image Distance

Objective

To demonstrate the effects of different OIDs on the sharpness of the recorded details.

Procedure

Using a dry bone and a resolution grid capable of resolving up to 10 LP/mm, four exposures will be taken. Set up your equipment to perform a tabletop procedure using a slow (detail) film-screen recording system cassette, 14" by 17" (36 cm × 43 cm), 40" (102 cm) SID, 50 kVp, and the large focus. All technical factors should be the same except for the OID employed.

Exposure 1: Minimal OID

1. Divide your 14" by 17" (36 cm × 43 cm) cassette into four equal sections using lead blocker dividers. Cover three sections, leaving one section uncovered. Mark the film section with a lead number 1.
2. Place a small bone (e.g., a talus, os calcis, or metacarpal) and the resolution grid adjacent to one another directly upon your cassette, centering the resolution grid to the midpoint of your cassette.
3. Using a perpendicular angle, center your central ray to the midpoint of the resolution grid, and collimate your beam to include the bone and the grid.
4. Select the appropriate exposure factors to produce a 1.0 to 1.2 bone density. (Use the same mAs identified in Experiment 9.)
5. Expose the film section.

Exposure 2: 2" (5 cm) OID

1. Adjust the lead dividers to cover the exposed film section and to uncover another section. Mark this section with a lead number 2.
2. Repeat Steps 2, 3, and 4 for Film 1, but place the bone and the resolution grid on a 2" thick radiolucent sponge or other suitable material.
3. Expose the film section.

Exposure 3: 4" (10 cm) OID

1. Adjust the lead dividers to cover the second exposed film section and to uncover a new section of the film. Mark this section with a lead number 3.
2. Repeat Steps 2, 3, and 4 for Film 1, but place the bone and the resolution grid on a 4" thick radiolucent sponge or other suitable material.
3. Expose the film section.

Exposure 4: 6" (15 cm) OID

1. Adjust the lead dividers to cover the third exposed film section and to uncover the final unexposed section of the film. Mark this section with a lead number 4.
2. Repeat Steps 2, 3, and 4 for Film 1, but place the bone and the resolution grid on a 6" thick radiolucent sponge or other suitable material.
3. Expose the film section and process the film.

Analysis

This experiment will enable you to observe and determine the geometric unsharpness and loss of recorded detail attributed to a minimal OID compared with an OID of 2", 4", and 6" (5 cm, 10 cm, and 15 cm). Review the recorded details of the bone and the resolution grid in each section of the radiograph. Determine the greatest number of line pairs per millimeter recorded in each radiographic image. Using the geometric unsharpness formula, solve for the geometric unsharpness of each image.

Conclusions

1. Review the radiographic images of the bone. Is there an obvious loss of recorded details from one image to the next? Which image demonstrates the better recorded detail? Why?

2. Indicate the greatest number of line pairs per millimeter that you can see clearly defined in each resolution grid image.

   Exposure 1: _______ minimal OID
   Exposure 2: _______ 2" (5 cm) OID
   Exposure 3: _______ 4" (10 cm) OID
   Exposure 4: _______ 6" (15 cm) OID

3. Identify the nominal focus size specification provided by the manufacturer for the large focus and the other geometric unsharpness factors employed in the production of the four exposures. Record this information in the appropriate spaces provided.

_______	mm, large focus	_______	SID all exposures
0.25"*	OID Exposure 1	_______	SOD Exposure 1
_______	OID Exposure 2	_______	SOD Exposure 2
_______	OID Exposure 3	_______	SOD Exposure 3
_______	OID Exposure 4	_______	SOD Exposure 4

*0.25" represents a nominal OID value for the purpose of the experiment (thickness of cassette front and intensifying screen). Using the geometric unsharpness formula, determine the unsharpness associated with the different OIDs using the factors identified above.

Focus Size × OID/SOD

Exposure 1: _______ geometric unsharpness
Exposure 2: _______ geometric unsharpness
Exposure 3: _______ geometric unsharpness
Exposure 4: _______ geometric unsharpness

4. Is the loss of sharpness as great in the images of the resolution grid as in the images of the bone? Explain the reasons for the difference, if any.

5. Compare the results of Experiment 9 examining the size of focus with the results of this experiment examining the OID. Which factor appears to have the greater influence on recorded detail? Provide statements to support your conclusion.

6. Considering the results of this experiment, describe the role of the OID in producing geometric unsharpness, and indicate how you would use this information to advantage to produce an image with minimal geometric unsharpness.

Name ______________________________ Date ______________

EXPERIMENT 11

RECORDED DETAIL—Geometric Unsharpness: Source-Image Distance

Objective

To demonstrate the effects of different SIDs on the sharpness of the recorded details.

Procedure

Using a dry bone and a resolution grid capable of resolving up to 10 LP/mm, a total of four exposures will be taken. Set up your equipment to perform a tabletop procedure using a slow (detail) film-screen recording system cassette, 14" by 17" (36 cm × 43 cm), 40" (102 cm) SID, 50 kVp, and the large focus. All technical factors should be the same except for the mAs adjustments necessary with the SID changes. It may be necessary to place your cassette on the floor in order to achieve the SID required for this experiment.

Exposure 1: 40" (102 cm) SID

1. Divide your 14" by 17" (36 cm × 43 cm) cassette into four equal sections using lead blocker dividers. Cover three sections, leaving one section uncovered. Mark the film section with a lead number 1.
2. Place a small bone (e.g., a talus, os calcis, or metacarpal) and the resolution grid adjacent to one another on a 2" thick radiolucent sponge. Place the sponge on your cassette, centering the resolution grid to the midpoint of your film section.
3. Using an SID of 40" (102 cm) and a perpendicular angle, center your central ray to the midpoint of the resolution grid, and collimate your beam to include the bone and the grid.
4. Select the appropriate exposure factors to produce a 1.0 to 1.2 bone density. (Check the exposure factors used for Experiment 10.)
5. Expose your divided film section.

Exposure 2: 20" (51 cm) SID

1. Set up your procedures similar to Film 1, but cover up the exposed film section, and uncover a new section, marking it with a lead number 2.

2. Decrease your SID to 20" (51 cm), and adjust your mAs according to the inverse square law in order to maintain the radiographic density.
3. Expose your divided film section.

Exposure 3: 60" (153 cm) SID

1. Set up your procedures similar to Film 1, but cover up the exposed film section, and uncover a new section, marking it with a lead number 3.
2. Increase your SID to 60" (153 cm), and adjust your mAs according to the inverse square law in order to maintain the radiographic density. You may have to place your cassette on the floor and use a measuring tape to adjust your SID accurately to the required distance.
3. Expose your divided film section.

Exposure 4: 80" (204 cm) SID

1. Set up your procedures similar to Film 1, but cover up the exposed film section, and uncover the final unexposed film section, marking it with a lead number 4.
2. Increase your SID to 80" (204 cm), and adjust your mAs according to the inverse square law in order to maintain the radiographic density. You may have to place your cassette on the floor and use a measuring tape to adjust your SID accurately to the required distance.
3. Expose your divided film section, and process your film.

Analysis

This experiment will enable you to observe and determine the geometric unsharpness and recorded detail loss attributed to the use of four different SIDs. Review the recorded details of the bone and the resolution grid in each radiographic image. Determine the greatest number of line pairs per millimeter recorded in each. Using the geometric unsharpness formula, solve for the geometric unsharpness of each recorded image.

Conclusions

1. Review the overall radiographic images of the bone. Is there an obvious loss of recorded details from one image to the next? Which image demonstrates the better recorded detail?

2. Indicate the greatest number of line pairs per millimeter that you can see clearly defined in each resolution grid's image.

   Exposure 1: _______ LP/mm, 40" (102 cm) SID
   Exposure 2: _______ LP/mm, 20" (51 cm) SID
   Exposure 3: _______ LP/mm, 60" (153 cm) SID
   Exposure 4: _______ LP/mm, 80" (204 cm) SID

3. Identify the nominal focus size specification provided by the manufacturer and the other geometric unsharpness factors employed in the production of the four exposures. Record this information in the appropriate spaces below:

   _______ mm, large focus   _______ OID all exposures
   _______ SID Exposure 1   _______ SOD Exposure 1
   _______ SID Exposure 2   _______ SOD Exposure 2
   _______ SID Exposure 3   _______ SOD Exposure 3
   _______ SID Exposure 4   _______ SOD Exposure 4

   Using the geometric unsharpness formula, determine the unsharpness associated with the different SIDs using the factors identified above.

   _______ Geometric unsharpness, Exposure 1
   _______ Geometric unsharpness, Exposure 2
   _______ Geometric unsharpness, Exposure 3
   _______ Geometric unsharpness, Exposure 4

4. Is the loss as great in the resolution grid's image as in the bone's image? Explain the reasons for the difference, if any.

5. Compare the results of Experiments 9 and 10 examining the size of focus and the OID and the results of this experiment examining the SID. Which factor appears to have the greatest influence on sharpness? Provide statements to support your conclusions. List the three factors affecting geometric unsharpness in order of increasing adverse effect on image sharpness.

6. Considering the results of this experiment, describe the role of the SID in producing geometric unsharpness, and indicate how you would use this information to advantage to produce an image with minimal geometric unsharpness.

Name ______________________________ Date ______________

EXPERIMENT 12
RECORDED DETAIL—Summary of Image Unsharpness

Objective

To demonstrate the cumulative influences of the factors contributing to geometric unsharpness and material unsharpness on the recorded details of the radiographic image and to observe the improved image details as these factors are adjusted one by one.

Procedure

Using a small, dry bone (e.g., a talus, os calcis, or metacarpal) and a resolution grid capable of resolving up to 10 LP/mm, a series of six radiographs will be taken using 10" by 12" (25 cm × 30 cm) cassettes as a tabletop procedure. Adjustments in materials, technical procedure, and exposure factors will be suggested between each exposure. It is important to maintain the recorded image's radiographic density on all radiographs produced.

Film 1: All Factors Abused

Use the large focus, a 30" (76 cm) SID, an OID of 8" (20 cm), and the most rapid speed film-screen recording system available.

1. Place the bone and the resolution grid adjacent to one another on radiolucent sponges so that they are located 8" (20 cm) above the film. Center the resolution grid to the middle of the cassette.
2. Using a 30" (76 cm) SID perpendicular angle, center your central ray to the midpoint of the resolution grid, and collimate your beam to include the bone and the grid.
3. Using the large focus and 50 kVp, select the appropriate exposure factors to produce a 1.0 to 1.2 bone density.
4. Expose and process your film.

Films 2 through 6

For the remaining films, repeat Steps 1 through 4 in Film 1, with the following changes.

Film 2: Change to a Small Focus

Change to the small focus. Use the same milliamperage station if at all possible as employed with the large focus. Depending on the calibration of your equipment, adjust for tube output differences, if any, to maintain radiographic density.

Film 3: Increase the SID

Using the small focus, also increase your SID to 60" (153 cm). Increase your mAs according to the inverse square law to compensate for the increased distance.

Film 4: Decrease the OID

Using the small focus and an SID of 60" (153 cm), decrease your OID to 1" (2.54 cm) by placing the bone and resolution grid on a radiolucent sponge so that they now lie 1" (2.54 cm) above the film surface.

Film 5: Change to a Slow Speed (Detail) Recording System

Using the small focus, an SID of 60" (153 cm), and a minimal OID of 1" (2.54 cm), change to a slow speed (detail) film-screen recording system. Adjust your technical factors to compensate for the difference in the speed of the imaging system in order to maintain the radiographic density of the recorded image.

Film 6: Change to Direct Exposure

Using the small focus, an SID of 60" (153 cm), and a minimal OID of 1" (2.54 cm), change to a direct exposure (cardboard) film holder. Adjust your technical factors to compensate for the loss of the image intensification factor of the intensifying screens in order to maintain the radiographic density of the recorded image.

Analysis

This experiment will enable you to review the recorded details when all of the factors affecting recorded detail except for film-screen contact and motion have been abused. It will also enable you to see the specific improvement in recorded detail associated with each factor as it is adjusted to improve the radiographic quality of the image.

Review the radiographic images of all of the films side by side on an illuminator. Compare the recorded image of Film 1 to Film 2, Film 2 to Film 3, and so forth. Finally, compare Film 1 with Film 6. You will be able to determine the influence of geometric and material unsharpness of each film's recorded image. You will also be able to determine which factors exhibit the greatest influence on recorded detail.

Conclusions

1. Identify the nominal focus size specification provided by the manufacturer and the other (geometric and material) unsharpness factors employed in the production of the six radiographs. Record this information in the appropriate spaces provided.

Film 1
_____ mm, Large focus
_____ SID
_____ OID
_____ SOD
_____ Rapid speed imaging system

Film 2
_____ mm, Small focus
_____ SID
_____ OID
_____ SOD
_____ Rapid speed imaging system

Film 3
_____ mm, Small focus
_____ SID
_____ OID
_____ SOD
_____ Rapid speed imaging system

Film 4
_____ mm, Small focus
_____ SID
_____ OID
_____ SOD
_____ Rapid speed imaging system

Film 5
_____ mm, Small focus
_____ SID
_____ OID
_____ SOD
_____ Slow speed imaging system

Film 6
_____ mm, Small focus
_____ SID
_____ OID
_____ SOD
_____ Direct exposure

2. Determine the **total unsharpness** (geometric unsharpness + screen unsharpness) using the geometric unsharpness formula and the known inherent unsharpness factor related to the screen speed employed for each of the six radiographs produced using the factors identified above.
   Film 1: _______ mm, total unsharpness
   Film 2: _______ mm, total unsharpness
   Film 3: _______ mm, total unsharpness
   Film 4: _______ mm, total unsharpness
   Film 5: _______ mm, total unsharpness
   Film 6: _______ mm, total unsharpness

3. Indicate the total unsharpness that would result if a poor film-screen contact of 1.5 mm were added to the unsharpness recorded in Film 1 above. Use the poor film-screen contact formula to determine your answer.
   _______ mm, total unsharpness

4. Examine each radiographic image, and indicate the greatest number of line pairs per millimeter that you can see resolved in each of the six radiographs.

   Film 1: _____ LP/mm Film 2: _____ LP/mm Film 3: _____ LP/mm

   Film 4: _____ LP/mm Film 5: _____ LP/mm Film 6: _____ LP/mm

5. Compare the radiographic image of the bone in Film 1 with Film 2. Does the change to a small focus produce a significant improvement in the recorded details of the image? Without comparing these films with any others, would you consider the recorded detail of Film 1 acceptable? Is the recorded detail of Film 2 acceptable?

6. Compare the radiographic image of the bone in Film 2 with Film 3. Does the increase in the SID produce a significant improvement in the recorded details of the image? Without comparing these films with any others, would you consider the recorded detail of Film 3 acceptable?

7. Compare the radiographic image of the bone in Film 3 with Film 4. Does the decrease in the OID produce a significant improvement in the recorded details of the image? Without comparing these films with any others, would you consider the recorded detail of Film 4 acceptable?

8. Compare the radiographic image of the bone in Film 4 with Film 5. Does the change to a slow speed imaging system produce a significant improvement in the recorded details of the image? Without comparing these films to any others, would you consider the recorded detail of Film 5 acceptable?

9. Compare the radiographic image of the bone in Film 5 with Film 6. Does the change to a direct exposure procedure produce a significant improvement in the recorded details of the image? Without comparing these films to any others, would you consider the recorded detail of Film 6 acceptable?

10. If all you had to evaluate was Film 3, would the unsharpness recorded in the image be as obvious as when you compared this image with the subsequent films taken in this series? Explain your answer.

11. Review each radiographic image separately. At what point (which film) do you believe the radiographic image becomes acceptable?

12. Arrange the following factors in decreasing order of influence related to the radiographic property of recorded detail. The factor that contributes the greatest image unsharpness should be numbered 1, and the factor that contributes the least image unsharpness should be numbered 8.

_____ SID
_____ Involuntary motion
_____ X-ray film type
_____ Intensifying screens
_____ Size of focus
_____ Voluntary motion
_____ Film-screen contact
_____ OID

Name ____________________ Date __________

EXPERIMENT 13
DISTORTION—Size Distortion: Object-Image Distance

Objective

To demonstrate the effect of different OIDs on the recording of the image size and at the same time demonstrate the interrelationship between distortion and recorded detail.

Procedure

A series of four exposures will be taken tabletop using a small, dry bone (e.g., talus, os calcis, or metacarpal). Use the small focus and 50 kVp. All technical factors should remain the same except for the changes in the OID.

Exposure 1: Minimal OID

1. Using a slow speed (detail) film-screen recording system with a 10" by 12" (25 cm × 30 cm) cassette, divide the film in four sections using lead blockers. Cover three sections of the film with lead blockers.
2. Place the dry bone directly on the surface of the cassette, centered to the uncovered section. Mark the section with a lead number 1.
3. Direct your central ray to the midpoint of the dry bone, using a perpendicular angle at a 40" (102 cm) SID.
4. Select the appropriate exposure factors to produce a 1.0 to 1.2 bone density.
5. Expose the film section.

Exposure 2: 2" (5 cm) OID

Set up the procedure the same as for Exposure 1; however, cover the exposed film section, adjust the lead blockers to a new section, mark it with a lead number 2, and raise your dry bone on a radiolucent sponge to create an OID of 2" (5 cm). Expose the film section.

Exposure 3: 4" (10 cm) OID

Set up the procedure the same as for Exposure 1; however, cover the exposed film section, adjust the lead blockers to a new section, mark it with a lead number 3, and raise your dry bone on a radiolucent sponge to create an OID of 4" (10 cm). Expose the film section.

Exposure 4: 8" (20 cm) OID

Set up the procedure the same as for Exposure 1; however, adjust the lead blockers to the final unexposed new section, mark it with a lead number 4, and raise your dry bone on a radiolucent sponge to create an OID of 8" (20 cm). Expose the final film section, and process your film.

Analysis

Your radiograph will have four recorded images of the dry bone, each demonstrating a different image size. This experiment will enable you to measure the magnification of the image attributed to OID and to determine the relationship of distortion to recorded detail. You can also determine the level at which the size distortion becomes unacceptable. You can also calculate the percentage of magnification related to the different OID of the bone.

Conclusions

1. Accurately measure the dry bone's length and width in millimeters. Measure the length and width of each of the recorded images on the radiograph in millimeters. Record this information in the spaces provided below:

   Dry bone _____ mm length _____ mm width _____ mm total area
   No OID _____ mm length _____ mm width _____ mm total area
   2" OID _____ mm length _____ mm width _____ mm total area
   4" OID _____ mm length _____ mm width _____ mm total area
   8" OID _____ mm length _____ mm width _____ mm total area

2. Using the information provided in Question 1, determine the percentage of magnification in the following problems using the formula provided.

   Image Width – Object Width/Object Width × 100
   2" (5 cm) OID _____ % magnification
   4" (10 cm) OID _____ % magnification
   8" (20 cm) OID _____ % magnification

3. As the OID is increased, the recorded size of the image (*circle one*) increases/ decreases.

4. What effect, if any, does size distortion (magnification) have on the recorded details of the image?

5. At what point does the loss of image details become unacceptable? _____ OID. Explain the reasons for your answer.

6. Considering the results of this experiment, describe the role of the OID in producing size distortion, and indicate how you would use this information to advantage to produce an image with minimal size distortion.

Name ______________________________ Date ______________

EXPERIMENT 14
DISTORTION—Size Distortion: Source-Image Distance

Objective

To demonstrate the effect of different SIDs on the recording of the image size and at the same time demonstrate the interrelationship between distortion and recorded detail.

Procedure

A series of four exposures will be taken tabletop using a small, dry bone (e.g., talus, os calcis, or metacarpal). Use the small focus and 50 kVp. The cassette should be placed on the floor in order to make all the SID adjustments. All technical factors should remain the same except for the changes in the SID.

Exposure 1: Minimal SID

1. Using a slow speed (detail) film-screen recording system with a 10" by 12" (25 cm × 30 cm) cassette, divide the film in four sections using lead blockers. Cover three sections of the film with lead blockers.
2. Place the dry bone on a 2" radiolucent sponge directly on the surface of the cassette, centered to the uncovered section. Mark it with a lead number 1.
3. Direct your central ray to the midpoint of the dry bone, using a perpendicular angle at a 20" (51 cm) SID. Use a tape measure to adjust the distance.
4. Select the appropriate exposure factors to produce a 1.0 to 1.2 bone density using the inverse square law as a guide.
5. Expose the film section.

Exposure 2: 40" (102 cm) SID

Set up the procedure the same as for Exposure 1; however, cover the exposed film section, adjust the lead blockers to a new section, mark it with a lead number 2, and raise your SID to 40" (102 cm). Expose the film section.

Exposure 3: 60" (153 cm) SID

Set up the procedure the same as for Exposure 1; however, cover the exposed film section, adjust the lead blockers to a new section, mark it with a lead number 3, and raise your SID to 60" (153 cm). Expose the film section.

Exposure 4: 80" (204 cm) SID

Set up the procedure the same as for Exposure 1; however, adjust the lead blockers to the final unexposed new section, mark it with a lead number 4, and raise your SID to 80" (204 cm). Expose the final film section, and process your film.

Analysis

Your radiograph will have four recorded images of the dry bone, each demonstrating a different image size. This experiment will enable you to measure the magnification of the image attributed to SID and to determine the relationship of distortion to recorded detail. You can determine the level at which the size distortion becomes unacceptable. In addition, you can calculate the percentage of magnification related to the different SIDs.

Conclusions

1. Accurately measure the dry bone's length in millimeters. Measure the length of each of the recorded images on the radiograph in millimeters. Record these measurements as well as the mAs employed for each radiographic image in the spaces provided below:

   Dry bone _____ mm length
   20" (51 cm) SID _____ mm length ________ mAs
   40" (102 cm) SID _____ mm length ________ mAs
   60" (153 cm) SID _____ mm length ________ mAs
   80" (204 cm) SID _____ mm length ________ mAs

2. Using the information provided in Question 1, determine the percentage of magnification in the following problems using the formula provided.

   Image Length – Object Length/Object Length × 100
   20" (51 cm) SID _____ % magnification
   40" (102 cm) SID _____ % magnification
   60" (153 cm) SID _____ % magnification
   80" (204 cm) SID _____ % magnification

3. As the SID is increased, the recorded size of the image (*circle one*) increases/ decreases.

4. At what point does the loss of image details become unacceptable? _____ SID. Explain the reasons for your answer.

5. If we continue to increase the SID, at what point will size distortion be totally eliminated? Explain your answer.

6. Compare the results of Experiment 13 examining the influence of OID and the results of this experiment examining the influence of SID on size distortion. Which factor produces the greatest influence on size distortion? Provide statements to support your conclusion.

7. Increases in the SID can potentially lead to motion unsharpness. Explain the possible reasons for this statement.

8. Considering the results of this experiment, describe the role of the SID in producing size distortion, and indicate how you would use this information to advantage to produce an image with minimal size distortion.

Name ______________________________ Date ______________

EXPERIMENT 15
DISTORTION—Shape Distortion: Structure-Film Relationship

Objective

To demonstrate the distortion produced when the part to be examined is not aligned properly with the film surface. Ideally, the longitudinal plane of the structure of interest is placed parallel with the film surface.

Procedure

Tape a small, dry bone (e.g., os calcis, scapula, or sacrum) to a radiolucent sponge of 4" (10 cm) thickness. Be sure the longitudinal axis of the bone is flat against the sponge and securely taped in position. A series of three radiographs will be taken (tabletop) using a 14" by 17" (36 cm × 43 cm) slow speed (detail) film-screen recording system divided in thirds with lead blockers. Be sure to include the entire bone on each exposure. If the bone is too large, it may be necessary to expose separate films. All technical factors should remain the same between exposures except for the change in the position of the bone.

Exposure 1: Parallel Relationship

1. Divide your 14" by 17" (36 cm × 43 cm) cassette into three equal sections along the length of the cassette using lead blockers. Cover two sections with lead blockers, leaving one section uncovered. Mark the film section with a lead number 1.
2. Place the sponge with the bone taped on it flat on the surface of the film so that its longitudinal axis is parallel with the film.
3. Direct your central ray to the midpoint of the dry bone using a perpendicular angle and an SID of 40" (102 cm). Collimate to the size of the bone.
4. Using the small focus and 50 kVp, select the appropriate exposure factors to produce a 1.0 to 1.2 bone density.
5. Expose this film section.

Exposure 2: 30° Angle Relationship

Arrange your procedure as you did with Exposure 1; however, raise one end of the length of the bone in order to achieve a 30° angle from a parallel relationship. Use a new section of your film, blocking off the rest of the film; mark it with a lead number 2. Expose this film section.

Exposure 3: 30° Angle Relationship

Arrange your procedure as you did with Exposure 1; however return the bone to the parallel relationship, and then raise the *opposite* end of the bone in order to achieve a 30° angle from a parallel relationship. Use the last unexposed film section, blocking off the rest; mark it with a lead number 3. Expose and process your film.

Analysis

Your radiograph will have three recorded images of the dry bone, each demonstrating a different image shape. Examine each radiographic image throughout its entire recorded surface. The correct parallel relationship will record the image with the least amount of shape distortion. Although magnified due to the 4" (10 cm) OID, it appears as an image with an overall shape similar to that of the dry bone.

Conclusions

1. Describe the differences in anatomical appearance between the recorded images in Exposure 1 and in Exposure 2. Which portions of the image appear elongated? Which portions appear foreshortened?

2. Describe the differences in anatomical appearance between the recorded images in Exposure 1 and in Exposure 3. Which portions of the image appear elongated? Which portions appear foreshortened?

3. Is the shape distortion created by Exposure 2 the same as that produced by Exposure 3? Describe their similarities and their differences.

4. In millimeters, measure the length of the bone and the radiographic image for each exposure, and record it in the space provided below:

   Bone: _____ mm length　　Exposure 1: _____ mm length
   Exposure 2: _____ mm length　　Exposure 3: _____ mm length

5. Can shape distortion caused by an improper part-film relationship also contribute to size distortion? Explain your answer.

6. Differentiate between the body part and the actual structures of interest related to the concerns for the production of shape distortion within the recorded image.

7. To minimize shape distortion, indicate the most ideal relationship between the structures of interest and the plane of the film.

Additional Analysis

This experiment demonstrates the shape distortion of a single structure. For a better understanding of shape distortion represented by the complexity of overlapping and superimposed body structures, *repeat this experiment* using a more representative phantom, such as the one described below.

Tape a scapula flat against a 2" (5 cm) radiolucent sponge. Cover this with another 2" (5 cm) sponge, and tape a clavicle in the middle of the second sponge. Cover this with a third 2" (5 cm) sponge, and tape a sacrum onto the third sponge centered over the bones below it. Tape the sponges together.

Other bones can be substituted, but they should all be of different shapes. This phantom will demonstrate the effects of multiple sizes, shapes, and OID in the production of shape distortion and the part-film relationship. Compare the results of this experiment with those obtained with the single bone phantom.

Name ______________________________ Date ______________

EXPERIMENT 16

DISTORTION—Shape Distortion: Central Ray–Structure–Film Relationship

Objective

To demonstrate the distortion produced when the central ray is improperly centered to the structures of interest within the body part being examined. Ideally, the structures of interest will be centered to the middle of the film and the central ray directed to this point.

Procedure

Tape a small, dry bone (e.g., lumbar vertebra, os calcis, scapula, or sacrum) to a radiolucent sponge of 4" (10 cm) thickness. Be sure the longitudinal axis of the bone is flat against the sponge and securely taped in position. A series of three radiographs will be taken (tabletop) using a 14" by 17" (36 cm × 43 cm) slow speed (detail) film-screen recording system divided in thirds with lead blockers. Be sure to include the entire bone on each exposure. If the bone is too large, it may be necessary to expose separate films. All technical factors should remain the same between exposures except for the change in the position of the bone.

Exposure 1: Central Ray Centered to Part

1. Divide your 14" by 17" (36 cm × 43 cm) cassette into three equal sections along the length of the cassette using lead blockers. Cover two sections with lead blockers, leaving one section uncovered. Mark the film section with a lead number 1.
2. Place the sponge with the bone taped on it flat on the surface of the film so that its longitudinal axis is parallel with the film, and the middle of the bone is centered to the midpoint of the divided film section.
3. Direct your central ray to the midpoint of the dry bone using a perpendicular angle and an SID of 40" (102 cm). Collimate to the size of the bone.
4. Using the small focus and 50 kVp, select the appropriate exposure factors to produce a 1.0 to 1.2 bone density.
5. Expose this film section.

Exposure 2: Central Ray Off-centered to Part

Arrange your procedure as you did with Exposure 1; however, move your central ray to the left of center along the longitudinal axis of the bone until it is 6" (15 cm) off-center from its original position. Open your collimator to include the entire bone. Use a new section of your film, blocking off the rest of the film; mark it with a lead number 2. Expose this film section.

Exposure 3: Central Ray Off-centered to the Part

Arrange your procedure as you did with Exposure 1; however, return the central ray to the center of the bone, and continue moving it to the right of center along the longitudinal axis of the bone until it is 6" (15 cm) off-center from the midpoint of the bone in the opposite direction. Open your collimator to include the entire bone. Use the last unexposed film section, blocking off the rest; mark it with a lead number 3. Expose and process your film.

Analysis

Your radiograph will have three recorded images of the dry bone, each demonstrating a different image shape. Examine each radiographic image throughout its entire recorded surface. The exposure taken with the central ray centered to the midpoint of the bone will record the least amount of shape distortion. Although magnified due to the 4" (10 cm) OID, it appears as an image with an overall shape similar to that of the dry bone.

Conclusions

1. Describe the differences in anatomical appearance between the recorded images in Exposure 1 and Exposure 2. Which portions of the image appear elongated? Which portions appear foreshortened?

2. Describe the differences in anatomical appearance between the recorded images in Exposure 1 and Exposure 3. Which portions of the image appear elongated? Which portions appear foreshortened?

3. Is the shape distortion created by Exposure 2 the same as that produced by Exposure 3? Describe their similarities and their differences.

4. In millimeters, measure the length of the bone and the radiographic image for each exposure, and record it in the space provided below:

   Bone: _____ mm length　　Exposure 1: _____ mm length

   Exposure 2: _____ mm length　　Exposure 3: _____ mm length

5. Can shape distortion caused by an improper central ray–part–film centering also contribute to size distortion? Explain your answer.

6. Compare the shape distortion produced in this experiment with the shape distortion produced in Experiment 15. Of the two causes of shape distortion identified in these experiments, which produces the more obvious misrepresentation of the structures of interest? Provide reasons for your answer.

7. The shape distortion caused by the off-centering of the central ray to the part is frequently more subtle in its appearance. Explain this statement related to the need to visualize joint spaces or the attempt to demonstrate nondisplaced fracture lines and bone fragments.

Additional Analysis

This experiment demonstrates the shape distortion of a single structure. For a better understanding of shape distortion represented by the complexity of overlapping and superimposed body structures, *repeat this experiment* using a more representative phantom, such as the one described in the additional analysis section of Experiment 15. Compare the results of this additional experiment with those obtained with a single bone phantom.

Name ______________________________ Date ______________

EXPERIMENT 17
DISTORTION—Shape Distortion: Central Ray Direction

Objective

To demonstrate the distortion produced when the central ray is angled improperly to the plane of the structures of interest within the body part being examined. Ideally, the central ray will be directed with a perpendicular angle to the plane of the structures of interest.

Procedure

Tape a small, dry bone (e.g., lumbar vertebra, os calcis, scapula, or sacrum) to a radiolucent sponge of 4" (10 cm) thickness. Be sure the longitudinal axis of the bone is flat against the sponge and securely taped in position. A series of three radiographs will be taken (tabletop) using a 14" by 17" (36 cm × 43 cm) slow speed (detail) film-screen recording system divided in thirds with lead blockers. Be sure to include the entire bone on each exposure. If the bone is too large, it may be necessary to expose separate films. All technical factors should remain the same between exposures except for the change in the position of the bone.

Exposure 1: Central Ray Perpendicular to the Structures of Interest

1. Divide your 14" by 17" (36 cm × 43 cm) cassette into three equal sections along the length of the cassette using lead blockers. Cover two sections with lead blockers, leaving one section uncovered. Mark the film section with a lead number 1.
2. Place the sponge with the bone taped on it flat on the surface of the film so that its longitudinal axis is parallel with the film, and the middle of the bone is centered to the midpoint of the divided film section.
3. Direct your central ray to the midpoint of the dry bone using a perpendicular angle and an SID of 40" (102 cm). Collimate to the size of the bone.
4. Using the small focus and 50 kVp, select the appropriate exposure factors to produce a 1.0 to 1.2 bone density.
5. Expose this film section.

Exposure 2: Central Ray Angled 25° to the Structures of Interest

Arrange your procedure as you did with Exposure 1; however, move your central ray to the left of center along the longitudinal axis of the bone, and angle the central ray back toward the center of the bone using a 25° angle. Reduce the vertical tube height 1" (2.54 cm) for each 5° of tube angle. Adjust your collimator to include the entire bone. Use a new section of your film, blocking off the rest of the film; mark it with a lead number 2. Expose this film section.

Exposure 3: Central Ray Angled 25° to the Structures of Interest

Arrange your procedure as you did with Exposure 1; however, move your central ray to the right of center along the longitudinal axis of the bone, and angle the central ray back in the opposite direction toward the center of the bone using a 25° angle. Reduce the vertical tube height 1" (2.54 cm) for each 5° of tube angle. Adjust your collimator to include the entire bone. Use the final unexposed section of your film, blocking off the rest of the film; mark it with a lead number 3. Expose this film section, and process your film.

Analysis

Your radiograph will have three recorded images of the dry bone, each demonstrating a different image shape. Examine each radiographic image throughout its entire recorded surface. The exposure taken with the central ray perpendicular to the midpoint of the bone will record the least amount of shape distortion. Although magnified due to the 4" (10 cm) OID, it appears as an image with an overall shape similar to that of the dry bone.

Conclusions

1. Describe the differences in anatomical appearance between the recorded images in Exposure 1 and Exposure 2. Which portions of the image appear elongated? Which portions appear foreshortened?

2. Describe the differences in anatomical appearance between the recorded images in Exposure 1 and Exposure 3. Which portions of the image appear elongated? Which portions appear foreshortened?

3. Is the shape distortion created by Exposure 2 the same as that produced by Exposure 3? Describe their similarities and their differences.

4. In millimeters, measure the length of the bone and the radiographic image for each exposure and record it in the space provided below:

   Bone: _____ mm length   Exposure 1: _____ mm length

   Exposure 2: _____ mm length   Exposure 3: _____ mm length

5. Can shape distortion caused by an improper central ray angle through the part also contribute to size distortion? Explain your answer.

6. Compare the shape distortion produced in this experiment with the shape distortion produced in Experiments 15 and 16. Which factor contributing to shape distortion appears to produce the more obvious misrepresentation of the structures of interest? Provide reasons for your answer.

7. Would it be more appropriate to direct the central ray perpendicular to the film or to the structures of interest within the body part in order to minimize shape distortion? Explain your answer, and provide practical examples that support your position.

Additional Analysis

This experiment demonstrates the shape distortion of a single structure. For a better understanding of shape distortion represented by the complexity of overlapping and superimposed body structures, *repeat this experiment* using a more representative phantom, such as the one described in the additional analysis section of Experiment 15. Compare the results of this additional experiment with those obtained with a single bone phantom.

Name ______________________________ Date ______________

EXPERIMENT 18
DISTORTION—Shape Distortion: Used to Advantage To Avoid Superimposition

Objective

To demonstrate that distortion can be used to advantage in a controlled application in order to clear the structures of interest from superimposed or overlapping structures.

Procedure

Select samples of the following examinations from the radiographic film file. If unavailable, produce these projections using a suitable phantom of the thorax or a skeleton.

1. **Chest examination for lungs.** Be sure to include a posteroanterior projection and an apical lordotic projection.
2. **Clavicle.** Be sure to include a posteroanterior projection and an angled (30°–45°) posteroanterior projection.

Analysis

Review the radiographs of both examinations carefully with specific attention to the anatomical relationships recorded on each film. Describe the appearance of the structures of interest in each projection.

Conclusions

1. Describe the position of the clavicles in the posteroanterior chest projection.

2. Describe how the anatomical appearance of the apical lordotic projection of the chest differs from the posteroanterior projection of the chest.

3. Describe how the projection in Question 2 can be considered an example of the use of controlled distortion to avoid superimposition.

4. Describe how the anatomical appearance of the angled posteroanterior projection of the clavicle differs from the true posteroanterior projection of the clavicle.

5. Describe how the projection in Question 4 can be considered an example of the use of controlled distortion to avoid superimposition.

6. Identify three other examinations/projections that use distortion to advantage in this manner, and describe how they accomplish this goal.

Name ______________________________ Date ______________

EXPERIMENT 19
DISTORTION—Shape Distortion: Used to Advantage To Demonstrate Anatomy

Objective

To demonstrate that distortion can be used to advantage in a controlled application in order to record properly the size, shape, and relationship of specific anatomical structures.

Procedure

Select samples of the following examinations from the radiographic file. If unavailable, produce these projections using a suitable phantom of the pelvic girdle and foot, or use a skeleton.

1. **Lumbosacral spine.** Be sure to include an anteroposterior projection of the lumbar spine and an anteroposterior projection of the lumbosacral articulation.
2. **Os calcis.** Be sure to include an axial projection.

Analysis

Review the radiographs of both examinations carefully with specific attention to the anatomical relationships recorded on each film.

Conclusions

1. Describe how the anatomical appearance of the lumbosacral articulation in the anteroposterior projection of the lumbar spine differs from the anteroposterior projection of the lumbosacral articulation.

2. Which of the two films records the lumbosacral joint space to advantage? Indicate how this is an example of the use of controlled distortion to demonstrate anatomy.

3. Describe the anatomical appearance of the os calcis in the axial projection compared to the true appearance of the bone.

4. Indicate how the axial projection serves as an example of controlled distortion to demonstrate anatomy.

5. Identify three other examinations/projections that use distortion to advantage in this manner and describe how they accomplish this goal.

Name ______________________________ Date ____________

EXPERIMENT 20
RADIOGRAPHIC DENSITY—Effect of Milliamperage

Objective

To demonstrate the effect of milliamperage on radiographic density.

Procedure

With the use of a phantom part (e.g., elbow or ankle) and an aluminum penetrometer (step wedge), a series of four exposures will be made using different milliamperage values. The exposures should be made tabletop using a 14" by 17" (36 cm × 43 cm) slow speed (detail) film-screen recording system cassette divided into four sections by lead blockers. It is important that the same tube focus be used for all exposures. The milliamperage values suggested may have to be adjusted to meet the requirements of the equipment you are using. A 40" (102 cm) SID and 60 kVp are selected. All technical factors should remain the same between exposures except for the change in the milliamperage.

Exposure 1: 100 mA

1. Cover three sections of the film, leaving one section uncovered. Place the phantom part in the middle of the divided film section, tape it securely into position, and place the penetrometer next to it. Mark the film section with a lead number 1.
2. Select the appropriate exposure time to produce a bone density of 1.0 to 1.2.
3. Direct your central ray to the midpoint of the phantom using a perpendicular angle. Collimate appropriately to include the phantom and the penetrometer.
4. Expose the film section.

Exposure 2: 50 mA

1. Cover the exposed film section, and uncover a new section, marking it with a lead number 2.
2. Using the same tube focus, change the milliamperage setting to 50.
3. Expose the film section.

Exposure 3: 200 mA

1. Cover the exposed film section, and uncover a new section, marking it with a lead number 3.

2. Using the same tube focus, change the milliamperage setting to 200.
3. Expose the film section.

Exposure 4: 400 mA

1. Cover the exposed film section, and uncover the final unexposed section, marking it with a lead number 4.
2. Using the same tube focus, change the milliamperage setting to 400.
3. Expose the film section, and process the film.

Analysis

Your radiograph will have four recorded images of the selected phantom part and the penetrometer, each demonstrating a different radiographic density. The radiographic images go from insufficient to excessive radiographic density. A measurement of the specific radiographic densities of a selected point within the images can be accurately assessed with the aid of a densitometer. The relationship of milliamperage to radiographic density can be examined and evaluated.

Conclusions

1. The tube current is measured in milliamperes. Describe exactly what is happening in the x-ray tube as the tube current is increased from 50 to 100, 200, and 400 mA.

2. As the milliamperage is increased, what effect does this have on the radiographic density of the radiograph?

3. Indicate the mAs selected in each of the four exposures produced in this experiment. Record these values in the spaces provided below:

   Exposure 1: _____ mA × _____ time of exposure = _____ mAs
   Exposure 2: _____ mA × _____ time of exposure = _____ mAs
   Exposure 3: _____ mA × _____ time of exposure = _____ mAs
   Exposure 4: _____ mA × _____ time of exposure = _____ mAs

4. Select a bone area within the radiographic image of the phantom and a specific step of the penetrometer, and circle the same area on each exposure. Measure the radiographic density represented by these selected areas for each exposure with a densitometer. Record these values in the spaces provided below:

   Exposure 1: _____ phantom density; _____ penetrometer density
   Exposure 2: _____ phantom density; _____ penetrometer density
   Exposure 3: _____ phantom density; _____ penetrometer density
   Exposure 4: _____ phantom density; _____ penetrometer density

5. Considering the radiographic density levels recorded above and the mAs increases represented by the milliamperage changes indicated in Question 3, can any specific relationship between the milliamperage and radiographic density be identified? Describe this relationship, if any.

6. Which radiographic density would represent the greatest visibility of the recorded details and thus be more representative of a film that is of radiographic quality? How would you describe the images of the other three exposures? Be specific.

Name ______________________________ Date ______________

EXPERIMENT 21
RADIOGRAPHIC DENSITY—Effect of Exposure Time

Objective

To demonstrate the effect of the time of exposure on radiographic density.

Procedure

With the use of a phantom part (e.g., elbow or ankle) and an aluminum penetrometer (step wedge), a series of four exposures will be made using different exposure times. The exposures should be made tabletop using a 14" by 17" (36 cm × 43 cm) slow speed (detail) film-screen recording system cassette divided into four sections by lead blockers. It is important that the same tube focus and the same 100 mA be used for all exposures. The exposure times suggested may have to be adjusted to meet the requirements of the equipment you are using. A 40" (102 cm) SID and 60 kVp are selected. All technical factors should remain the same between exposures except for the change in the milliamperage.

Exposure 1: Exposure Time To Produce 1.0 to 1.2 Bone Density

1. Cover three sections of the film, leaving one section uncovered. Place the phantom part in the middle of the divided film section, tape it securely into position, and place the penetrometer next to it. Mark the film section with a lead number 1.
2. Select the appropriate exposure time to produce a bone density of 1.0 to 1.2. (See Experiment 20, Exposure 1.)
3. Direct your central ray to the midpoint of the phantom using a perpendicular angle. Collimate appropriately to include the phantom and the penetrometer.
4. Expose the film section.

Exposure 2: One Half the Exposure Time

1. Cover the exposed film section, and uncover a new section, marking it with a lead number 2.
2. Using the same tube focus and 100 mA, change the time of exposure to one half the value used in Exposure 1.
3. Expose the film section.

Exposure 3: Double the Exposure Time

1. Cover the exposed film section, and uncover a new section, marking it with a lead number 3.
2. Using the same tube focus and 100 mA, change the time of exposure to double the value used in Exposure 1.
3. Expose the film section.

Exposure 4: Quadruple the Exposure Time

1. Cover the exposed film section, and uncover the final unexposed section, marking it with a lead number 4.
2. Using the same tube focus and 100 mA, change the time of exposure to four times the value used in Exposure 1.
3. Expose the film section, and process the film.

Analysis

Your radiograph will have four recorded images of the selected phantom part and the penetrometer, each demonstrating a different radiographic density. The radiographic images go from insufficient to excessive radiographic density. A measurement of the specific radiographic densities of a selected point within the images can be accurately assessed with the aid of a densitometer. The relationship of exposure time to radiographic density can be examined and evaluated.

Conclusions

1. As the time of exposure is increased, what effect is there on the radiographic density of the recorded image?

2. Indicate the mAs selected in each of the four exposures produced in this experiment. Record these values in the spaces provided below:

   Exposure 1: _____ mA × _____ time of exposure = _____ mAs
   Exposure 2: _____ mA × _____ time of exposure = _____ mAs
   Exposure 3: _____ mA × _____ time of exposure = _____ mAs
   Exposure 4: _____ mA × _____ time of exposure = _____ mAs

3. Select a bone area within the radiographic image of the phantom and a specific step of the penetrometer, and circle the same area on each exposure. Measure the radiographic density represented by these selected areas for each exposure with a densitometer. Record these values in the spaces provided below.

   Exposure 1: _____ phantom density; _____ penetrometer density
   Exposure 2: _____ phantom density; _____ penetrometer density
   Exposure 3: _____ phantom density; _____ penetrometer density
   Exposure 4: _____ phantom density; _____ penetrometer density

4. Considering the radiographic density levels recorded above and the mAs increases represented by the time of exposure changes indicated in Question 2, can any specific relationship between the milliamperage and radiographic density be identified? Describe this relationship, if any.

5. Which radiographic density would represent the greatest visibility of the recorded details and thus be more representative of a film that is of radiographic quality? How would you describe the images of the other three exposures? Be specific.

Name ______________________________ Date ______________

EXPERIMENT 22
RADIOGRAPHIC DENSITY—Milliampere-Seconds (mAs)

Objective

To demonstrate how radiographic density can be maintained, although different milliamperage and times of exposure are employed, as long as the mAs value remains constant.

Procedure

With the use of a phantom part (e.g., elbow or ankle) and an aluminum penetrometer (step wedge), a series of four exposures will be made using different milliamperage and times of exposure in order to produce the same mAs values. The exposures should be made tabletop using a 14" by 17" (36 cm × 43 cm) slow speed (detail) film-screen recording system cassette divided into four sections by lead blockers. If possible, the same tube focus should be used for all exposures. The mAs values suggested as well as the specific mA/time relationships may have to be adjusted to meet the requirements of the equipment you are using, but the principle will remain the same. A 40" (102 cm) SID and 60 kVp are selected. All technical factors should remain the same between exposures except for the manipulation of the mA/time relationship to maintain the mAs.

Exposure 1: 50 mA

1. Cover three sections of the film, leaving one section uncovered. Place the phantom part in the middle of the divided film section, tape it securely into position, and place the penetrometer next to it. Mark the film section with a lead number 1.
2. Using 50 mA, select an appropriate time of exposure that will produce a 1.0 to 1.2 bone density for the mAs employed. (Review the exposures employed in Experiments 20 and 21 to identify a suggested starting exposure.)
3. Direct your central ray to the midpoint of the phantom using a perpendicular angle. Collimate appropriately to include the phantom and the penetrometer.
4. Expose the film section.

Exposure 2: 100 mA

1. Cover the exposed film section, and uncover a new section, marking it with a lead number 2.

2. Using the same tube focus, select 100 mA, and adjust your time of exposure to produce the same mAs as indicated in Exposure 1.
3. Expose the film section.

Exposure 3: 200 mA

1. Cover the exposed film section, and uncover a new section, marking it with a lead number 3.
2. Using the same tube focus, select 200 mA, and adjust your time of exposure to produce the same mAs as indicated in Exposure 1.
3. Expose the film section.

Exposure 4: 400 mA

1. Cover the exposed film section, and uncover the final unexposed section, marking it with a lead number 4.
2. Using the same tube focus, select 400 mA, and adjust your time of exposure to produce the same mAs as indicated in Exposure 1.
3. Expose the film section, and process the film.

Analysis

Your radiograph will have four recorded images of the selected phantom part and the penetrometer. You will be able to investigate the relationship of the mAs in maintaining radiographic density when different milliamperage and times of exposure are selected. The relationship of mAs to radiographic density can be examined and evaluated by measuring the specific radiographic densities of both the phantom and the penetrometer with the aid of a densitometer.

Conclusions

1. How would you describe the relationship between the milliamperage and the time of exposure as it relates to mAs and radiographic density?

2. Indicate the milliamperage and time of exposure required to maintain the same mAs for each of the four exposures. Record this information in the spaces provided below:

Exposure 1: _____ mA × _____ time of exposure = _____ mAs
Exposure 2: _____ mA × _____ time of exposure = _____ mAs
Exposure 3: _____ mA × _____ time of exposure = _____ mAs
Exposure 4: _____ mA × _____ time of exposure = _____ mAs

3. Select a bone area within the radiographic image of the phantom and a specific step of the penetrometer, and circle the same area on each exposure. Measure the radiographic density represented by these selected areas for each exposure with a densitometer. Record these values in the spaces provided below:

   Exposure 1: _____ phantom density; _____ penetrometer density
   Exposure 2: _____ phantom density; _____ penetrometer density
   Exposure 3: _____ phantom density; _____ penetrometer density
   Exposure 4: _____ phantom density; _____ penetrometer density

4. Identify any major differences in the radiographic density between the four radiographic images. Analyze the differences, if any, and provide reasons this may have occurred.

5. What importance does the manipulation of the milliamperage and time of exposure to maintain mAs have to the performance of general radiographic procedures?

6. Identify a radiographic situation where this relationship could be applied to advantage.

7. Of what significance is the calibration of your x-ray generator output to the maintenance of radiographic density and to an overall program of quality assurance?

Name ______________________________ Date ____________

EXPERIMENT 23
RADIOGRAPHIC DENSITY—Evaluation of Tube Current (Milliamperage)

Objective

To evaluate the tube current (mA) for the various milliamperage settings on the control panel to determine whether the output is correct.

Procedure

Radiographic exposures will be made for each milliamperage setting available on your control panel for the small and large focus of the tube. The exposures should be made tabletop using a 14" by 17" (36 cm × 43 cm) direct exposure (cardboard) film holder or using the largest size direct exposure (cardboard) film holder available. The film should be divided into as many sections as are needed to demonstrate each available milliamperage setting on the control panel for the small focus. A second film will be divided into as many sections as are needed to demonstrate each available milliamperage setting on the control panel for the large focus. All exposures are taken using 40" (102 cm) SID and 40 kVp.

Film 1, Exposure 1: Small Focus

1. Determine the number of available milliamperage settings on your control panel for the small focus, and divide your film into that many sections using lead blockers. Block off all but one section, and mark that section with a lead number 1.
2. Select the smallest milliamperage setting for the small focus, and choose a 2-sec time of exposure.
3. Direct your central ray to the middle of your divided film section using a perpendicular angle. Collimate to an area 4" (10 cm) square.
4. Calculate and record the mAs utilized for the exposure. Observe the milliamperage meter or ballistic mAs meter while exposing the film section, and record the reading. Expose the film section.

Film 1: Small Focus, Additional Exposures

1. Perform the same procedure as indicated above for each of the milliamperage settings available for the small focus. Cover the exposed film sections, and uncover a new, unexposed section for each additional exposure. Mark each additional exposure with a progressive lead number 2, 3, 4, and so forth.

2. For each increase in the milliamperage setting, reduce your time of exposure in order to maintain the same mAs as used with Exposure 1.
3. Observe your milliamperage meter or ballistic mAs meter while exposing each additional film section, and record the reading.
4. When exposures of all milliamperage settings have been taken, process your film.

Film 2, Exposure 1: Large Focus

1. Determine the number of available milliamperage settings on your control panel for the large focus, and divide your film into that many sections using lead blockers. Block off all but one section, and mark that section with a lead number 1.
2. Select the smallest milliamperage setting for the large focus and *choose a time of exposure equal to the time of exposure chosen for that milliamperage setting for Film 1 using the small focus.*
3. Direct your central ray to the middle of your divided film section using a perpendicular angle. Collimate to an area 4" (10 cm) square.
4. Calculate and record the mAs utilized for the exposure. Observe the milliamperage meter or ballistic mAs meter while exposing the film section and record the reading. Expose the film section.

Film 2: Large Focus, Additional Exposures

1. Perform the same procedure as indicated above for each of the milliamperage settings available for the large focus. Cover the exposed film sections, and uncover a new, unexposed section for each additional exposure. Mark each additional exposure with a progressive lead number 2, 3, 4, and so forth.
2. For each increase in the milliamperage setting, reduce your time of exposure in order to maintain the same mAs as used with Exposure 1.
3. Observe your milliamperage meter or ballistic mAs meter while exposing each additional film section, and record the reading.
4. When exposures of all milliamperage settings have been taken, process your film.

Analysis

You will now have at least two radiographs with multiple exposures: one representing exposures using the small focus, the other representing exposures using the large focus. The mAs used for all the exposures were constant. If the x-ray equipment is properly calibrated, the radiographic density of all the radiographic images should be the same. The milliamperage meter or ballistic mAs meter readings for all the exposures also should have been constant.

Conclusions

1. Record the information requested for each of the milliamperage settings employed with the small focus.
   Exposure 1: _____ mA, _____ sec, _____ mAs, _____ mA/mAs meter reading
   Exposure 2: _____ mA, _____ sec, _____ mAs, _____ mA/mAs meter reading
   Exposure 3: _____ mA, _____ sec, _____ mAs, _____ mA/mAs meter reading
   Exposure 4: _____ mA, _____ sec, _____ mAs, _____ mA/mAs meter reading
   Exposure 5: _____ mA, _____ sec, _____ mAs, _____ mA/mAs meter reading

2. Describe any differences in the mA/mAs meter readings for each of the milliamperage settings. Explain how this may have occurred. **Note:** Report any differences to your instructor, supervisor, or quality control technologist.

3. Record the radiographic density for each of the milliamperage settings employed with the small focus.
   Exposure 1: _______ radiographic density
   Exposure 2: _______ radiographic density
   Exposure 3: _______ radiographic density
   Exposure 4: _______ radiographic density
   Exposure 5: _______ radiographic density

4. Describe any differences in the radiographic density recorded between the different milliamperage settings. Explain how this may have occurred. **Note:** Report any differences to your instructor, supervisor, or quality control technologist.

5. Record the information requested for each of the milliamperage settings employed with the large focus.
   Exposure 1: _____ mA, _____ sec, _____ mAs, _____ mA/mAs meter reading
   Exposure 2: _____ mA, _____ sec, _____ mAs, _____ mA/mAs meter reading
   Exposure 3: _____ mA, _____ sec, _____ mAs, _____ mA/mAs meter reading
   Exposure 4: _____ mA, _____ sec, _____ mAs, _____ mA/mAs meter reading
   Exposure 5: _____ mA, _____ sec, _____ mAs, _____ mA/mAs meter reading

6. Describe any differences in the mA/mAs meter readings for each of the milliamperage settings. Explain how this may have occurred. **Note:** Report any differences to your instructor, supervisor, or quality control technologist.

7. Record the radiographic density for each of the milliamperage settings employed with the large focus.

   Exposure 1: _______ radiographic density
   Exposure 2: _______ radiographic density
   Exposure 3: _______ radiographic density
   Exposure 4: _______ radiographic density
   Exposure 5: _______ radiographic density

8. Describe any differences in the radiographic density recorded between the different milliamperage settings. Explain how this may have occurred. **Note:** Report any differences to your instructor, supervisor, or quality control technologist.

9. Compare the results of the same milliamperage settings for the small and large focus on Film 1 and Film 2. Describe the similarities and/or differences noted.

10. Would it be correct to state that the x-ray equipment that you performed this experiment on is calibrated correctly and that the milliamperage settings are true and correct for both the small and large focus? Explain your answer.

11. Of what significance is the accuracy of the tube current output to the production of radiographic quality?

12. Discuss the need for an evaluation of the tube current output within the scope of a quality assurance program and how this demonstrates the application of the principles of ALARA.

Name ______________________ Date ____________

EXPERIMENT 24
RADIOGRAPHIC DENSITY—Comparison of Tube Current (Milliamperage) between Different X-Ray Units

Objective

To evaluate the tube current (mA) output between two different x-ray units.

Procedure

Review the procedure described in Experiment 23. Select a different x-ray unit having similar milliamperage settings for both the small and large focus as the equipment used for Experiment 23. Set up and perform the same procedure using the new x-ray equipment, choosing and exposing only those milliamperage stations that have a comparable setting on the x-ray equipment used for Experiment 23. Use the same technical factors employed for Experiment 23.

Film 1: Small Focus

Determine the number of available milliamperage selections on your control panel that match up with the milliamperage settings for the small focus on the equipment used for Experiment 23. Use the *same mAs settings* for your exposures. Expose the appropriate film sections, and process your film.

Film 2: Large Focus

Determine the number of available milliamperage selections on your control panel that match up with the milliamperage settings for the large focus on the equipment used for Experiment 23. Use the *same mAs settings* for your exposures. Expose the appropriate film sections, and process your film.

Analysis

You will now have two radiographs with multiple exposures, one representing exposures using the small focus, the other representing exposures using the large focus. The mAs used for all the exposures were constant and the same as those employed in Experiment 23. If the x-ray equipment is properly calibrated, the radiographic density of all the radiographic images should be the same, the milliamperage meter or ballistic mAs meter readings for all the exposures also should have been constant, and the radiographic densities and milliamperage meter or ballistic mAs meter readings between the two pieces of equipment should also be the same.

Conclusions

1. Record the information requested for each of the milliamperage settings employed with the small focus for this experiment.
   Exposure 1: _____ mA, _____ sec, _____ mAs, _____ mA/mAs meter reading
   Exposure 2: _____ mA, _____ sec, _____ mAs, _____ mA/mAs meter reading
   Exposure 3: _____ mA, _____ sec, _____ mAs, _____ mA/mAs meter reading
   Exposure 4: _____ mA, _____ sec, _____ mAs, _____ mA/mAs meter reading
   Exposure 5: _____ mA, _____ sec, _____ mAs, _____ mA/mAs meter reading

2. Describe any differences in the mA/mAs meter readings for each of the milliamperage settings. Explain how this may have occurred. **Note:** Report any differences to your instructor, supervisor, or quality control technologist.

3. Record the radiographic density for each of the milliamperage settings employed with the small focus for this experiment.
   Exposure 1: _______ radiographic density
   Exposure 2: _______ radiographic density
   Exposure 3: _______ radiographic density
   Exposure 4: _______ radiographic density
   Exposure 5: _______ radiographic density

4. Describe any differences in the radiographic density recorded between the different milliamperage settings. Explain how this may have occurred. **Note:** Report any differences to your instructor, supervisor, or quality control technologist.

5. Record the information requested for each of the milliamperage settings employed with the large focus for this experiment.
   Exposure 1: _____ mA, _____ sec, _____ mAs, _____ mA/mAs meter reading
   Exposure 2: _____ mA, _____ sec, _____ mAs, _____ mA/mAs meter reading
   Exposure 3: _____ mA, _____ sec, _____ mAs, _____ mA/mAs meter reading
   Exposure 4: _____ mA, _____ sec, _____ mAs, _____ mA/mAs meter reading
   Exposure 5: _____ mA, _____ sec, _____ mAs, _____ mA/mAs meter reading

6. Describe any differences in the mA/mAs meter readings for each of the milliamperage settings. Explain how this may have occurred. **Note:** Report any differences to your instructor, supervisor, or quality control technologist.

7. Record the radiographic density for each of the milliamperage settings employed with the large focus for this experiment.
   Exposure 1: _______ radiographic density
   Exposure 2: _______ radiographic density
   Exposure 3: _______ radiographic density
   Exposure 4: _______ radiographic density
   Exposure 5: _______ radiographic density

8. Describe any differences in the radiographic density recorded between the different milliamperage settings. Explain how this may have occurred. **Note:** Report any differences to your instructor, supervisor, or quality control technologist.

9. Compare the results of the same milliamperage settings for the small and large focus on Film 1 and Film 2 for Experiment 24. Describe the similarities and/or differences noted.

10. Would it be correct to state that the x-ray equipment with which you performed Experiment 24 is calibrated correctly and that the milliamperage settings are true and correct for both the small and large focus? Explain your answer.

11. Compare the mA/mAs meter readings for the same milliamperage settings for Experiment 24 with those demonstrated in Experiment 23. Describe any differences. **Note:** Report any differences between the two x-ray units to your instructor, supervisor, or quality control technologist.

12. Compare the radiographic densities for the same milliamperage settings for Experiment 24 with those demonstrated in Experiment 23. Describe any differences. **Note:** Report any differences between the two x-ray units to your instructor, supervisor, or quality control technologist.

13. Discuss the importance of the accuracy of the tube current output between the various x-ray units contained within a radiology department.

Name ______________________________ Date ______________

EXPERIMENT 25

RADIOGRAPHIC DENSITY—Evaluation of X-Ray Timing Mechanism: Spinning Top Test

Objective

To determine the accuracy of the timing mechanism in a two-phase generating x-ray unit.

Procedure

With the use of a spinning top device, a series of four exposures will be made using different exposure times. The exposures should be made tabletop using a 10" by 12" (25 cm × 30 cm) slow speed (detail) film-screen recording system cassette divided into four sections by lead blockers. A 40" (102 cm) SID, 100 mA, and 60 kVp are selected. All technical factors should remain the same between exposures except for the changes in the time of exposure.

Exposure 1: 1/60 (.0166) Sec Time of Exposure

1. Cover three sections of the film, leaving one section uncovered. Place the spinning top in the middle of the uncovered divided film section. Tape the base to the surface of the cassette front so that it will not move during the exposure. Mark this section with a lead number 1.
2. Direct your central ray to the midpoint of the spinning top using a perpendicular angle, and collimate appropriately to include the spinning top and the lead number.
3. Select a 1/60 (.0166) exposure time, and begin the spinning top by rotating it between your thumb and forefinger.
4. Expose the film section while the spinning top is in motion.

Exposure 2: 1/30 (.0333) Sec Time of Exposure

1. Cover the exposed film section, and uncover a new section, marking it with a lead number 2.
2. Change the time of exposure to 1/30 (.0333) sec and perform your procedure as you did in Exposure 1.
3. Expose the film section while the spinning top is in motion.

Exposure 3: 1/20 (.050) Sec Time of Exposure

1. Cover the exposed film section, and uncover a new section, marking it with a lead number 3.
2. Change the time of exposure to 1/20 (.050) sec, and perform your procedure as you did in Exposure 1.
3. Expose the film section while the spinning top is in motion.

Exposure 4: 1/10 (.10) Sec Time of Exposure

1. Cover the exposed film section, and uncover the final unexposed section, marking it with a lead number 4.
2. Change the time of exposure to 1/10 (.10) sec, and perform your procedure as you did in Exposure 1.
3. Expose the film section while the spinning top is in motion.

Analysis

Your radiograph will have four recorded images of the spinning top, each demonstrating a different number of "blips" or density dashes within the recorded image. A review of the number of "blips" will enable you to evaluate whether the timing mechanism for your x-ray unit is operating properly. A 60-cycle, two-phase generating x-ray unit will produce a total of 120 density "blips" in 1 sec.

Conclusions

1. Record the requested information in the spaces provided below:
   Exposure 1: _____ time of exposure, _____ density "blips"
   Exposure 2: _____ time of exposure, _____ density "blips"
   Exposure 3: _____ time of exposure, _____ density "blips"
   Exposure 4: _____ time of exposure, _____ density "blips"

2. How many "blips" or density dashes will be demonstrated in a spinning top test employing an exposure of 1/40 (.025) sec? 1/15 (.066) sec?

3. In Question 1, are the density dashes or "blips" demonstrated in Exposures 1 through 4 correct? If not, describe the differences and explain how this may have occurred. **Note:** If any differences from the correct number occur, report this information to your instructor, supervisor, or quality control technologist.

4. Of what significance is the accuracy of the timer mechanism to the production of radiographic quality?

5. Discuss the need for an evaluation of the timer mechanism within the scope of a quality assurance program and how this demonstrates the application of the principles of ALARA.

Name ____________________ Date ____________

EXPERIMENT 26
RADIOGRAPHIC DENSITY—Evaluation of X-Ray Timing Mechanism: Synchronous Spinning Top Test

Objective

To determine the accuracy of the timing mechanism in a three-phase generating x-ray unit.

Procedure

With the use of a synchronous spinning top device, a series of four exposures will be made using different exposure times. The exposures should be made tabletop using a 10" by 12" (25 cm × 30 cm) slow speed (detail) film-screen recording system cassette divided into four sections by lead blockers. A 40" (102 cm) SID, 100 mA, and 60 kVp are selected. All technical factors should remain the same between exposures except for the changes in the time of exposure.

Exposure 1: 1/2 (.50) Sec Time of Exposure

1. Cover three sections of the film, leaving one section uncovered. Place the spinning top in the middle of the uncovered divided film section. Tape the base to the surface of the cassette front so that it will not move during the exposure. Mark this section with a lead number 1.
2. Direct your central ray to the midpoint of the synchronous spinning top using a perpendicular angle, and collimate appropriately to include the spinning top and the lead number.
3. Select a 1/2 (.50) exposure time and begin the spinning top by flicking the switch that turns on the synchronous motor. The synchronous spinning top will make a complete revolution in 1 sec.
4. Expose the film section while the spinning top is in motion.

Exposure 2: 1/4 (.25) Sec Time of Exposure

1. Cover the exposed film section, and uncover a new section, marking it with a lead number 2.
2. Change the time of exposure to 1/4 (.25) sec, and perform your procedure as you did in Exposure 1.
3. Expose the film section while the spinning top is in motion.

Exposure 3: 1/8 (.125) Sec Time of Exposure

1. Cover the exposed film section, and uncover a new section, marking it with a lead number 3.
2. Change the time of exposure to 1/8 (.125) sec, and perform your procedure as you did in Exposure 1.
3. Expose the film section while the spinning top is in motion.

Exposure 4: 1/12 (.083) Sec Time of Exposure

1. Cover the exposed film section, and uncover the final unexposed section, marking it with a lead number 4.
2. Change the time of exposure to 1/12 (.083) sec, and perform your procedure as you did in Exposure 1.
3. Expose the film section while the spinning top is in motion.

Analysis

Your radiograph will have four recorded images of the spinning top, each demonstrating a different portion of a circle represented by an arc of radiographic density. A measurement of the arc representing a portion of a circle will enable you to evaluate whether the timing mechanism for your x-ray unit is operating properly. A three-phase generating x-ray unit will produce a recording of a complete circle (360°) in 1 sec. A protractor identifying different degrees of a circle from 0° to 180° would be helpful for this analysis.

Conclusions

1. Record the requested information in the spaces provided below:

   Exposure 1: _____ time of exposure, _____ ° arc
   Exposure 2: _____ time of exposure, _____ ° arc
   Exposure 3: _____ time of exposure, _____ ° arc
   Exposure 4: _____ time of exposure, _____ ° arc

2. How many degrees of a circle will be demonstrated in a synchronous spinning top test employing an exposure of 3/4 (.75) sec? 1/24 (.04) sec?

3. In Question 1, are the arcs of the circle demonstrated in Exposures 1 through 4 correct? If not, describe the differences, and explain how this may have occurred. **Note:** If any differences from the correct number of degrees occur, report this information to your instructor, supervisor, or quality control technologist.

4. Is the synchronous spinning top test useful for the examination of short exposure times, such as 1/60 sec, 1/120 sec, and 1/250 sec? Explain the reasons for your choice.

Name ______________________________ Date ____________

EXPERIMENT 27
RADIOGRAPHIC DENSITY—Image Visibility and Radiographic Density

Objective

To demonstrate the need for proper radiographic density by comparing the image visibility loss produced by insufficient and by excessive radiographic density.

Procedure

With the use of a phantom part (e.g., elbow or ankle) and an aluminum penetrometer (step wedge), three exposures are made using different mAs values. The exposures are made tabletop using a 14" by 17" (36 cm × 43 cm) slow speed (detail) film-screen recording system cassette divided into three sections. It is important that the same tube focus and the same 100 mA station be used for all three exposures. The time of exposure is adjusted to produce the different mAs required for all three recorded images. A 40" (102 cm) SID and 50 kVp are selected. All technical factors should remain the same between exposures except for the change in the time of exposure.

Exposure 1: Appropriate Image Density

1. Cover two sections of the film, leaving one section uncovered. Place the phantom part in the middle of the divided film section, tape it securely, and place the penetrometer next to it. Mark the film section with a lead number 1.
2. Using 100 mA, select the appropriate time of exposure to produce a bone density of 1.0 to 1.2.
3. Direct your central ray to the midpoint of the phantom using a perpendicular beam, and collimate appropriately to include the phantom and the penetrometer.
4. Expose the film section.

Exposure 2: Insufficient Image Density

1. Cover the exposed film section, and uncover a new section, marking it with a lead number 2.
2. Decrease your mAs by 50% from the mAs used in Exposure 1.
3. Expose the film section.

Exposure 3: Excessive Image Density

1. Cover the exposed film section, and uncover the final unexposed section, marking it with a lead number 3.
2. Increase your mAs by 50% from the mAs used with Exposure 1.
3. Expose the final film section and process your film.

Analysis

Your radiograph will have three recorded images of the selected phantom part and the penetrometer, each demonstrating a different radiographic density. The recorded images go from insufficient to excessive radiographic density. Review each image from the standpoint of the visibility of the structures contained within the body part. Compare the second and third images with the first in order to analyze the quality of the recorded images.

Conclusions

1. Select a specific area within the radiographic image and a middle step of the penetrometer, and circle the same area on each film. Measure the radiographic density of these selected areas for each radiograph with a densitometer. Record these values in the spaces provided below:

   Exposure 1: _____ phantom density; _____ penetrometer density
   Exposure 2: _____ phantom density; _____ penetrometer density
   Exposure 3: _____ phantom density; _____ penetrometer density

2. Indicate the number of penetrometer steps that are distinctly visible as separate densities in each penetrometer image. How do the images of the penetrometer differ from Exposure 1 to Exposure 2? From Exposure 1 to Exposure 3?

   Exposure 1: _____ penetrometer steps visible
   Exposure 2: _____ penetrometer steps visible
   Exposure 3: _____ penetrometer steps visible

3. Review the radiographic image in Exposure 2. Is the loss of image visibility more apparent in the less thick, less dense tissues or in the thicker, denser tissues? Explain why this occurred.

4. Review the radiographic image in Exposure 3. Is the loss of image visibility more apparent in the less thick, less dense tissues or in the thicker, denser tissues? Explain why this occurred.

5. Comparing Exposures 2 and 3 with Exposure 1, which image appears to have produced the greatest loss of image visibility? Provide reasons for your choice.

6. Review the radiographic image in Film 3 with a bright light. Discuss the image visibility demonstrated using this procedure as compared with viewing the image with a regular illuminator. Is this a satisfactory method of producing images of radiographic quality?

7. Summarize the effect of insufficient or excessive radiographic density on the visibility of your recorded image and on the photographic properties of radiographic quality.

Name ______________________________ Date ______________

EXPERIMENT 28
RADIOGRAPHIC DENSITY—Control of Radiographic Density

Objective

To demonstrate the exposure (mAs) increase required to produce a visible increase in radiographic density.

Procedure

As requested by the instructor, either or both of the following procedures will be performed.

Film 1: Tabletop Procedure

With the use of a phantom part (e.g., elbow or ankle), four exposures are made using different mAs values. The exposures should be made tabletop using a 14" by 17" (36 cm × 43 cm) slow speed (detail) film-screen recording system cassette divided into four sections by lead blockers. It is important that the same tube focus and the same 100 mA setting be used for all exposures. A 40" (102 cm) SID and 50 kVp are selected. All technical factors should remain the same between exposures except for the change in the time of exposure.

Exposure 1: Proper Radiographic Density

1. Cover three sections of the film, leaving one section uncovered. Place the phantom part in the middle of the divided film section, tape it securely, and mark the film section with a lead number 1.
2. Using 100 mA, select the appropriate time of exposure to produce a bone density of 1.0 to 1.2.
3. Direct your central ray to the midpoint of the phantom using a perpendicular beam, and collimate appropriately to include the phantom.
4. Expose the film section.

Exposures 2, 3, 4: Tabletop, Increase the mAs

Move to a new section of the film, and take Exposures 2, 3, and 4 by increasing the mAs from Exposure 1 by 10% for each exposure by adjusting your time of exposure. Mark each new film with a lead number 2, 3, and 4. Process your film.

Film 2: Bucky Procedure

With the use of a phantom part (e.g., pelvis), four exposures are made of the hip using different mAs values. The exposures should be made using the table Bucky with four 10" by 12" (25 cm × 30 cm) slow speed (detail) film-screen recording system cassettes. It is important that the same tube focus and the same 100 mA setting be used for all exposures. A 40" (102 cm) SID and 70 kVp are selected. All technical factors should remain the same between exposures except for the change in the time of exposure.

Exposure 1: Proper Radiographic Density

1. Center the hip joint to the midline of your table, and collimate to a size slightly smaller than the film size. Place your cassette in the Bucky tray, and center it to the phantom part. Mark your film with a lead number 1.
2. Using 100 mA, select the appropriate time of exposure to produce a bone density of 1.0 to 1.2.
3. Direct your central ray to the middle of the phantom part using a perpendicular beam.
4. Expose and process the film.

Exposures 2, 3, 4: Bucky, Increase the mAs

1. Reuse the same cassette, and take Exposures 2, 3, and 4 by increasing the mAs by 10% for each exposure by adjusting your time of exposure from Exposure 1. Mark each film with a lead number 2, 3, and 4. Process your film between each exposure.

Analysis

The object of performing both procedures is to demonstrate that the principle of controlled density increase operates for tabletop and Bucky examinations. By comparing the radiographic densities of the four images in each procedure side by side, you will be able to determine the percentage of exposure (mAs) increase necessary to demonstrate a visible radiographic density increase.

Conclusions

1. Indicate the mAs selected in each of the exposures produced with the tabletop procedure. Record these values in the spaces provided below:

   Exposure 1: _____ mA × _____ time of exposure = _____ mAs
   Exposure 2: _____ mA × _____ time of exposure = _____ mAs
   Exposure 3: _____ mA × _____ time of exposure = _____ mAs
   Exposure 4: _____ mA × _____ time of exposure = _____ mAs

2. Select a bone area within the radiographic image of the phantom, and circle the same area on each exposure. Measure the radiographic density represented by this area for each exposure with a densitometer. Record these values in the spaces provided below:

   Exposure 1: _____ phantom density
   Exposure 2: _____ phantom density
   Exposure 3: _____ phantom density
   Exposure 4: _____ phantom density

3. Review the radiographic images produced using the tabletop procedure. At what point does the radiographic density of the image *obviously* increase? What percentage of mAs increase does this represent?

4. Indicate the mAs selected in each of the exposures produced with the Bucky procedure. Record these values in the spaces provided below:

   Exposure 1: _____ mA × _____ time of exposure = _____ mAs
   Exposure 2: _____ mA × _____ time of exposure = _____ mAs
   Exposure 3: _____ mA × _____ time of exposure = _____ mAs
   Exposure 4: _____ mA × _____ time of exposure = _____ mAs

5. Select a bone area (head of the femur) within the radiographic image of the phantom, and circle the same area on each exposure. Measure the radiographic density represented by this area for each exposure with a densitometer. Record these values in the spaces provided below:

   Exposure 1: _____ phantom density
   Exposure 2: _____ phantom density
   Exposure 3: _____ phantom density
   Exposure 4: _____ phantom density

6. Review the radiographic images produced using the Bucky mechanism. At what point does the radiographic density of the image *obviously* increase? What percentage of mAs increase does this represent?

7. Of what practical value is this relationship when you are setting up your original exposure factors for an examination?

8. Of what significance is this relationship when you are adjusting your exposure factors to produce an additional or repeat film within the same examination if your original exposure produced an image with insufficient radiographic density?

Name ______________________________ Date ______________

EXPERIMENT 29
RADIOGRAPHIC DENSITY—Source-Image Distance and Exposure Rate

Objective

To demonstrate the influence of SID on the exposure rate of the x-ray beam and to demonstrate the difference in the area covered by a beam of radiation at different SIDs.

Procedure

Select a 14" by 17" (36 cm × 43 cm) slow speed (detail) film-screen recording system cassette, and place it on the tabletop. It should be divided into two sections with lead blockers. No phantoms are required for this procedure.

Exposure 1: 40" (102 cm) SID

Cover one half of your film with lead blockers. Center your central ray over the uncovered film section using an SID of 40" (102 cm). Using a perpendicular beam, collimate to produce a 4" (10 cm) square image of light upon the surface of your cassette. Measure with a ruler or measuring tape. Using 50 kVp and 100 mA, adjust your time of exposure to produce a 1.0 to 1.2 radiographic density. Expose your film section.

Exposure 2: 20" (51 cm) SID

Cover the exposed film section, and uncover the other half of the film. Center your central ray over the uncovered film section, and change your SID to 20" (51 cm) using a perpendicular beam. Do not adjust your collimation or change your exposure factors from Exposure 1. Expose your film section, and process the film.

Analysis

A review of the two radiographic images will demonstrate the effect of different SIDs on the area covered by the beam of radiation and the intensity of the beam of radiation at each SID.

Conclusion

1. What is the total area of the recorded image produced with the 40" (102 cm) SID? The 20" (51 cm) SID? (Area = length × width.)

2. How much larger (expressed as a ratio) is the recorded image produced at the 40" (102 cm) SID compared with the recorded image produced at the 20" (51 cm) SID?

   a. 2:1 c. 6:1

   b. 4:1 d. 8:1

3. Describe the reason(s) for this observed increase in the size of the recorded image when the SID was changed from 20" (51 cm) to 40" (102 cm).

4. Using a densitometer, record the radiographic density of each image. Explain the relationship between the two densities and how this change in radiation intensity came about.

   Exposure 1: _____ density Exposure 2: _____ density

Name ______________________________ Date ______________

EXPERIMENT 30
RADIOGRAPHIC DENSITY—Influence of Source-Image Distance

Objective

To demonstrate the influence of different SIDs on radiographic density and the relationship between the SID and radiographic density.

Procedure

With the use of a phantom part (e.g., elbow or ankle) and an aluminum penetrometer (step wedge), a series of four exposures will be made using different SIDs. The exposures should be made tabletop using a 14" by 17" (36 cm × 43 cm) slow speed (detail) film-screen recording system cassette divided into four sections by lead blockers. It is important that the same tube focus, the same 100 mA and total mAs, and 50 kVp be used for all exposures. The time of exposure should be selected to produce a 1.0 to 1.2 bone density. All technical factors should remain the same except for the change in the SID.

Exposure 1: 40" (102 cm) SID

1. Center the phantom and penetrometer to the first film section, and secure it in position. Center your central ray to the phantom using a perpendicular angle. Mark the film section with a lead number 1.
2. Using a 40" (102 cm) SID, collimate the beam to include the phantom and penetrometer.
3. Expose the film section.

Exposure 2: 30" (76 cm) SID

1. Set up everything the same as for Exposure 1; however, change to a new film section, decrease your SID to 30" (76 cm), and adjust your collimation for the distance change. Mark your film section with a lead number 2.
2. Expose the film section.

Exposure 3: 20" (51 cm) SID

1. Set up everything the same as for Exposure 1; however, change to a new film section, decrease your SID to 20" (51 cm), and adjust your collimation for the distance change. Mark your film section with a lead number 3.
2. Expose the film section.

Exposure 4: 60" (153 cm) SID

1. Set up everything the same as for Exposure 1; however, change to the final unexposed film section, increase your SID to 60" (153 cm), and adjust your collimation for the distance change. Mark your film section with a lead number 4.
2. Expose the final film section, and process your film.

Analysis

Your radiograph will demonstrate a total of four images of the phantom and penetrometer. Review the recorded image of each related to the radiographic density and the overall visibility of the images recorded.

Conclusions

1. Describe the influence of different SID increases/decreases on the radiographic density of your recorded images.

2. Of the four recorded images, which SID represents the recording with the best image visibility? Explain the reasons for your choice.

3. Select and circle a specific area within the radiographic image of the phantom and one of the middle steps of the penetrometer for each of the four exposures, and measure the radiographic density of each with a densitometer. Record the radiographic density in the spaces provided below:
   Exposure 1: 40" (102 cm) SID _____ phantom density; _____ penetrometer
   Exposure 2: 30" (76 cm) SID _____ phantom density; _____ penetrometer
   Exposure 3: 20" (51 cm) SID _____ phantom density; _____ penetrometer
   Exposure 4: 60" (153 cm) SID _____ phantom density; _____ penetrometer

4. Which SID produces the greatest radiographic density? The least radiographic density? Why?

5. Compare the radiographic density of the image produced at the 20" (51 cm) SID with the image produced at the 40" (102 cm) SID. Describe the relationship, if any, between the densities of the two images, and identify the radiological principle involved.

6. Compare the radiographic density of the image produced at 40" (102 cm) SID with the image produced at 60" (153 cm) SID. Is the relationship comparable to that described in Question 5? Explain your answer.

7. Compare the radiographic density of the image produced at 30" (76 cm) SID with the image produced at 60" (153 cm) SID. Is the relationship produced comparable to that described in Question 5? Explain your answer.

8. What conclusions can you make related to the influence of the SID on the radiographic density of the recorded image?

Name ______________________________ Date ______________

EXPERIMENT 31

RADIOGRAPHIC DENSITY—Source-Image Distance and the Inverse Square Law

Objective

To apply the principles of the inverse square law to a practical situation. To demonstrate that the radiographic density of the recorded image can be maintained at different SIDs by adjusting the mAs according to the principles of the inverse square law.

Principle

The inverse square law states that the intensity of an x-ray beam is inversely proportional to the square of the distance. A practical formula based on the principles of the inverse square law has been developed, enabling us to adjust accurately our exposure factors to maintain our radiographic density at different SIDs.

$$\text{New mAs} = \text{old mAs} \times \text{new distance}^2/\text{old distance}^2$$

Procedure

With the use of a phantom part (e.g., elbow or ankle) and an aluminum penetrometer (step wedge), a series of four exposures will be made using different SIDs and different exposure factors. The exposures should be made tabletop using a 14" by 17" (36 cm × 43 cm) slow speed (detail) film-screen recording system cassette divided into four sections by lead blockers. It is important that the same tube focus, the same 100 mA setting, and the same 50 kVp be used for all exposures.

Exposure 1: 40" (102 cm) SID

1. Center the phantom and penetrometer to the first film section, and secure it in position. Center your central ray to the phantom using a perpendicular angle. Mark the film section with a lead number 1.
2. Using a 40" (102 cm) SID, collimate the beam to include the phantom and penetrometer.
3. Select an appropriate time of exposure to produce a 1.0 to 1.2 bone density.
4. Expose the film section.

Exposure 2: 30" (76 cm) SID

1. Set up everything the same as for Exposure 1; however, change to a new film section, decrease your SID to 30" (76 cm), and adjust your collimation for the distance change. Mark your film section with a lead number 2.
2. Using the inverse square law formula, adjust your mAs by changing the time of exposure to maintain your radiographic density at the new distance.
3. Expose the film section.

Exposure 3: 20" (51 cm) SID

1. Set up everything the same as for Exposure 1; however, change to a new film section, decrease your SID to 20" (51 cm), and adjust your collimation for the distance change. Mark your film section with a lead number 3.
2. Using the inverse square law formula, adjust your mAs by changing the time of exposure to maintain your radiographic density at the new distance.
3. Expose the film section.

Exposure 4: 60" (153 cm) SID

1. Set up everything the same as for Exposure 1; however, change to the final unexposed film section, increase your SID to 60" (153 cm), and adjust your collimation for the distance change. Mark your film section with a lead number 4.
2. Using the inverse square law formula, adjust your mAs by changing the time of exposure to maintain your radiographic density at the new distance.
3. Expose the final film section, and process your film.

Analysis

Your radiograph should demonstrate a total of four images of the phantom and penetrometer. The overall radiographic density of each recorded image should be comparable.

Conclusions

1. Using the inverse square law formula, calculate the proper mAs to use for the new SID when your original mAs was selected for a 40" (102 cm) SID. Show all work.

Exposure 1: _______ mAs at 40" (102 cm) SID
Exposure 2: _______ mAs at 30" (76 cm) SID
Exposure 3: _______ mAs at 20" (51 cm) SID
Exposure 4: _______ mAs at 60" (153 cm) SID

2. Select and circle a specific area within the radiographic image of the phantom and one of the middle steps of the penetrometer for each of the four exposures, and measure the radiographic density of each with a densitometer. Record the radiographic density in the spaces provided below:

   Exposure 1: 40" (102 cm) SID _____ phantom density; _____ penetrometer
   Exposure 2: 30" (76 cm) SID _____ phantom density; _____ penetrometer
   Exposure 3: 20" (51 cm) SID _____ phantom density; _____ penetrometer
   Exposure 4: 60" (153 cm) SID _____ phantom density; _____ penetrometer

3. Review the radiographic appearance of each of the recorded images. Does the overall visibility of the recorded images appear similar? Do the radiographic densities of the images compare satisfactorily with each other on the densitometer readings? Are all the images of radiographic quality?

4. Describe the differences, if any, in the recorded density between the four recorded images, and explain the differences noted.

5. Is it possible to maintain radiographic density while adjusting your SID? Discuss the practical application of the inverse square law formula to general radiography.

6. Describe how the radiographic quality factors of recorded detail and distortion have been affected by your change from 40" (102 cm) SID to 20" (51 cm) SID and by your change from 40" (102 cm) SID to 60" (153 cm) SID. Be specific.

7. Discuss the relationship between the radiographic exposure required at 20" (51 cm) SID and the exposure required at 40" (102 cm) SID; the radiographic exposure required at 60" (153 cm) SID and the exposure required at 30" (76 cm) SID.

Name ______________________________ Date ______________

EXPERIMENT 32
RADIOGRAPHIC DENSITY—Kilovoltage

Objective

To demonstrate the effect of kilovoltage on the radiographic density of the recorded image.

Procedure

Use of a slow speed (detail) film-screen recording system cassette (14" by 17"; 36 cm × 43 cm) divided into four sections by lead blockers to record four images of a phantom elbow or ankle and an aluminum penetrometer (step wedge) as a tabletop procedure. The exposures will be made using different kilovoltage, and all other factors remain the same. Use the same tube focus and milliamperage setting for all exposures.

Exposure 1: 50 kVp

1. Divide the film with lead strips, cover three film sections, and center the phantom to the uncovered section, taping it into place. Place the penetrometer adjacent to it, and collimate your beam to the divided film section to include both the phantom and the penetrometer.
2. Arrange your central ray at a perpendicular angle to the midpoint of the phantom using 40" (102 cm) SID. Mark the film section with a lead number 1.
3. Using 50 kVp, select an appropriate mAs to produce a bone density of 1.0 to 1.2, and expose the film section.

Exposure 2: 40 kVp

Cover the exposed film section, and uncover a new section. Set up as for Exposure 1, decrease your kilovoltage to 40, mark the film section with a lead number 2, and expose the film section.

Exposure 3: 60 kVp

Cover the exposed film section, and uncover a new section. Set up as for Exposure 1, increase your kilovoltage to 60, mark the film section with a lead number 3, and expose the film section.

Exposure 4: 80 kVp

Cover the exposed film section, and uncover the final unexposed section. Set up as for Exposure 1, increase your kilovoltage to 80, mark the film section with a lead number 4, expose the film section, and process your film.

Analysis

Your radiograph will have four recorded images of the selected phantom part and the penetrometer, each demonstrating a different radiographic density. The radiographic densities go from an insufficient to an excessive value. A measurement of the specific radiographic densities of a selected point within the recorded images can be made with the aid of a densitometer.

Conclusions

1. Carefully review the radiographic visibility of each image. Describe the effect of kilovoltage on radiographic density.

2. Consider the effect on radiographic density of increasing the kilovoltage from 40 to 60 (50% increase) and the effect of increasing the mAs by the same percentage. Which of the factors do you believe would have a greater influence on radiographic density? Explain.

3. Choose a specific area within the radiographic images of the phantom and a middle step of the penetrometer, and circle them on each exposure's image. Using a densitometer, measure the radiographic density represented by these selected areas for each image. Record them below:
   Exposure 1: _____ phantom density; _____ penetrometer step density
   Exposure 2: _____ phantom density; _____ penetrometer step density
   Exposure 3: _____ phantom density; _____ penetrometer step density
   Exposure 4: _____ phantom density; _____ penetrometer step density

4. Which of the four exposures would be considered closest to a recorded image that is of radiographic quality? Give reasons for your selection.

5. Are the kilovoltage increases the only factor that influenced the radiographic density increases observed between the four radiographic images? To what extent has the intensification factor of the intensifying screens contributed to the change?

Name ____ Date ________________

EXPERIMENT 33
RADIOGRAPHIC DENSITY—Kilovoltage: Influence of Low versus High Kilovoltage

Objective

To demonstrate that the influence of kilovoltage on radiographic density is not the same throughout the diagnostic range of kilovoltage levels.

Procedure

Two sets of radiographs will be produced: one representing a series of exposures in the 50-kVp range, and the other a series of exposures in the 80-kVp range.

Procedure 1: Tabletop Procedure, 50-kVp Range

With the use of a phantom part (e.g., elbow or ankle), a series of four exposures are made using different kilovoltage. The exposures should be made tabletop using a 14" by 17" (36 cm × 43 cm) slow speed (detail) film-screen recording system cassette divided into four sections using lead dividers.

Film 1, Exposure 1

1. Cover three sections of the divided film with lead blockers, leaving one section uncovered. Place the phantom part and the penetrometer onto the uncovered film section with the phantom part centered to the section. Mark the film section with a lead number 1.
2. Using 50 kVp, select an appropriate mAs to produce a 1.0 to 1.2 bone density. (See Experiment 32, Exposure 1.)
3. Direct your central ray to the midpoint of the phantom using a perpendicular beam; collimate appropriately to cover the phantom and the penetrometer. Expose the film section.

Exposures 2, 3, and 4

For your second, third, and fourth exposures, move to a new film section for each exposure, blocking the exposed sections and marking the successive exposures with a lead number 2, 3, and 4 and increasing your kilovoltage to 52, 54, and 56, respectively. All other factors remain the same. After completing your final exposure, process your film.

Procedure 2: Bucky Procedure, 80-kVp Range

With the use of a phantom part (e.g., pelvis), a series of four radiographs are made using different kilovoltage. The exposures should be made with a Bucky using a 10" by 12" (25 cm × 30 cm) slow speed (detail) film-screen recording system cassette.

Film 1

1. Place the pelvis phantom and the penetrometer onto the tabletop, and center the hip to the midline of the table, placing the penetrometer adjacent to the side of the phantom. **Note:** It may be necessary to off-center the pelvis phantom in order to include the penetrometer on the film. Mark the film with a lead number 1.
2. Using 80 kVp, select an appropriate mAs to produce a 1.0 to 1.2 bone density.
3. Direct your central ray to the midpoint of the tabletop and centered to enter at the joint space of the hip using a perpendicular beam. Collimate appropriately to cover the phantom and the penetrometer (film size). Expose and process the film.

Films 2, 3, and 4

For your second, third, and fourth films, use the same cassette for each new radiograph, marking the successive exposures with a lead number 2, 3, and 4 and increasing your kVp to 82, 84, and 86, respectively. All other factors remain the same. Expose and process each film.

Analysis

Your first radiograph will have four recorded images representing exposures of 50, 52, 54, and 56 kVp. The second series of exposures will have four separate radiographs of the same image representing 80, 82, 84, and 86 kVp. Review the radiographic densities of each procedure, measure the density increases with a densitometer, and compare the kVp/density relationship differences between the application of low and high kilovoltage levels.

Conclusions

1. Review the radiographic appearance of the recorded images of the tabletop procedure. Identify the kilovoltage level that demonstrates an *obvious* radiographic density change from the first exposure taken at 50 kVp:

   _____ 52 kVp _____ 54 kVp _____ 56 kVp

   Review the radiographic appearance of the Bucky procedure of the hip. Identify the kilovoltage level that demonstrates an *obvious* radiographic density change from that produced at 80 kVp:

   _____ 82 kVp _____ 84 kVp _____ 86 kVp

2. Circle a specific point within the radiographic images of each of the series of exposures. Circle the middle step of the penetrometer. Measure the radiographic density with a densitometer, and record the value below:

   Tabletop Procedure, Low 50-kVp Range

   Exposure 1: _____ image density; _____ penetrometer step density
   Exposure 2: _____ image density; _____ penetrometer step density
   Exposure 3: _____ image density; _____ penetrometer step density
   Exposure 4: _____ image density; _____ penetrometer step density

   Bucky Procedure, High 80-kVp Range

   Film 1: _____ image density; _____ penetrometer step density
   Film 2: _____ image density; _____ penetrometer step density
   Film 3: _____ image density; _____ penetrometer step density
   Film 4: _____ image density; _____ penetrometer step density

3. Comparing the radiographic density changes produced at the 80-kVp range to those produced at the 50-kVp range (each representing a 2-kVp increment increase), what is your opinion of adjusting the kilovoltage to produce and maintain a desired radiographic density? Discuss the accuracy and practical application of the relationship of kilovoltage adjustments to radiographic density.

4. Review the images of the penetrometer for each group of exposures. Identify the total number of radiographic densities visible within the radiographic image of the penetrometer:

   Low 50-kVp Range

   Exposure 1: _____ visible penetrometer steps
   Exposure 2: _____ visible penetrometer steps
   Exposure 3: _____ visible penetrometer steps
   Exposure 4: _____ visible penetrometer steps

   High 80-kVp Range

   Film 1: _____ visible penetrometer steps
   Film 2: _____ visible penetrometer steps
   Film 3: _____ visible penetrometer steps
   Film 4: _____ visible penetrometer steps

5. Describe what happens to the overall visibility of the recorded image as you increase your kilovoltage from 50 kVp in the first radiograph and as you increase your kilovoltage from 80 kVp in the second series of radiographs.

6. Do you believe the radiographic density increases would be similar for 2-kVp changes at 40 kVp rather than 50 kVp? At 100 kVp rather than 80 kVp? Explain your answers.

7. From this experiment, do you believe that the use of kilovoltage manipulations for the maintenance and control of radiographic density is an accurate and acceptable procedure? Explain.

Name ______________________________ Date ______________

EXPERIMENT 34
RADIOGRAPHIC DENSITY—Kilovoltage: Intensifying Screen versus Nonscreen

Objective

To demonstrate that the influence of kilovoltage on radiographic density differs according to the materials selected to record the image. The same kilovoltage increase will have a different effect on radiographic density with intensifying screen radiography than with nonscreen radiography.

Procedure

Two sets of radiographs will be produced: one representing a series of exposures taken as a nonscreen procedure, and the other a series of exposures taken with a slow speed (detail) film-screen recording system cassette. The same phantom can be used with both films, and both are performed tabletop. The bone density of the recorded image for the initial exposure of both films should have a 1.0 to 1.2 density. The suggested mAs in the experiments may have to be adjusted in order to achieve this.

Film 1: Intensifying Screen Procedure

Note: If Experiment 33 has been performed, you may use the radiograph from *Procedure 1: Tabletop Procedure, 50-kVp Range* from Experiment 33 for *Film 1: Intensifying Screen Procedure* of this experiment and begin your experiment with *Film 2: Nonscreen Procedure.* If not, follow the instructions provided below.

With the use of a phantom part (e.g., elbow or ankle), a series of four exposures are made using different kilovoltage. The exposures should be made tabletop using a 14" by 17" (36 cm × 43 cm) slow speed (detail) film-screen recording system cassette divided into four sections using lead dividers.

Exposure 1

1. Cover three sections of the divided film with lead blockers, leaving one section uncovered. Place the phantom part and the penetrometer onto the uncovered film section with the phantom part centered to the section. Mark the film section with a lead number 1.
2. Using 50 kVp, select an appropriate mAs to produce a 1.0 to 1.2 bone density. (See Experiment 32, Exposure 1.)

3. Direct your central ray to the midpoint of the phantom using a perpendicular beam; collimate appropriately to cover the phantom and the penetrometer. Expose the film section.

Exposures 2, 3, and 4

For your second, third, and fourth exposures, move to a new film section for each exposure, blocking the exposed sections and marking the successive exposures with a lead number 2, 3, and 4 and increasing your kilovoltage to 52, 54, and 56, respectively. All other factors remain the same. After completing your final exposure, process your film.

Film 2: Nonscreen Procedure

1. Using a 14" by 17" (36 cm × 43 cm) direct exposure (cardboard) film holder, divide it into four sections using lead blockers. Cover three sections, and center the same phantom and penetrometer used for the intensifying screen procedure to the uncovered film section. Mark the film section with a lead number 1.
2. Using a 40" (102 cm) SID, direct your central ray to the midpoint of the phantom using a perpendicular angle. Collimate appropriately to include the phantom and penetrometer.
3. Using 50 kVp, select the appropriate mAs to produce a bone density of 1.0 to 1.2. Identify the image intensification factor of screens at 50 kVp as a starting point to determine your exposure. Several exposures may have to be taken to achieve the proper starting mAs for the nonscreen procedure. Expose the first film section.
4. For your second, third, and fourth exposures, adjust your lead blockers, move to a new unexposed film section for each exposure, and mark the section with a lead number 2, 3, and 4. Increase your kilovoltage to 52, 54, and 56, respectively, for each additional exposure. When your final film section is exposed, process your film.

Analysis

Your two radiographs will each have four recorded images. The recorded images employing the 50 kVp on each film should possess a similar *overall* radiographic density. (The radiographic contrast between the two radiographs will differ considerably, but the overall radiographic density for the first exposure on each film should possess a similar level.) You can now assess the kVp/density relationship of different recording materials.

Conclusions

1. Review the appearance of the recorded images produced with the intensifying screens related to the radiographic density differences produced as you increased your kilovoltage from 50 to 56. Circle a specific point within the recorded images of the phantom and a middle step of the penetrometer, and measure the density for each of the four exposures using a densitometer. Record your findings in the spaces provided below:

   Intensifying Screen Procedure

   Exposure 1: _____ phantom density; _____ penetrometer step density
   Exposure 2: _____ phantom density; _____ penetrometer step density
   Exposure 3: _____ phantom density; _____ penetrometer step density
   Exposure 4: _____ phantom density; _____ penetrometer step density

2. Review the appearance of the recorded images produced as a nonscreen procedure related to the radiographic density differences produced as you increased your kilovoltage from 50 to 56. Circle a specific point within the recorded images of the phantom and a middle step of the penetrometer, and measure the density for each of the four exposures using a densitometer. Record your findings in the spaces provided below:

   Nonscreen Procedure

   Exposure 1: _____ phantom density; _____ penetrometer step density
   Exposure 2: _____ phantom density; _____ penetrometer step density
   Exposure 3: _____ phantom density; _____ penetrometer step density
   Exposure 4: _____ phantom density; _____ penetrometer step density

3. Compare the radiographic density increase between the recorded images produced on the two radiographic films. Is the radiographic density increase produced by the increase in the kilovoltage in the nonscreen procedure the same as the increase produced in the intensifying screen procedure? Describe any difference noted, and discuss how this difference was produced.

4. Does this experiment support or oppose the statement that kilovoltage changes are not an accurate method of assessing or controlling radiographic density? Explain your answer.

5. Do you believe that you would have greater exposure latitude in the selection of exposure factors (margin for error) when you are performing a procedure with intensifying screens or with nonscreen holders? Explain the reasons for your choice.

6. How does the intensification factor of intensifying screens influence the radiographic density of the recorded image related to the kilovoltage selected?

7. Which series of radiographic images possesses overall greater image visibility throughout the series of exposures? Why?

Name ____ Date ____________________

EXPERIMENT 35
RADIOGRAPHIC DENSITY—Kilovoltage: Necessary Kilovoltage To Penetrate

Objective

To demonstrate that the kilovoltage necessary to penetrate must be employed in order to produce a satisfactory radiographic density.

Procedure

With the use of a 10" by 12" (25 cm × 30 cm) slow speed (detail) film-screen recording system cassette collimated to the film size, a series of exposures of the pelvis phantom and an aluminum penetrometer (step wedge) is taken using a Bucky mechanism. If possible, use the same cassette for each exposure, and employ the same size focus and milliamperage setting, adjusting your time of exposure in order to adjust your mAs. **Note:** If Experiment 33 has been performed, you may use the radiograph from *Procedure 2: Bucky Procedure, 80-kVp Range* from Experiment 33 for *Film 1: 80 kVp, Sufficient To Penetrate* of this experiment and begin your experiment with *Film 2: 40 kVp, Insufficient To Penetrate, Same mAs.* If not, follow the instructions provided below.

Film 1: 80 kVp, Sufficient To Penetrate

1. Place the pelvis phantom and the penetrometer onto the tabletop, and center the hip to the midline of the table, placing the penetrometer adjacent to the side of the phantom. **Note:** It may be necessary to off-center the pelvis phantom in order to include the penetrometer on the film. Mark the film with a lead number 1.
2. Using 80 kVp, select an appropriate mAs to produce a 1.0 to 1.2 bone density.
3. Direct your central ray to the midpoint of the tabletop, and ensure it is centered to enter at the joint space of the hip using a perpendicular beam. Collimate appropriately to cover the phantom and the penetrometer (film size). Expose and process the film.

Film 2: 40 kVp, Insufficient To Penetrate, Same mAs

Set up your procedure as you did for Film 1, but reduce the kilovoltage to 40. Using the same mAs, expose and process your film. Mark the film with a lead number 2.

Film 3: 40 kVp, Insufficient To Penetrate, Increased mAs

Set up your procedure as you did for Film 1, but reduce the kilovoltage to 40. Increase your mAs five times (5 ×) from the mAs used for Film 1; expose and process your film. Mark the film with a lead number 3.

Film 4: 40 kVp, Insufficient To Penetrate, Increased mAs

Set up your procedure as you did for Film 1, but reduce the kilovoltage to 40. Increase your mAs ten times (10 ×) from the mAs used for Film 1; expose and process your film. Mark the film with a lead number 4.

Analysis

Review the four radiographic films side by side, comparing the radiographic density of the phantom and the penetrometer. This experiment will enable you to determine whether any mAs increase will compensate for an insufficient kilovoltage to penetrate the part.

Conclusions

1. Circle a specific point within the radiographic images of each of the radiographs. Circle the middle step of the penetrometer. Measure the radiographic density with a densitometer, and record the value below:

   Exposure 1: _____ image density; _____ penetrometer step density
   Exposure 2: _____ image density; _____ penetrometer step density
   Exposure 3: _____ image density; _____ penetrometer step density
   Exposure 4: _____ image density; _____ penetrometer step density

2. Identify the exposure factors employed for each of the radiographs, and record them in the spaces provided below:

   Film 1: _____ kVp, _____ mA, _____ time of exposure, _____ mAs
   Film 2: _____ kVp, _____ mA, _____ time of exposure, _____ mAs
   Film 3: _____ kVp, _____ mA, _____ time of exposure, _____ mAs
   Film 4: _____ kVp, _____ mA, _____ time of exposure, _____ mAs

3. Using 80 kVp, has a satisfactory radiographic density been achieved? Explain your answer.

4. In Film 2, using 40 kVp and the same mAs as Film 1, has a satisfactory density been achieved? Explain your answer.

5. In Film 3, using 40 kVp and five times (5 ×) the mAs, has a satisfactory density been achieved? Explain your answer.

6. In Film 4, using 40 kVp and ten times (10 ×) the mAs, has a satisfactory density been achieved? Explain your answer.

7. Do you believe that a satisfactory density level of the film would be achieved if you increased your mAs to perhaps 20 times (20 ×) or 30 times (30 ×)? Explain the reason for your answer.

8. Describe the principle of radiographic quality that this experiment is designed to prove.

Name ________________________________ Date ______________

EXPERIMENT 36
RADIOGRAPHIC DENSITY—kVp/mAs Relationship

Objective

To demonstrate the relationship between the kilovoltage and mAs and to demonstrate that the radiographic density can be maintained with a manipulation of the kVp/mAs relationship as long as the original exposure employed a kilovoltage sufficient to penetrate the part. The kVp/mAs 15% rule is used.

Procedure

Using a 10" by 12" (25 cm × 30 cm) slow speed (detail) film-screen recording system cassette collimated to the size of the film, a series of four radiographs of the pelvis phantom and an aluminum penetrometer (step wedge) is taken using a Bucky mechanism. If possible use the same size focus, milliamperage setting, and cassette for each radiograph.

Film 1: 70 kVp, Sufficient To Penetrate

1. Center the hip joint of the pelvis phantom to the midline of the tabletop, and direct your central ray to this point using a perpendicular angle at a 40" (102 cm) SID. **Note:** You may have to off-center the phantom to allow your penetrometer lying adjacent to the lateral edge of the phantom to be included on the film. Mark your film with a lead number 1.
2. Using 70 kVp and a 100-mA setting, select the appropriate time of exposure for the mAs required to produce a bone density of 1.0 to 1.2.
3. Expose and process your film.

Film 2: Increase the Kilovoltage by 15%, Reduce the mAs by 50%

1. Set up your procedure the same as for Film 1. Increase your kilovoltage to 80 (15%), and reduce your mAs by 50% from your original mAs. Mark your film with a lead number 2.
2. Expose and process your film.

Film 3: Increase the Kilovoltage by 15%, Reduce the mAs by 50%

1. Set up your procedure the same as for Film 1. Adjust your exposure factors from those used for Film 2. Increase your kilovoltage to 92 (15%), and reduce

your mAs by 50% from the mAs used for Film 2. Mark your film with a lead number 3.

2. Expose and process your film.

Film 4: Increase the Kilovoltage by 15%, Reduce the mAs by 50%

1. Set up your procedure the same as for Film 1. Adjust your exposure factors from those used for Film 3. Increase your kilovoltage to 106 (15%), and reduce your mAs by 50% from the mAs used for Film 3. Mark your film with a lead number 4.
2. Expose and process your film.

Analysis

Review the radiographic images and penetrometer of the four radiographs side by side, comparing the radiographic density of each. The overall radiographic density of the four radiographs should be similar, although the radiographic contrast produced between the four recorded images may have changed.

Conclusions

1. Select a point within the image of each film, circle it, and select a middle step of the image of the penetrometer. Using a densitometer, measure the radiographic density of each, and record it in the space provided below:

   Film 1: _____ image density; _____ penetrometer step density
   Film 2: _____ image density; _____ penetrometer step density
   Film 3: _____ image density; _____ penetrometer step density
   Film 4: _____ image density; _____ penetrometer step density

2. Identify the exposure factors used for each of the four radiographs in the spaces provided below:

   Exposure 1: _____ kVp, _____ mA, _____ time of exposure, _____ mAs
   Exposure 2: _____ kVp, _____ mA, _____ time of exposure, _____ mAs
   Exposure 3: _____ kVp, _____ mA, _____ time of exposure, _____ mAs
   Exposure 4: _____ kVp, _____ mA, _____ time of exposure, _____ mAs

3. Are the radiographic images of the hip in each film comparable in their overall radiographic density? Explain.

4. What, if any, difference in the visibility of the image is observable between the four recorded images of the hip? Describe in detail.

5. Compare the radiographic contrast of Film 1, taken with 70 kVp, with that of Film 4, taken with 106 kVp. Describe the differences, if any, in the photographic properties of the two images.

6. Identify the *number* of penetrometer steps visible for each of the four radiographs. Describe what has happened to change the relationship between the different radiographic densities recorded.

7. Is the kVp/mAs 15% rule applicable for radiographic density control? Describe its practical application related to the control of motion. Discuss the limitations of its practical application in routine radiographic procedures.

8. Can the kVp/mAs 15% rule be utilized to manipulate the radiographic image in order to produce a desired scale of radiographic contrast? If so, describe an example of its application for this purpose.

Name ______________________________ Date ______________

EXPERIMENT 37
RADIOGRAPHIC DENSITY—Intensifying Screens: Intensification Factor

Objective

To demonstrate the light emission properties of intensifying screens related to different levels of kilovoltage.

Procedure

Take a 14" by 17" (36 cm × 43 cm) slow speed (detail) film-screen recording system cassette into the darkroom, and remove the radiographic film. Close the cassette, and bring it into an x-ray examination room. Place the cassette on the tabletop, and open it up so that the intensifying screens are facing the x-ray tube. Center the central ray to the middle of the front screen of the cassette, and collimate to an area that measures 10" (25 cm) square using a 40" (102 cm) SID.

Go behind your control panel, and observe the screen through the glass panel. Set up your machine as follows, and turn out all the lights in the room. Give yourself a few minutes (10 min minimum) to accommodate to the darkness.

Exposure 1

Using your lowest possible milliamperage selection, select a 1-sec time of exposure at 40 kVp. Expose the screen, and observe the light emission from the intensifying screen during the exposure. Record your observations.

Exposure 2

Keeping everything the same as in Exposure 1, change your kilovoltage to 50. Expose, observe, and record your observations.

Exposure 3

Keeping everything the same as in Exposure l, change your kilovoltage to 60. Expose, observe, and record your observations.

Exposure 4

Keeping everything the same as in Exposure 1, change your kilovoltage to 70. Expose, observe, and record your observations.

Analysis

You should have been able to observe different light emissions from the intensifying screen as you changed the kilovoltage. Record your observations of the intensification factor of intensifying screens in the conclusions section below.

Conclusions

1. Describe the phenomenon that occurred when the intensifying screen absorbed radiation during an exposure.

2. How would you describe the color of the light emission?

3. Describe the differences in intensifying screen light emission observed between Exposures 1, 2, 3, and 4. Were the differences observed the same between 40 and 50 kVp as between 50 and 60 kVp? Between 60 and 70 kVp?

4 Discuss the influence of kilovoltage on the light emission properties of the intensifying screen and its application to radiographic procedures.

Additional Analysis

If materials are available, repeat this experiment using other intensifying screen phosphor materials: calcium tungstate versus rare earth intensifying screens and/or different rare earth phosphors. Compare your visual observations of these other phosphor materials with those observed from your original experiment. Provide answers to the conclusion questions for these additional materials.

Name ______________________________ Date ______________

EXPERIMENT 38
RADIOGRAPHIC DENSITY—Intensifying Screens versus Direct Exposure

Objective

To demonstrate that the radiographic density of an image can be maintained when going from direct exposure radiography to intensifying screen radiography by using the intensification factor appropriate for the selected kilovoltage range.

Procedure

Two sets of radiographs are produced, one using a slow speed (detail) film-screen recording system cassette, the other performed as a direct exposure (cardboard) film holder procedure. Both procedures are performed tabletop at a 40" (102 cm) SID with screen-type film using a perpendicular angle collimated to the area of interest or less. An elbow or ankle phantom and an aluminum penetrometer (step wedge) are used for all exposures. Using 50 kVp, the same focus size using a 100-mA station should be employed for all exposures, adjusting the time of exposure to achieve the necessary mAs.

Film 1: Intensifying Screen Procedure

1. Using a 10" by 12" (25 cm × 30 cm) cassette, place your phantom and penetrometer on the cassette, and center your central ray to the middle of the structures to be recorded. Using 50 kVp and a 100-mA station, select the time of exposure necessary to produce the mAs needed to produce a 1.0 to 1.2 bone density.
2. Expose and process your film.

Film 2: Direct Exposure Procedure

Exposure 1: Same Exposure Factors

1. If available, select a 14" by 17" (36 cm × 43 cm) cardboard holder, and divide it in four sections using lead blockers. Cover three sections, leaving one section uncovered, and mark this section with a lead number 1. If necessary, use smaller film sizes, and produce separate radiographs.
2. Set up your procedure the same as for the intensifying screen procedure, and using the same exposure factors used for Film 1, expose your film section.

Exposure 2: Increase Your mAs by a Factor of 50

1. Cover the exposed film section, uncover a new section, and mark this section with a lead number 2.
2. Set up your procedure the same as for the intensifying screen procedure, but increase your mAs by a factor of 50 from the exposure used for Film 1; expose your film section.

Exposure 3: Increase Your mAs by a Factor of 60

1. Cover the exposed film section, uncover a new section, and mark this section with a lead number 3.
2. Set up your procedure the same as for the intensifying screen procedure, but increase your mAs by a factor of 60 from the exposure used for Film 1; expose your film section.

Exposure 4: Increase Your mAs by a Factor of 80

1. Cover the exposed film section, uncover the final unexposed section, and mark this section with a lead number 4.
2. Set up your procedure the same as for the intensifying screen procedure, but increase your mAs by a factor of 80 from the exposure used for Film 1; expose your film section, and process your film.

Analysis

You can review the radiographic densities produced for an intensifying screen procedure and a direct exposure procedure that adjusted the mAs from the original exposure using different screen intensification factors. You can determine the intensification factor for 50 kVp. You can also analyze the effectiveness of changing from intensifying screen radiography to direct exposure radiography using the intensification factor to maintain radiographic density. **Note:** The intensification factor can also be used to change from direct exposure radiography to intensifying screen radiography.

Conclusions

1. Select Film 1, and, using a densitometer, measure the radiographic density of the penetrometer's image to find the penetrometer step that comes nearest to a 1.0 to 1.2 density. Record that level and its reading in the spaces provided below:

   Film 1: ______ penetrometer density reading, step # _____

2. Select the *same* penetrometer step on Film 2 for each of the four exposures, and, using a densitometer, measure the radiographic density of the penetrometer's image. Record that level and its reading in the spaces provided below:

   Film 2, Exposure 1: _______ penetrometer density reading
   Film 2, Exposure 2: _______ penetrometer density reading
   Film 2, Exposure 3: _______ penetrometer density reading
   Film 2, Exposure 4: _______ penetrometer density reading

3. Record the exposure factors employed for all exposures in the spaces provided below:

   Film 1: _____ kVp, _____ mA, _____ time of exposure, _____ mAs
   Film 2:
   Exposure 1: _____ kVp, _____ mA, _____ time of exposure, _____ mAs
   Exposure 2: _____ kVp, _____ mA, _____ time of exposure, _____ mAs
   Exposure 3: _____ kVp, _____ mA, _____ time of exposure, _____ mAs
   Exposure 4: _____ kVp, _____ mA, _____ time of exposure, _____ mAs

4. Compare the radiographic density readings for the four exposures of Film 2 with that of Film 1. Which exposure of Film 2 approximates the radiographic density of Film 1? What is the intensification factor for 50 kVp? (If the density readings are similar, proceed to Question 6.)

5. If none of the radiographic density readings for the four exposures of Film 2 approximate Film 1, *repeat the experiment for Film 2* adjusting the mAs until the same penetrometer step matches that of Film 1. Determine the correct intensification factor for 50 kVp by comparing the new mAs for the repeat radiograph with the original mAs selected for Film 1. Record the intensification factor and new mAs in the spaces provided below:

   Repeat Experiment:
   _____ kVp, _____ mA, _____ time of exposure, _____ mAs
   _____ intensification factor for 50 kVp

6. Compare the overall appearance of the recorded images of the knee in Film 1 and the exposure in Film 2 or the repeat exposure in Question 5 that approximates the same radiographic density and represents the proper intensification factor for 50 kVp. Are the overall radiographic densities of the recorded images of the knee similar? Explain as needed. Compare the visibility of the

recorded details as it relates to the radiographic contrast between the two comparable recorded images, and describe your findings.

7. Indicate the significance of the intensification factor to the maintenance of radiographic density when changing from direct exposure to intensifying screen radiography or vice versa.

Name ______________________________ Date ______________

EXPERIMENT 39
RADIOGRAPHIC DENSITY—Intensifying Screens and Screen Speed

Objective

To demonstrate the effect of different speed intensifying screens on the light emission of the screens when exposed to x-radiation.

Procedure

If available, in the darkroom, remove the radiographic film from four cassettes with different intensifying screen speeds. If possible, choose cassettes of the same size (e.g., 10" by 12" [25 cm × 30 cm]). Close the cassettes, and bring them into an x-ray examination room. If four different screen speeds are unavailable, select cassettes representing those screen speeds that are available.

Open each cassette, and place the cassettes on the table screen-side up. Arrange the cassettes so that one corner of each is aligned adjacent to and touching the others, so you will have four different speed screens aligned with one another meeting together at their corners. Identify the cassettes by the numbers 1, 2, 3, and 4 beforehand on the basis of their speed values, going from lowest to highest.

Center your central ray over the junction point where the four cassettes meet, and using a perpendicular angle, collimate your beam to cover an area 14" (36 cm) square using a 40" (102 cm) SID.

Go behind your control panel, and observe the screens through the glass panel. Set up your machine as follows, turn out all the lights in the room, and allow yourself a minimum of 10 min to accommodate to the darkness.

Using 60 kVp, select the 100-mA station, and energize your tube for 1 sec. Observe the light emission pattern of the different speed intensifying screens during the exposure.

Analysis

Record your observations of the light emission of the different speeds of intensifying screens in the spaces provided.

Conclusions

1. Describe the phenomenon observed when the exposure was made.

2. Was the color of the light emission for each screen approximately the same? Explain as needed.

3. Record the speed value of the intensifying screen for each of the four cassettes in the space provided below:

   Cassette 1: _______ intensifying screen speed value
   Cassette 2: _______ intensifying screen speed value
   Cassette 3: _______ intensifying screen speed value
   Cassette 4: _______ intensifying screen speed value

4. Describe the differences in light intensity emitted related to the speed of the intensifying screens.

5. Discuss the influence of intensifying screen speed on the light emission properties of the screens and its application to radiographic procedures.

6. Identify the intrinsic factors that influence the speed of an intensifying screen.

Name ______________________________ Date ______________

EXPERIMENT 40
RADIOGRAPHIC DENSITY—Different Speed Intensifying Screens

Objective

To demonstrate the influence of different intensifying screen speeds on the radiographic density of the image. To demonstrate that radiographic density can be controlled and maintained with exposure factor adjustments to compensate for the different speeds of the screens.

Procedure

With four different 10" by 12" (25 cm × 30 cm) cassettes representing four different intensifying screen speeds, a total of seven radiographs will be taken. All radiographs will be taken tabletop using an aluminum penetrometer (step wedge) and a phantom elbow or ankle in the anteroposterior projection taped securely and centered to the middle of the cassette. The central ray is centered to the phantom/penetrometer arrangement and directed in a perpendicular angle. The beam is collimated to include an area 1/2" beyond the phantom/penetrometer arrangement. Use the same size focus and milliamperage station for all films. Adjust your technical exposure factors by changing the time of exposure. Mark your recording system speed in the space provided below. **Note:** If fewer than four different recording system speeds are available, use what you have at your disposal, and adjust the procedure accordingly.

Film 1: 100 Speed Film-Screen Recording System

1. Using 50 kVp, select the appropriate mAs required to produce a 1.0 to 1.2 bone density within the phantom part.
2. Select the slowest speed system available, and arrange your cassette as indicated above.
3. Expose and process the film, marking it with a lead number 1.

Film 2: _______ Speed Film-Screen Recording System

1. Using the same technical factors and setup as indicated above for Film 1, select the next higher speed recording system available, and mark it with a lead number 2.
2. Expose and process your film.

Film 3: _______ Speed Film-Screen Recording System

1. Using the same technical factors and setup as indicated above for Film 1, select the next higher speed recording system available, and mark it with a lead number 3.
2. Expose and process your film.

Film 4: _______ Speed Film-Screen Recording System

1. Using the same technical factors and setup as indicated above for Film 1, select the highest speed recording system available, and mark it with a lead number 4.
2. Expose and process your film.

Using the 100 speed film-screen recording system film as your standard, adjust your technical factors according to the intensifying screen speeds and exposure values in Chapter 4, Table 4-3.

Repeat Film 2: _______ Speed Film-Screen Recording System

1. Adjust your mAs by decreasing your time of exposure according to the recording system speed in order to produce the same radiographic density as Film 1.
2. Set up your procedure as for Film 1, and mark it with the lead number 2R.
3. Expose and process your film.

Repeat Film 3: _______ Speed Film-Screen Recording System

1. Adjust your mAs by decreasing your time of exposure according to the recording system speed in order to produce the same radiographic density as Film 1.
2. Set up your procedure as for Film 1, and mark it with the lead number 3R.
3. Expose and process your film.

Repeat Film 4: _______ Speed Film-Screen Recording System

1. Adjust your mAs by decreasing your time of exposure according to the recording system speed in order to produce the same radiographic density as Film 1.
2. Set up your procedure as for Film 1, and mark it with the lead number 4R.
3. Expose and process your film.

Analysis

Your series of radiographs will demonstrate the effect of intensifying screen speeds on the radiographic density of the image when no compensation in exposure is made. It will also demonstrate that radiographic density can be controlled and maintained when employing different intensifying screen speeds if the appropriate adjustments in the exposure factors are made.

Conclusions

1. Select the same specific area within the radiographic image of the phantom and a middle step of the penetrometer, and circle it on each of the seven radiographs. Measure the radiographic density of this area for each radiograph, and record your findings in the spaces provided below:

   Film 1: _______ phantom density; _______ penetrometer density
   Film 2: _______ phantom density; _______ penetrometer density
   Film 3: _______ phantom density; _______ penetrometer density
   Film 4: _______ phantom density; _______ penetrometer density
   Film 2R: _______ phantom density; _______ penetrometer density
   Film 3R: _______ phantom density; _______ penetrometer density
   Film 4R: _______ phantom density; _______ penetrometer density

2. Compare the radiographic visibility of the recorded image produced by Film 1 (100 speed system) with the radiographic images produced by Films 2, 3, and 4. Using Film 1 as your standard, discuss each recorded image in terms of radiographic quality.

3. Compare the radiographic visibility of the recorded image produced by Film 1 (100 speed system) with the radiographic images produced by Films 2R, 3R, and 4R. Using Film 1 as your standard, discuss each recorded image in terms of radiographic quality.

4. Describe the influence of screen speed on the radiographic density of the radiographic image.

5. Discuss the importance of knowing the screen speed values when performing routine radiographic procedures.

6. Describe a practical application of the principles related to intensifying screen speeds and their influence on radiographic density.

7. Are the overall radiographic densities of Films 1, 2R, 3R, and 4R approximately the same? Describe any major differences, and indicate reasons for differences (i.e., other influences that may have affected the results).

8. How is the availability of different film-screen recording system speeds an example of the application of the principles of ALARA?

Name ______________________________ Date ____________

EXPERIMENT 41
RADIOGRAPHIC DENSITY—Angle of the Central Ray

Objective

To demonstrate the effect of different central ray angulation on the radiographic density of the recorded image and the need for exposure factor adjustments when using increased central ray angulation. This experiment will suggest adjustments in the vertical tube height to maintain the SID rather than adjustments in the mAs to maintain the radiographic density of the recorded image. Review Table 4-5.

Procedure

Using a 10" by 12" (25 cm × 30 cm) slow speed (detail) film-screen recording system cassette, placed in the Bucky tray, a series of seven radiographs will be produced. All the radiographs are taken of a lateral phantom skull using a 40" (102 cm) SID. If possible, the same size focus and a 100-mA setting should be employed for all exposures. Select 70 kVp and the time of exposure to produce a 1.0 to 1.2 bone density of the skull.

Film 1: Perpendicular Central Ray

1. Center your skull phantom to the midline of the table in the lateral position, and tape it in place. Center your central ray to the area of the sella turcica using a perpendicular angle. Center your cassette in the Bucky tray to your central ray. Mark your film with a lead number 1.
2. Expose and process your film.

Film 2: 15° Caudal Angle

1. Set up your procedure as for Film 1; however, adjust your central ray to employ a 15° caudal angle, and move the tube stand so that it is centered to emerge through the sella turcica. Do not adjust the height of your tube stand. Measure the *actual* SID employed with a measuring tape, and write in your measurement in the space provided. __________ in/cm. Mark your film with a lead number 2.
2. Move your Bucky tray in a caudal direction to accommodate for the tube angle so that the central ray passes through the center of the film.
3. Using the same technical exposure factors, expose and process your film.

Film 3: 30° Caudal Angle

1. Set up your procedure as for Film 1; however, adjust your central ray to employ a 30° caudal angle, and move the tube stand so that it is centered to emerge through the sella turcica. Do not adjust the height of your tube stand. Measure the *actual* SID employed with a measuring tape and write in your measurement in the space provided. __________ in/cm. Mark your film with a lead number 3.
2. Move your Bucky tray in a caudal direction to accommodate for the tube angle so that the central ray passes through the center of the film.
3. Using the same technical exposure factors, expose and process your film.

Film 4: 45° Caudal Angle

1. Set up your procedure as for Film 1; however, adjust your central ray to employ a 45° caudal angle, and move the tube stand so that it is centered to emerge through the sella turcica. Do not adjust the height of your tube stand. Measure the *actual* SID employed with a measuring tape, and write in your measurement in the space provided. __________ in/cm. Mark your film with a lead number 4.
2. Move your Bucky tray in a caudal direction to accommodate for the tube angle so that the central ray passes through the center of the film.
3. Using the same technical exposure factors, expose and process your film.

Films 2, 3, and 4 will be repeated, adjusting the vertical tube height of the tube stand in order to maintain the 40" (102 cm) SID according to the guidelines identified in Chapter 4, Table 4-5.

Repeat Film 2: 15° Caudal Angle Vertical Tube Height Reduction

1. Set up your procedure as for Film 1; however, adjust your central ray to employ a 15° caudal angle, and move the tube stand so that it is centered to emerge through the sella turcica. Adjust (reduce) the vertical height of your tube stand in order to maintain a 40" (102 cm) SID. Measure the *actual* vertical height adjustment with a measuring tape, and write in your distance reduction in the space provided. __________ in/cm. Mark your film with the lead number 2R.
2. Adjust the position of your Bucky tray so that the central ray passes through the center of the film.
3. Using the same technical exposure factors, expose and process your film.

Repeat Film 3: 30° Caudal Angle Vertical Tube Height Reduction

1. Set up your procedure as for Film 1; however, adjust your central ray to employ a 30° caudal angle, and move the tube stand so that it is centered to emerge through the sella turcica. Adjust (reduce) the vertical height of your tube stand in order to maintain a 40" (102 cm) SID. Measure the *actual* vertical

height adjustment with a measuring tape, and write in your distance reduction in the space provided. _________ in/cm. Mark your film with the lead number 3R.

2. Adjust the position of your Bucky tray so that the central ray passes through the center of the film.
3. Using the same technical exposure factors, expose and process your film.

Repeat Film 4: 45° Caudal Angle Vertical Tube Height Reduction

1. Set up your procedure as for Film 1; however, adjust your central ray to employ a 45° caudal angle, and move the tube stand so that it is centered to emerge through the sella turcica. Adjust (reduce) the vertical height of your tube stand in order to maintain a 40" (102 cm) SID. Measure the *actual* vertical height adjustment with a measuring tape, and write in your distance reduction in the space provided. _________ in/cm. Mark your film with the lead number 4R.
2. Adjust the position of your Bucky tray so that the central ray passes through the center of the film.
3. Using the same technical exposure factors, expose and process your film.

Analysis

Your series of radiographs will demonstrate the effect of central ray angulation on the radiographic density of the recorded image when no adjustment in the vertical tube height is made to maintain the SID and how the radiographic density can be maintained by an adjustment of the vertical tube height.

Conclusions

1. Select the same specific area within the radiographic image of the skull, circle it on each of the seven radiographs. Measure the radiographic density of this area for each radiograph and record your findings in the spaces provided below:

   Film 1: _______ phantom density
   Film 2: _______ phantom density Film 2: _______ phantom density
   Film 3: _______ phantom density Film 3: _______ phantom density
   Film 4: _______ phantom density Film 4: _______ phantom density

2. Identify the actual SID employed for Films 2, 3, and 4, and indicate the vertical height reduction required for Repeat Films 2, 3, and 4 in order to maintain the desired 40" (102 cm) SID in the spaces provided below:

   Film 1: __40"__ SID
   Film 2: _______ actual SID Repeat Film 2: _______ vertical height reduction

Film 3: _______ actual SID Repeat Film 3: _______ vertical height reduction
Film 4: _______ actual SID Repeat Film 4: _______ vertical height reduction

3. Compare the overall visibility of the image produced by Film 1 with the radiographic images produced by Films 2, 3, and 4. Discuss the differences, if any.

4. What happened to the radiographic density when the angle of the central ray increased?

5. Compare the overall visibility of the image produced by Film 1 with the radiographic images produced by Repeat Films 2R, 3R, and 4R. Discuss the differences, if any.

6. Does the central ray angle have an influence on the radiographic density of the recorded image? Is it possible to maintain the radiographic density by maintaining the SID by adjusting the vertical tube height?

7. Is the radiographic density of the image influenced the same regardless of the angle of the central ray employed? Explain your answer.

Note: Radiographic density can also be maintained by the adjustment of the exposure factors rather than adjustment of the vertical tube height. A second experiment can be produced to demonstrate this influence. The exposure factor influence is also identified in Chapter 4, Table 4-5.

Name ______________________________ Date ______________

EXPERIMENT 42

RADIOGRAPHIC DENSITY—Radiographic Fog: Darkroom Light Leaks

Objective

To investigate the presence of light leaks into the darkroom and to repair and/or report your findings.

Procedure

Enter the darkroom, and close the entrance. Turn off any white lights, and adapt to the darkened environment of the safelights for a period not less than 15 min. Turn off the safelights so that the darkroom is in total darkness, and check for any light leaks into the darkroom.

1. Check all around the ceiling where it meets with the walls.
2. Check all around the edges where the passboxes are constructed into the walls. Have someone open the passbox doors in the x-ray examining rooms with the overhead white lights on in the examining rooms, and check for light leaks around the passbox doors.
3. Check all around the electrical and plumbing fixtures.
4. Check all around the film processing unit(s) where they meet with the walls or the construction areas where they enter into the darkroom.
5. Finally, be especially vigilant in checking all around the darkroom entrance including the area where it meets with the floor.

Analysis

After having adapted to the darkened environment of the darkroom, you can determine whether there are any small or large extraneous light leaks into the darkroom from the outside. Carefully identify any areas of light leaks by placing small Post-it paper or similar markers in every area where a light leak has been detected.

Conclusions

1. Discuss the significance of light leaks into the darkroom to the overall quality of the radiographic images produced in the radiology department.

2. List the locations of any light leaks found during your inspection. Report your findings to your instructor, supervisor, or quality control technologist.

3. Should routine checks for darkroom light leaks be included in an overall program of quality assurance? Explain the reason(s) for your answer.

Name ______________________________ Date ______________

EXPERIMENT 43

RADIOGRAPHIC DENSITY—Radiographic Fog: Darkroom Safelight Test

Objective

To investigate the safety of the direct "safelight" system employed in the darkroom and to demonstrate how to perform a safelight test.

Procedure

Using a 14" by 17" (36 cm × 43 cm) 100 speed film-screen recording system, make a light exposure with the beam restriction open to cover the entire film size. The exposure should use a low 40 kVp and an mAs sufficient to produce an overall radiographic density of 1.5. Take the exposed cassette to the darkroom. Enter the darkroom and close the entrance. Turn off any white lights, and adapt to the darkened environment of the safelights for a period not less than 15 min.

1. Prepare your exposure timer for a 1-min interval.
2. Have a completely exposed and processed 14" by 17" (36 cm × 43 cm) film available for use during the safelight test. The film should be totally black.
3. Open the cassette, and place the exposed film on the tabletop working surface directly under the area of the safelights.
4. Immediately cover four fifths of your exposed film with the totally black film prepared in advance. With a pencil or pen, write the number 5 on the uncovered one-fifth section of the film. Start your timer.
5. When 1 min is up, quickly move your black film covering back another one fifth of the film length. (A total of two fifths of the film is now exposed to the safelights.) Reset your timer, and mark the new uncovered film section with a number 4.
6. Continue the procedure moving your black film covering back one fifth of the film length until the entire film is uncovered, and a total of five sections have been identified, marked in turn from 5 to 1.
7. Process your film.

Analysis

You have a radiograph that has been exposed to the safelight system of the darkroom for varying periods of time, from 1 min to 5 min. You can examine the film for any differences in radiographic density from one end to the other. Differences in radiographic density indicate that the amount of light reaching the film at some point during the test was unsafe and caused additional radiographic density (fog) to be recorded on the film.

Conclusions

1. Examine the film for any differences in radiographic density from one end to the other. Are there any differences visible? Describe the differences, if any. Report any differences to your instructor, supervisor, or quality control technologist.

2. At what point in time does the difference in radiographic density appear? What does this mean to you related to the processing of your films?

3. If the difference is noted at 3 min or longer, is there a significant problem with the safelight system? Explain your answer.

4. If the difference is noted earlier than 3 min, is there a potential problem with the safelight system? If your answer is yes, what steps can be undertaken to eliminate the problem?

5. Should routine tests for the safety of the darkroom safelight system be included in an overall program of quality assurance? Explain the reason(s) for your answer.

6. Of what importance are tests for the safety of the darkroom safelight system to the overall operation of the radiology department? Explain the reason(s) for your answer.

Name ______________________________ Date ______________

EXPERIMENT 44
RADIOGRAPHIC DENSITY—Manual Film Processing

Objective

To demonstrate that film processing is a major influence on the radiographic density of the image and to establish the necessity of a proper time and temperature control in the processing of radiographic film. Manual film processing normally employs a time/temperature control for development of 3 min at 68° F. To perform this experiment, manual processing equipment is required.

Procedure

A series of three radiographs will be taken and processed using a 10" by 12" (25 cm × 30 cm) slow speed (detail) film-screen recording system at 40" (102 cm) SID. A small phantom part (e.g., elbow or ankle) and an aluminum penetrometer (step wedge) are used for all three exposures. The same size focus and milliamperage setting are used at 50 kVp. The only change will be in time/temperature control of the development process.

Film 1

1. Place the cassette on the tabletop, place the phantom part upon the cassette centered to the film, and tape it into position. The penetrometer is placed adjacent to the phantom. The film is marked with a lead number 1.
2. The central ray is centered over the phantom part and collimated to a size to include the phantom and the penetrometer using a perpendicular angle.
3. Using a 100-mA setting, select a time of exposure sufficient to produce a 1.0 to 1.2 bone density within the recorded image of the phantom.
4. Expose the film.

Films 2 and 3

Set up exactly as for Film 1, and mark the films with lead numbers 2 and 3, respectively. Expose and process all three films together.

Processing

1. Hang all three films on film holders, and place them into the developing solution in the following sequence: Film 2 should be placed to the left side of

the tank, Film 1 should be placed in the middle, and Film 3 should be placed to the right side of the tank. Set your timer mechanism for 1½ min.

2. Pull Film 2 out of the developer at the end of 1½ min, reset your timer, and continue the processing steps for Film 2.
3. Pull Film 1 out of the developer at the end of the second timing cycle (3 min), reset your timer, and continue the processing steps for Film 1 and Film 2.
4. Pull Film 3 out of the developer at the end of the third timing cycle (4½ min), and continue the processing steps for all three films.

Analysis

You can now compare the three radiographic images developed for 1½ min, 3 min, and 4½ min for differences in radiographic density.

Conclusions

1. Using a densitometer, measure the density of the same selected area within the radiographic image of each film and the middle step of the penetrometer, and record the radiographic densities in the spaces provided below:

   Film 1: _______ phantom density; _______ penetrometer density
   Film 2: _______ phantom density; _______ penetrometer density
   Film 3: _______ phantom density; _______ penetrometer density

2. Describe the radiographic density and visibility of details of the film produced at the decreased development time. Compare this with the properly processed Film 1.

3. Describe the radiographic density and visibility of details of the film produced at the increased development time. Compare this with the properly processed Film 1.

4. Discuss the importance of the control of the time of development and its influence on radiographic density related to the processing of film.

Note: A similar experiment can be performed to examine the effect of chemical activity on the development of the film by changing the temperature of the developing solution from 68°F to 60°F and 76°F.

Name ______________________________ Date ______________

EXPERIMENT 45
RADIOGRAPHIC DENSITY—Automated Film Processing

Objective

To demonstrate that film processing is a major influence on the radiographic density of the image and to establish the necessity of a proper time and temperature control in the processing of radiographic film. Automated film processing normally employs a time/temperature control for development. The time is controlled by means of a regulated roller system, and the temperatures of the active chemical solution are kept above 90°F. Temperature control is more critical with automated film processing than with manual processing. To perform this experiment, manual processing equipment is required.

Procedure

A series of three radiographs will be taken and processed using a 10" by 12" (25 cm × 30 cm) slow speed (detail) film-screen recording system at 40" (102 cm) SID. A small phantom part (e.g., elbow or ankle) and an aluminum penetrometer (step wedge) are used for all three exposures. The same size focus and milliamperage setting is used at 50 kVp. The only change will be in the time/temperature control of the development process.

Film 1

1. Place the cassette on the tabletop, place the phantom part upon the cassette centered to the film, and tape it into position. The penetrometer is placed adjacent to the phantom. The film is marked with a lead number 1.
2. The central ray is centered over the phantom part and collimated to a size to include the phantom and the penetrometer using a perpendicular angle.
3. Using a 100-mA setting, select a time of exposure sufficient to produce a 1.0 to 1.2 bone density within the recorded image of the phantom.
4. Expose the film.

Films 2 and 3

Set up exactly as for Film 1, and mark the films with lead numbers 2 and 3. Expose and process all three films together.

Processing

1. Process Film 1 using the recommended temperature for the developing solution.
2. Process Film 2 after having increased the temperature of your developing solution by 5°F from the recommended temperature.
3. Process Film 3 after having decreased the temperature of your developing solution by 5°F from the recommended temperature.

Note: Be sure to readjust your solution temperature and run test strips for quality control prior to using the processing unit for the processing of radiographic films from patient examinations or additional experiments.

Analysis

You can now compare the three radiographic images developed at the proper temperature, at a reduced temperature, and at an increased temperature of the development solution for differences in radiographic density.

Conclusions

1. Using a densitometer, measure the density of the same selected area within the radiographic image of each film and the middle step of the penetrometer, and record the radiographic densities in the spaces provided below:

   Film 1: ______ phantom density; ______ penetrometer density
   Film 2: ______ phantom density; ______ penetrometer density
   Film 3: ______ phantom density; ______ penetrometer density

2. Describe the radiographic density and visibility of details of the film produced at the lowered development temperature. Compare this with the properly processed Film 1.

3. Describe the radiographic density and visibility of details of the film produced at the increased development temperature. Compare this with the properly processed Film 1.

4. Discuss the importance of temperature control and its influence on radiographic density related to the processing of film.

5. Why would the temperature control of the developing solution be more critical with automated processing compared with manual processing?

Name ____ Date ________________

EXPERIMENT 46
RADIOGRAPHIC DENSITY—Anode Heel Effect

Objective

To demonstrate the lack of uniformity of x-ray intensity known as the **anode heel effect.**

Procedure

If possible, remove the beam limiting device completely so that only the bottom of the tube housing is visible. Arrange your tube in a perpendicular angle with the tabletop, and reduce your SID to 30" (76 cm). If you are unable to remove your beam limitation device, open the collimator to a beam coverage of 4" wide and as much length as the device will permit. Place a 14" by 17" (36 cm × 43 cm) slow speed (detail) film-screen recording system cassette lengthwise on the tabletop, and center the central ray to the middle of the cassette. Block the cassette with lead strips so that an image 4" wide and the full length of the cassette will be permitted. Use a lead marker to identify the left side of the beam. Using 50 kVp, select an exposure to produce a radiographic density of 1.8 to 2.0; expose and process the film.

Analysis

Place the film on an illuminator, and review the radiographic density along the length of the exposed strip. You should be able to observe a difference in radiographic density between the two extreme ends of the film.

Conclusions

1. Encircle a small area at the center of the film, at a point adjacent to the extreme left side of the film, and at a point adjacent to the extreme right side of the film. Using a densitometer, measure the density of the three selected areas, and record the radiographic density in the spaces provided below:

   _______ Radiographic density, extreme left side of the film
   _______ Radiographic density, middle of the film
   _______ Radiographic density, extreme right side of the film

2. Is the radiographic density uniform throughout the exposed strip? If not, describe any difference noted.

3. Locate the position of your tube cathode (–) and tube anode (+) on the equipment used for this experiment (left or right side), and record them below:

   Tube cathode (–) __________ side of the tube

   Tube anode (+) __________ side of the tube

4. Is the intensity of the beam greater toward the cathode or anode of the tube? Provide reasons for the difference noted.

5. Of what significance would this effect be for the overall radiographic appearance of an image?

6. Indicate a procedure where the anode heel effect must be taken into consideration before setting up the patient–tube–film relationship.

Name ______________________________ Date ____________

EXPERIMENT 47
RADIOGRAPHIC DENSITY—Application of the Anode Heel Effect

Objective

To demonstrate the anode heel effect used to advantage to produce a more uniform density throughout the image.

Procedure

Film 1: Cervical Region of Thorax Placed Toward the Cathode

Using 36" (91 cm) SID, open your collimator sufficient to cover the entire length of a 14" by 17" (36 cm × 43 cm) slow speed (detail) film-screen recording system cassette placed in the Bucky tray. Place the thorax phantom on the tabletop, centering the spine to the midline of the table. Center your film in the Bucky tray to the middle of the phantom. Adjust the width of the collimator to a 4" (10 cm) recorded width and a full 17" length on the film. Place the upper portion of the thorax toward the cathode end of the tube (mark the area near the neck with a lead number 1), and center the central ray to the midpoint of the spine and to the center of the film using a perpendicular beam. Using 70 kVp, select an exposure to produce a bone density of 1.0 to 1.2 within the thoracic spine of the phantom. Expose and process the film.

Film 2: Cervical Region of Thorax Placed Toward the Anode

Arrange your procedure exactly as you did for Film 1; however, reverse the position of the thorax phantom. Place the upper cervical area of the thorax phantom toward the anode end of the tube and mark the area near the neck with a lead number 2. Expose and process your film using the same technical factors.

Analysis

A comparison of the two radiographic images will reveal the proper method to arrange your procedure to take advantage of the anode heel effect.

Conclusions

1. Is there a difference in the visibility of the recorded details between the two radiographic images? Describe the difference, if any.

2. Which of the two radiographs demonstrates the spine with better visibility of the recorded details throughout its entire length? Why?

3. In structures of the body of similar thickness and density throughout their lengths, can the anode heel effect be used to advantage? Explain your answer.

4. Why is the length of the structure important in determining whether the anode heel effect can be used to advantage?

5. When using the anode heel effect to advantage, what basic consideration must be made in setting up the arrangement of the part to be examined with the x-ray tube?

Name ______________________________ Date ______________

EXPERIMENT 48
RADIOGRAPHIC DENSITY—Orthopedic Cast Materials

Objective

To demonstrate the influence of orthopedic plaster cast materials on the radiographic density of the image and to prove that overall radiographic density can be maintained through the adjustment of the technical exposure factors.

Procedure

You will need access to the materials used to apply an orthopedic cast for this experiment. A series of five radiographic images of a phantom ankle in the anteroposterior projection and an aluminum penetrometer (step wedge) will be taken using 10" by 12" (36 cm × 43 cm) slow speed (detail) film-speed recording system cassettes at a 40" (102 cm) SID.

Film 1: Standard Radiographic Image

1. Place the cassette on the tabletop, and place the phantom part and the penetrometer upon the cassette centering the arrangement to the film and taping the phantom in place.
2. Using 60 kVp, select the 100-mA small focus setting on your control panel, and choose the time of exposure necessary to produce a 1.0 to 1.2 bone density. Mark your film with a lead number 1.
3. Center your central ray to the ankle phantom using a perpendicular angle, and collimate your beam to include the phantom and penetrometer.
4. Expose and process your film.

The phantom is now prepared for the next series of films by applying a plaster orthopedic cast using the same cotton wadding and plaster materials of the same design and thickness as would be applied to an injured foot and ankle.

Film 2: Orthopedic Cast, Same Exposure Factors

Set up Film 2 the same as for Film 1, mark it with a lead number 2, and expose and process your film.

Film 3: Orthopedic Cast, Increased mAs

Film 3 is set up as above and is marked with a lead number 3, but the mAs is doubled. The kilovoltage remains at 60. Expose and process your film.

Film 4: Orthopedic Cast, Increased Kilovoltage

Film 4 is set up as above and is marked with a lead number 4, but the kilovoltage is increased by 10 to 70 kVp. The mAs remains at the original value used for Film 1. Expose and process your film.

Film 5: Orthopedic Cast, Increased mAs and Kilovoltage

Film 5 is set up as above and is marked with a lead number 5, but the mAs is doubled, and the kilovoltage is increased by 10% to 66 kVp. Expose and process your film.

Analysis

By comparing the films taken with the orthopedic casting materials with the original film taken without the cast, you can determine the influence of the casting materials on the visibility of the image as well as the necessary adjustments required to maintain the radiographic density of the recorded image.

Conclusions

1. Select the same bone density within the image of the phantom for each of the radiographic images and a middle step of the penetrometer, and record the radiographic density in the spaces provided below:

   Film 1: ______ phantom density; ______ penetrometer density
   Film 2: ______ phantom density; ______ penetrometer density
   Film 3: ______ phantom density; ______ penetrometer density
   Film 4: ______ phantom density; ______ penetrometer density
   Film 5: ______ phantom density; ______ penetrometer density

2. Compare the radiographic density and visibility of the image of Film 2 with those of Film 1. Are the radiographic densities equal? Is the radiographic contrast similar? Describe the differences, if any.

3. Compare the density and visibility of the image of Film 3 with those of Film 1 and Film 2. Has the increase in the mAs improved the image compared with Film 2? Is the radiographic density of the image equal to Film 1? Is the radiographic contrast similar to Film 1? Describe the differences, if any.

4. Compare the density and visibility of the image of Film 4 with those of Film 1 and Film 2. Has the increase in the kilovoltage improved the image compared with Film 2? Is the radiographic density of the image equal to Film 1? Is the radiographic contrast similar to Film 1? Describe the differences, if any.

5. Compare the density and visibility of the image of Film 5 with those of Film 1 and Film 2. Has the increase in the mAs and kilovoltage improved the image compared with Film 2? Is the overall radiographic density of the image equal to Film 1? Is the overall radiographic contrast similar to Film 1? Describe the differences, if any.

6. Which film most closely approximates the total image visibility demonstrated by Film 1? Explain the reasons for this similarity related to factors of thickness and opacity of part and the technical exposure factor adjustments utilized to produce this change.

Name ______________________________ Date ____________

EXPERIMENT 49
RADIOGRAPHIC DENSITY—Beam Filtration

Objective

To demonstrate the influence of different filter thicknesses on the radiographic density of the recorded image. The appropriate total filtration normally employed for beam filtration in the diagnostic range is 2.5 mm aluminum, consisting of 0.5 mm inherent filtration and 2.0 mm added filtration.

Procedure

Using a 10" by 12" (25 cm × 30 cm) slow speed (detail) film-screen recording system cassette collimated to a size just smaller than the film, a series of four radiographs of a phantom ankle and an aluminum penetrometer (step wedge) are taken tabletop using a perpendicular beam centered to the part at 40" (102 cm) SID. The same technical exposure factors are used for all four exposures. The same size focus, milliamperage, and 50 kVp are selected. The time of exposure is chosen to produce a 1.0 to 1.2 bone density within the recorded image of the phantom.

Film 1: Normal Filtration, 2.5 mm Aluminum

Set up the procedure as described above, mark the film with a lead number 1, and expose and process the film.

Film 2: Remove All Added Filtration, 0.5 mm Inherent Filtration

Remove all added filtration from your tube housing. Set up your procedure the same as for Film 1; mark the film with a lead number 2; and using the same exposure factors, expose and process your film.

Film 3: Add 1.0 mm Aluminum Filtration

Add 1.0 mm aluminum filtration to the tube housing. Set up your procedure the same as for Film 1; mark the film with a lead number 3; and using the same exposure factors, expose and process your film.

Film 4: Add 3.0 mm Aluminum Filtration

Add 3.0 mm aluminum filtration to the tube housing. Set up your procedure the same as for Film 1; mark the film with a lead number 4; and using the same exposure factors, expose and process your film.

Note: At the completion of the procedure, be sure to replace the proper 2.0 mm aluminum filtration to the tube housing.

Analysis

You can now review a series of radiographic images that demonstrate the effect of beam filtration on the radiographic density of the recorded image.

Conclusions

1. Circle the same area within the recorded image of the phantom in each radiograph and the middle step of the penetrometer, and record the radiographic density in the spaces provided below:

   Film 1: _______ phantom density; _______ penetrometer density
   Film 2: _______ phantom density; _______ penetrometer density
   Film 3: _______ phantom density; _______ penetrometer density
   Film 4: _______ phantom density; _______ penetrometer density

2. What happened to the radiographic density when all the added filtration was removed (Film 2)? Describe the appearance of the radiographic image.

3. What happened to the radiographic density when 1.0 mm aluminum filtration was added (Film 3)? Describe the appearance of the radiographic image.

4. What happened to the radiographic density when 3.0 mm aluminum filtration was added (Film 4)? Describe the appearance of the radiographic image.

5. Review the radiographic appearance of the penetrometer's image in each of the radiographs, and describe how the changes in filtration affected the overall scale of radiographic contrast.

6. Would the use of filtration beyond the recommended 2.5 mm aluminum serve a useful purpose in radiography? To produce a similar radiographic density with increased filtration, what would happen to the patient exposure?

Name ______________________________ Date ______________

EXPERIMENT 50
RADIOGRAPHIC DENSITY—Compensating Filters

Objective

To demonstrate the application of compensating filters to improve the appearance of the radiographic image and the influence of compensating filters on radiographic density.

Procedure

A series of three radiographs will be performed tabletop using 10" by 12" (25 cm × 30 cm) slow speed (detail) film-screen recording system cassettes. The procedure will use a foot phantom placed into the anteroposterior projection for all three radiographs. A 40" (102 cm) SID is employed, and the central ray is angled 15° from the perpendicular in a direction toward the thicker part of the foot and centered to enter at the location of the proximal metatarsals. The central ray is collimated to include the phantom part.

Film 1: Anteroposterior Projection of the Foot

Set up your procedure as described above. Using 60 kVp, select the mAs necessary to produce a 1.0 to 1.2 bone density at the proximal metatarsal level of the bone phantom. Mark your film with a lead number 1. Expose and process your film.

Film 2: Anteroposterior Projection of the Foot Using a Compensating Filter

1. Set up your procedure as for Film 1 using the same exposure factors. Mark your film with a lead number 2.
2. Select a wedge-type compensating filter, and attach it to the collimator or tube housing, according to the equipment used. Be sure that the thicker end of the wedge is aligned to the toes of the foot, and the thinner portion of the wedge is facing the body of the foot. Expose and process your film.

Film 3: Anteroposterior Projection of the Foot Using a Compensating Filter after an Adjustment in the Exposure Factors

Set up your procedure as for Film 2; however, adjust your exposure factors by increasing your kilovoltage by 10 to 70 kVp. Expose and process your film.

Analysis

You can compare the radiographic quality of the recorded image of the foot as performed routinely with the images produced when a compensating filter is employed. You can analyze the influence of the compensating filter on radiographic density and its influence when it is used to advantage and the exposure factors are adjusted.

Conclusions

1. Select a bony area in the phalangeal (toe) region of the foot and the tarsal region of the foot for each radiographic image. Using a densitometer, determine the radiographic density, and record it in the space provided below:

   Film 1: ______ phalangeal density; ______ tarsal density
   Film 2: ______ phalangeal density; ______ tarsal density
   Film 3: ______ phalangeal density; ______ tarsal density

2. Describe the overall radiographic appearance and visibility of the recorded details in Film 1. Is the radiographic density uniform throughout the entire recorded image? Are all of the structures of interest fully visible?

3. Describe the overall radiographic appearance and visibility of the recorded details in Film 2. Compare the appearance of the recorded image with Film 1. Discuss the influence of the compensating filter on the overall radiographic density of the recorded image.

4. Describe the overall radiographic appearance and visibility of the recorded details in Film 3. Compare the appearance of the recorded image with Film 1. Discuss the influence of the compensating filter on the overall radiographic density of the recorded image when the exposure factors have been adjusted.

5. Which radiographic image provides the greatest visibility of all of the structures of interest? Explain your answer.

Name ______________________________ Date ______________

EXPERIMENT 51
RADIOGRAPHIC CONTRAST—Subject Contrast

Objective

To identify the different naturally occurring tissue densities of the body recorded within a number of radiographic images.

Procedure

The following tissue densities can be identified within a radiographic image either naturally, as a result of pathology, or as a result of induced subject contrast by means of a contrast agent:

1. Gas, normal
2. Gas, abnormal and induced
3. Fat
4. Cartilage
5. Hollow organs, empty
6. Muscles
7. Solid organs
8. Hollow organs, filled
9. Bone
10. Metal, foreign body or orthopedic prosthesis

Secure the following radiographs from the teaching file or the department's radiological files. Select films of radiographic quality demonstrating proper density and appropriate contrast for this experiment. Only one radiograph of each type of examination is necessary.

1. Routine, posteroanterior chest for lungs
2. Emphysema, posteroanterior chest
3. Routine, posteroanterior chest for ribs
4. Routine, supine abdomen
5. Intestinal obstruction, (supine, erect, or decubitus) abdomen
6. Barium enema with air, (supine or decubitus) abdomen
7. Small bowel series (1-hr progress), supine abdomen
8. Intravenous pyelogram (15-min progress), supine abdomen

9. Pelvic pneumogram (gynogram), anteroposterior pelvis, or pneumoencephalogram, lateral skull
10. Routine, lateral skull
11. Routine, Caldwell or Waters projection for sinuses
12. Sinusitis, Caldwell or Waters projection for sinuses
13. Routine, anteroposterior forearm
14. Arthrogram, anteroposterior knee
15. Postsurgical hip pinning, anteroposterior hip

Analysis

Reviewing the radiographic examples identified above, it is possible to visualize multiple tissue densities in each radiograph as well as a variety of different subject contrasts within the radiographic scale of contrast in each radiographic image. For each of the 15 radiographs, you should identify as many of the 10 naturally occurring or artificially produced subject contrasts visible within each of the radiographic images.

Conclusions

1. Review each of the 15 radiographs you have selected. Identify as many of the 10 tissue densities and subject contrasts as you can within the recorded image of each radiograph. Make a list, and record the identified subject contrasts you have been able to visualize, together with the specific anatomical structures that they represent, for each of the 15 radiographs.

    Example: Film 1, Normal supine abdomen

    1. Gas—stomach

    7. Solid organ—liver

2. Review your list and the radiographic films with your instructor. Most of the ten subject contrast structures will be present within each of the selected radiographs. Were there additional tissue densities that you overlooked? Add them to your list.

3. Discuss the importance of a thorough knowledge of anatomy, an understanding of physiology, and a recognition of the routine pathological conditions as they relate to the following:

   Subject contrast:

   Selection of exposure factors:

   Radiographic contrast:

4. Compare the selection of exposure factors, subject contrast, and radiographic contrast in a chest examination taken for possible rib fractures with those in a chest examination taken for lung pathology. Describe your findings.

5. Review the anteroposterior projection of the forearm, and examine the radiographic contrast recorded by a single structure of interest, the radius. Describe the different thickness and opacity of the radius that contributed to the recording of different radiographic densities of the bone.

Name ______________________________ Date ______________

EXPERIMENT 52
RADIOGRAPHIC CONTRAST—Film Speed and Latitude

Objective

To demonstrate the influence of different film speeds and latitude (recording properties) on the radiographic contrast and visibility of the recorded image.

Procedure

Select different speed and latitude films for this experiment. If unavailable, secure a selection of radiographic films of different speeds and contrast recording ability (latitude; e.g., slow speed–medium contrast; medium speed–high contrast and extended contrast; fast speed, high contrast and extended contrast) from a radiographic film manufacturer's technical representative.

Using a 10" by 12" (25 cm × 30 cm) slow speed (detail) film-screen recording system cassette collimated to a size just smaller than the film, exposures of a phantom ankle and an aluminum penetrometer (step wedge) are taken tabletop using a perpendicular beam at 40" (102 cm) SID for each of the different film speeds and latitudes that you have available. Using 50 kVp, you should employ the same exposure factors for each film used. Select an appropriate mAs to produce a 1.0 to 1.2 bone density within the radiographic image of the phantom. The films should be numbered consecutively, exposed, and processed.

Analysis

With a densitometer, each density level of the penetrometer is recorded for each radiograph taken. Each step of the penetrometer represents a doubling of the exposure. Each 0.3 increase in the log relative exposure represents a doubling of the exposure. These densities should now be plotted against the relative log exposure in order to demonstrate the sensitometric characteristic curve of each film's response. The chart provided in this experiment can be used for this purpose. Use a different color of ink for each curve, and number the curves consecutively.

Conclusions

1. Describe the purpose of having different film speeds available within the radiology department.

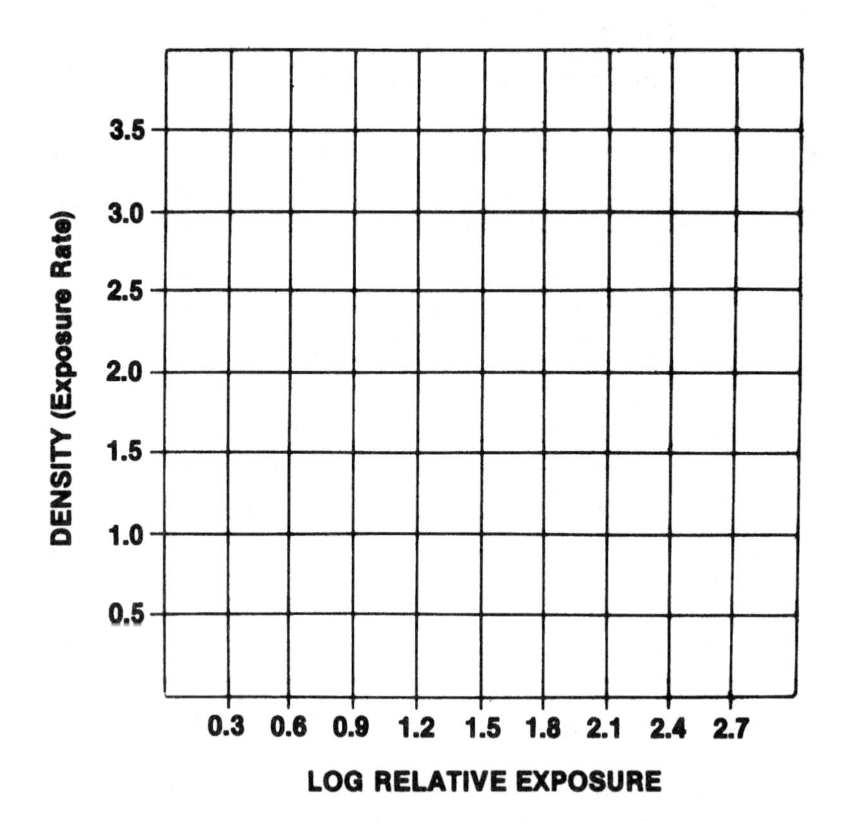

2. Describe the use of a *high-contrast film* for a shoulder examination versus an abdomen examination. Describe the radiographic contrast scale produced within each radiographic image. In which of these two examinations would this type of film be used to advantage? Explain.

3. Compare the radiographic images produced and identify the film with the highest sensitivity from the evaluation of the characteristic curves produced. Explain the reason for your choice.

4. Which film's characteristic curve represents the film with the highest inherent contrast?

5. Which film's characteristic curve represents a film with extended (response) latitude?

6. Of all the radiographic images, which film represents the recorded image possessing the greatest radiographic quality?

Name ____ Date ______________

EXPERIMENT 53
RADIOGRAPHIC CONTRAST—Exposure Latitude

Objective

To demonstrate that the maximum exposure latitude is available when the kilovoltage necessary for penetration is selected. This means that the technologist will have the greatest margin for error in the selection of the mAs values that will produce a diagnostic radiographic image.

Procedure

With a 10" by 12" (25 cm × 30 cm) slow speed (detail) film-screen recording system cassette collimated to a size just smaller than the film, a series of exposures of the hip phantom is taken using a Bucky mechanism. Center the hip joint of the pelvis phantom to the midline of the table, and adjust its position so that an aluminum penetrometer (step wedge) placed at the lateral aspect of the hip will be included on the film. Direct the central ray perpendicularly to the hip joint using a 40" (102 cm) SID, and center the film in the Bucky tray to the central ray.

Film 1: 80 kVp

1. Using 80 kVp and a small focus, select the appropriate mAs to produce a bone density of 1.0 to 1.2 in the area of the acetabulum. Mark the film with a lead number 1.
2. Expose and process your film.

Film 2: 80 kVp, Reduced mAs

1. Set up your procedure the same as for Film 1. Using 80 kVp, reduce the mAs 25% from the exposure used for Film 1. Mark the film with a lead number 2.
2. Expose and process your film.

Film 3: 80 kVp, Increased mAs

1. Set up your procedure the same as for Film 1. Using 80 kVp, increase the mAs 25% from the exposure used for Film 1. Mark the film with a lead number 2.
2. Expose and process your film.

Another series of exposures is made similar to those taken above; however, the kilovoltage is reduced to 60 kVp.

Film 4: 60 kVp

1. Using 60 kVp and a small focus, select the appropriate mAs to produce a bone density of 1.0 to 1.2 in the area of the acetabulum. Mark the film with a lead number 4. **Note:** The overall radiographic density of this film should be similar to Film 1.
2. Expose and process your film.

Film 5: 60 kVp, Reduced mAs

1. Set up your procedure the same as for Film 1. Using 60 kVp, reduce the mAs 25% from the exposure used for Film 4. Mark the film with a lead number 5.
2. Expose and process your film.

Film 6: 60 kVp, Increased mAs

1. Set up your procedure the same as for Film 1. Using 60 kVp, increase the mAs 25% from the exposure used for Film 4. Mark the film with a lead number 6.
2. Expose and process your film.

Analysis

Each of the two series of three radiographs starts with an mAs value appropriate for the kilovoltage selected in order to produce a proper radiographic density. From the proper 1.0 to 1.2 density of Film 1 and Film 4, the mAs is manipulated above and below the original mAs in 25% increments. You can now compare the exposure latitude for a hip joint when employing 80 kVp and 60 kVp.

Conclusions

1. Circle an area of the acetabulum of the hip in each of the radiographic images, and select the middle step of the penetrometer. Using a densitometer, identify the density of each image, and record the density in the space provided below:

   Film 1: ______ phantom density; ______ penetrometer density
   Film 2: ______ phantom density; ______ penetrometer density
   Film 3: ______ phantom density; ______ penetrometer density
   Film 4: ______ phantom density; ______ penetrometer density
   Film 5: ______ phantom density; ______ penetrometer density
   Film 6: ______ phantom density; ______ penetrometer density

2. Review the images produced by Films 1 through 3. Although all three films may not be of radiographic quality, would they be acceptable as diagnostic quality images overall? Is the radiographic contrast sufficient to visualize the details of the recorded image in all three radiographs? Explain your answer.

3. Review the images produced by Films 4 through 6. Are all three radiographs acceptable as diagnostic quality images overall? Is the radiographic contrast sufficient to visualize the details of the recorded image in all three radiographs? Explain your answer.

4. Which series of films demonstrates the greatest exposure latitude? Why?

5. Which kilovoltage represents the more appropriate penetration level for the hip examination? Explain your answer.

6. Is the greater image visibility loss created by an increased or a decreased mAs when the kilovoltage used is 60 kVp? Why?

7. Is the greater image visibility loss created by an increased or a decreased mAs when the kilovoltage is 80 kVp? Why?

Name ____ Date ________________

EXPERIMENT 54
RADIOGRAPHIC CONTRAST—Scale of Contrast: kVp/mAs

Objective

To demonstrate the influence that the kVp/mAs relationship has on the scale of contrast within the recorded image.

Procedure

Using a 10" by 12" (25 cm × 30 cm) slow speed (detail) film-screen recording system cassette collimated to a size just smaller than the film, a series of exposures of a pelvis phantom and an aluminum penetrometer (step wedge) is taken using a Bucky mechanism. The hip is centered in such a way that the penetrometer placed adjacent to the phantom will be included within the recorded image. The central ray is directed perpendicular and centered to the phantom using a 40" (102 cm) SID. The film is centered to the central ray.

Film 1: 70 kVp

1. Using 70 kVp, select the appropriate mAs to produce a 1.0 to 1.2 bone density in the region of the acetabulum of the hip. Mark the film with a lead number 1.
2. Expose and process the film.

Film 2: Apply the kVp/mAs 15% Rule

1. Increase your kilovoltage to 80; reduce your mAs by 50% from the mAs used for Film 1. Mark your film with the lead number 2.
2. Expose and process your film.

Film 3: Apply the kVp/mAs 15% Rule

1. Increase your kilovoltage to 92; reduce your mAs by 50% from the mAs used for Film 2. Mark your film with the lead number 3.
2. Expose and process your film.

Analysis

You can compare the three radiographs with regard to the overall density of the recorded images. You can also observe the changes in the scale of contrast produced.

Conclusions

1. Circle the acetabulum region for each radiographic image, and using a densitometer, record the density in the spaces provided below:

   Film 1: _______ phantom density
   Film 2: _______ phantom density
   Film 3: _______ phantom density

2. Select the middle step of the penetrometer and the step immediately above and below it. Using a densitometer, record the densities for each radiographic image in the spaces provided below:

   Film 1: _______ step below _______ midstep _______ step above
   Film 2: _______ step below _______ midstep _______ step above
   Film 3: _______ step below _______ midstep _______ step above

3. Compare the *overall* radiographic density of each of the three recorded images. Are they comparable? Explain.

4. Describe the differences in the scale of radiographic contrast demonstrated between Films 1, 2, and 3 based on your visual observation of the recorded images and the densitometer readings of the penetrometer recorded in Question 2.

5. As the kilovoltage in Films 2 and 3 increased, the scale of radiographic contrast (*circle one*) increased/decreased.

6. Describe the diagnostic quality of each recorded image. Would all three radiographic images be considered of diagnostic quality? Explain.

7. Would the kVp/mAs 15% rule operate successfully in the opposite direction, that is, if you were to decrease the kilovoltage by 15% and increase the mAs by 50%? Explain your answer.

Name ______________________________ Date ______________

EXPERIMENT 55
RADIOGRAPHIC CONTRAST—Penetration

Objective

To demonstrate that the proper relationship between the kilovoltage and the mAs is necessary if the radiographic contrast of the image is to be appropriate to visualize the structures of interest. This can be demonstrated by showing that insufficient kilovoltage will prevent radiation from passing through the part and that no practical increase in mAs will correct for a severely underpenetrated part.

Procedure

The effect of kilovoltage levels insufficient to penetrate the part was demonstrated in its relationship to the production and control of radiographic density in Experiment 35. (If Experiment 35 has not been performed, produce the four radiographs from that experiment at this time.)

Analysis

Review the radiographs produced in Experiment 35 with regard to the loss of image visibility and the influence on the radiographic scale of contrast as related to inadequate kilovoltage levels. Compare the radiographic images side by side in order to analyze the visibility of the details from the point of view of proper radiographic contrast.

Conclusions

1. In analyzing the visibility of a radiographic image, why must the first consideration be whether or not proper penetration of the part has been achieved?

2. What is meant by the statement that the greatest exposure latitude will be achieved when the kilovoltage necessary for penetration has been selected?

3. Can proper radiographic density and, therefore, an appropriate scale of contrast occur in the absence of proper penetration? Explain.

4. What happened to the radiographic contrast of the recorded images when the kilovoltage was reduced to 40 kVp? Why has this occurred?

5. Discuss the significance of proper penetration of the part to radiographic contrast and the visibility of the details of the recorded image.

Name ______________________________ Date ______________

EXPERIMENT 56

RADIOGRAPHIC CONTRAST—Radiographic Fog: Object-Image Distance

Objective

To demonstrate the influence of OID on the production of radiographic fog and the visibility of the recorded image.

Procedure

Use a 10" by 12" (25 cm × 30 cm) slow speed (detail) film-screen recording system cassette divided in two by lead strips. A block of paraffin, 3" (7.6 cm) thick and at least 5" (12.7 cm) in length, is placed on top of an elbow phantom and secured in that position. A 40" (102 cm) SID is employed using a perpendicular angle of the central ray. Using 60 kVp, the same size focus and milliamperage setting are employed for both exposures. Collimation is adjusted to include the phantom and paraffin block.

Exposure 1

1. Center the phantom part with the paraffin block attached to it to the uncovered film section. The paraffin block must be away from the film and on top of the phantom. Center your central ray to the divided film section. Mark the film section with a lead number 1.
2. Select the appropriate mAs necessary to produce a 1.0 to 1.2 bone density within the elbow phantom.
3. Expose your film section.

Exposure 2

1. Remove the phantom, and cover the exposed film section with a lead blocker.
2. Return the phantom and paraffin block to the unexposed film section. However, reverse the position of the phantom, placing the paraffin block directly upon the film surface with the phantom part on top of it.
3. Set up as for Exposure 1, and mark the film section with a lead number 2.
4. Using the same technical factors, expose and process your film.

Analysis

A comparison of the two radiographic images will enable you to visualize the influence of secondary and scattered radiation (radiographic fog) produced by the increased thickness of part and the greater OID of the structures of interest. (See Chapter 5, Figure 5-8.)

Conclusions

1. Using a densitometer, determine and record the radiographic density of the distal humerus of the elbow joint for each radiographic image in the space provided below:

   Exposure 1: _______ phantom density
   Exposure 2: _______ phantom density

2. Describe the radiographic contrast of each radiographic image. Is it different? Indicate the reasons for the difference in the radiographic contrast between Exposures 1 and 2, if any.

3. In what manner does scatter radiation affect the visibility of the image details?

4. Is there a noticeable difference in the size of the recorded images? Which image appears larger? Why?

5. Describe the influence of the thickness of the part and therefore of different OIDs of structures within the part on the radiographic quality of the image related to radiographic contrast and the visibility of the recorded details.

6. In the case of an obese patient, describe how this influence (Question 5) would affect your decision to perform a clavicle examination (i.e., whether anteroposterior or posteroanterior), assuming there is no contraindication to performing the examination using either projection.

Name ______________________________ Date ______________

EXPERIMENT 57
RADIOGRAPHIC CONTRAST—Radiographic Fog: Air Gap Technique

Objective

To demonstrate the improvement in radiographic contrast attributed to the use of the air gap technique.

Procedure

Experiment 56 should be completed before attempting this experiment. Use the same materials from Experiment 56 for this procedure. Set up your procedure with the paraffin placed closest to the film surface and the phantom part on top as for Exposure 2 in Experiment 56.

Two radiographs will be produced using two 8" by 10" (20 cm × 25 cm) slow speed (detail) film-screen recording system cassettes. The radiographs will be identified as Exposures 3 and 4 since the overall analysis of this experiment will include the radiographic images produced in Experiment 56.

Exposure 3, Increased OID Air Gap

1. Raise the paraffin/phantom unit on radiolucent sponges to produce an air gap of an additional 6" to 10" (15 cm × 25 cm) of OID and place it on an 8" by 10" (20 cm × 25 cm) cassette. (See Chapter 5, Figure 36.) Mark the film with the lead number 3.
2. Using 60 kVp and the same mAs employed in Experiment 56, mark this film with the lead number 3.
3. Expose and process your film.

Exposure 4, Increased OID Air Gap and Increased Exposure

1. Using another 8" by 10" (20 cm × 25 cm) cassette, set up as for Film 1, Exposure 3; however, using the same kilovoltage, increase your mAs by 50%.
2. Mark your film with a lead number 4.
3. Expose and process your film.

Analysis

The analysis for Experiment 57 will include the two radiographic images produced in Experiment 56 as well as the additional two exposures produced for this experiment. Place all four radiographic images side by side on film illuminators in order to compare and analyze the contrast improvement attributed to the use of the air gap technique.

Conclusions

1. Using a densitometer, determine and record the radiographic density of the distal humerus of the elbow joint for each radiographic image in the space provided below:

   Exposure 1: _______ phantom density, Experiment 56
   Exposure 2: _______ phantom density, Experiment 56
   Exposure 3: _______ phantom density, Experiment 57
   Exposure 4: _______ phantom density, Experiment 57

2. Compare Exposure 3 to Exposure 2 from Experiment 56. Describe the differences, if any, in the radiographic density of the images. Why have these changes occurred, since the technical factors employed for both radiographic images are the same?

3. Compare Exposure 3 to Exposure 2 from Experiment 56. Which radiographic image appears to be affected by radiographic fog and the loss of radiographic contrast? Why has this occurred, since the technical factors employed are the same?

4. Compare Exposure 3 to Exposure 4. Describe the differences, if any, in radiographic density and image contrast.

5. Compare Exposure 4 to Exposure 1 from Experiment 56. Are the overall radiographic density and contrast comparable? Explain the principles involved in the use of the air gap technique.

6. Is the air gap technique applicable to most radiographic procedures? Consider the geometric properties of the recorded image when formulating your answer.

Name ______________________________ Date ______________

EXPERIMENT 58

RADIOGRAPHIC CONTRAST—Measurement of Beam Limitation Devices

Objective

To determine the actual beam limitation properties of various beam limitation devices.

Procedure

The ability to perform this experiment depends on the availability of x-ray equipment that employs aperture diaphragms and cones as beam limiting devices.

Film 1: Aperture Diaphragm

1. Select a diaphragm, and measure the diameter of the opening in inches/centimeters. Measure the distance from the tube focus to the aperture opening where the diaphragm fits into the tube housing. Adjust your SID to 40" (102 cm), or identify the SID utilized by the x-ray unit if the distance cannot be adjusted.
2. Center a slow speed (detail) film-screen recording system cassette large enough to include the entire recorded image below the middle of the aperture opening. Mark the middle of the film with a lead number 1.
3. Expose the film using 50 kVp and an mAs sufficient to produce a 1.8 to 2.0 radiographic density. Process your film.

Film 2: Cone (Expose Film 3 If an Extension Cone Is Available)

1. Select a straight design cone, and measure the diameter of the lower opening in inches/centimeters. Measure the distance from the tube focus to the bottom of the cone with the cone mounted on the tube housing. Adjust your SID to 40" (102 cm). If you have an extension cone, use it, and take two measurements and two exposures, one with the cone collapsed and one with the cone extended.
2. Center a slow speed (detail) film-screen recording system cassette large enough to include the entire recorded image below the middle of the cone opening. Mark the middle of the film with a lead number 2. If using an extension cone, Film 2 is performed with the cone collapsed. Film 3 is exposed with the cone extended and marked with a lead number 3.

3. Expose the film(s) using 50 kVp and an mAs sufficient to produce a 1.8 to 2.0 radiographic density. Process your film(s).

Analysis

The radiographic images will demonstrate the exact beam limitation properties of your aperture diaphragm and cone devices when employed at a specific SID. You can determine whether the formulas for assessing the coverage of the x-ray beam are accurate and applicable.

Conclusions

1. Using the diaphragm formula provided below and the measurements you have taken of the aperture diaphragm used in this experiment, determine the projected image size.

$$\frac{\text{anode-diaphragm distance}}{\text{diameter of aperture}} = \frac{\text{source-image distance}}{\text{size of projected image}}$$

2. Measure the recorded image diameter on the radiograph taken with the aperture diaphragm. Does it coincide with the size indicated in Question 1? Explain any differences.

3. Using the cone formula provided below and the measurements you have taken of the cone used in this experiment, solve for the projected image. (If an extension cone was used, solve for both the collapsed and extended measurements.)

$$\frac{\text{source-image distance} \times \text{lower diameter of cone}}{\text{distance from tube focus to the bottom of cone}}$$

4. Measure the recorded image diameter on the radiograph taken with the cone. Does it coincide with the size indicated in Question 3? (If an extension cone was used, measure both images produced.) Explain any differences.

5. Examine the edges of the recorded image for the aperture diaphragm. Describe any unsharpness (penumbra) or undercutting of the recorded image that you can observe.

6. Examine the edges of the recorded image for the cone. Describe any unsharpness (penumbra) or undercutting of the recorded image that you can observe. If an extension cone was used, is there a difference between the penumbra created in the collapsed versus the extended image? Explain.

7. Is the unsharpness (penumbra) associated with the recorded image edge worse with a diaphragm (question 5) or with a cone (question 6)? Explain.

Name ______________________________ Date ______________

EXPERIMENT 59

RADIOGRAPHIC CONTRAST—Assessing the Alignment Accuracy of a Collimator

Objective

To demonstrate a simple method to investigate the beam limitation alignment accuracy of a collimator.

Procedure

Center a 10" by 12" (25 cm × 30 cm) slow speed (detail) film-screen recording system cassette directly below a collimator device that has been placed at an SID of 40" (102 cm). Be sure that the x-ray tube is arranged in a position that will project a perpendicular x-ray beam toward the film. Open the lead shutters of the collimator until its light or laser beam apparatus projects an image that measures 6" (15 cm) square upon the cassette. Use a ruler to measure this value accurately. Straighten out four new paper clips and bend them in half until each one forms a 90° angle. Place one paper clip in each corner of the projected image of light, and accurately align it with the edge of the light beam, so that the metal wire is even with the edge of the light beam. Tape the metal clips in place to prevent their movement. Expose your film using 50 kVp and an mAs sufficient to produce a 1.8 to 2.0 density, and process it.

Analysis

A review of the recorded image produced will enable you to evaluate the accuracy of the alignment of the beam limitation of your collimator with the beam of light projected toward your patient and used for the centering of the part to be examined and the centering of your film in the Bucky tray. The beam limitation of the collimator device and the beam of light projected toward the film should be accurately aligned.

Conclusions

1. Review the radiographic image produced with the collimator device. Measure the exact limits of the recorded image of the paper clips and the actual radiographic image of the beam limitation device, and record the total area of coverage (length × width) in the space provided below:

   Radiographic image: _______ length × _______ width = _______ total area

   Paper clip image: _______ length × _______ width = _______ total area

2. Does the image of the borders of the recorded paper clip coincide exactly with the radiographic image of the beam coverage? If not, describe the difference between the two recorded images. **Note:** Report any differences to your instructor, supervisor, or quality control technologist.

3. Review the edges of the radiographic image produced for evidence of unsharpness (penumbra) or undercutting caused by the collimator. Describe any evidence of penumbra present.

4. Compare the recorded image with the images produced in Experiment 58 using the aperture diaphragm and cone. Which image appears to produce the least edge undercutting (penumbra)? Why?

Name ______________________________ Date ______________

EXPERIMENT 60
RADIOGRAPHIC CONTRAST—Radiographic Fog: Beam Restriction

Objective

To demonstrate the influence of scattered radiation, the ability to reduce the radiographic fog, and the improvement of the image visibility attributed to the proper application of beam limitation.

Procedure

Using the pelvis phantom, a series of two radiographs will be taken with the hip centered to the midline of the x-ray table. The procedure will use the Bucky mechanism and employ an 8" by 10" (20 cm × 25 cm) slow speed (detail) film-screen recording system cassette. Using a 40" (102 cm) SID and 80 kVp, center your film within the Bucky tray to the hip joint. Align your central ray using a perpendicular angle to the center of the film.

Film 1: Proper Beam Restriction

1. Set up your procedure as described above. Select the appropriate mAs to produce a 1.0 to 1.2 bone density of the acetabulum.
2. Collimate your beam restriction to a size slightly smaller than the film size. Ideally, evidence of beam limitation will be recorded within the radiographic image.
3. Mark your film with a lead number 1. Expose and process your film.

Film 2: Improper Beam Restriction

1. Set up your procedure as described above for Film 1. Use the same technical exposure factors.
2. Open your collimator to its largest possible limits.
3. Mark your film with a lead number 2. Expose and process your film.

Analysis

Place the two radiographs side by side on an illuminator, and review them for overall density and radiographic contrast.

Conclusions

1. Select the same three areas within the radiographic images of the hip (the acetabulum, the femoral neck where it meets with the trochanter, and the wing of the ilium), and circle them on each radiograph. Using a densitometer, measure and record the radiographic densities in the spaces provided below:

   Film 1: _______ acetabulum density _______ femoral neck density

   _______ ilium wing density

   Film 2: _______ acetabulum density _______ femoral neck density

   _______ ilium wing density

2. Describe the differences in radiographic density and the visibility of the recorded details, if any, between the two radiographic images, and explain the origins of these differences.

3. Considering the overall recorded densities, which radiographic image provides for greater visibility of details and thus better radiographic contrast? Provide reasons for your choice.

4. Describe the influence of beam limitation on the production of a film of radiographic quality.

5. Describe the proper application of beam limitation in all radiographic examinations.

Name ______________________________ Date ______________

EXPERIMENT 61

RADIOGRAPHIC CONTRAST—Radiographic Fog: Beam Restriction and Density Control

Objective

To demonstrate the influence of radiographic fog on the radiographic density of the recorded image when significant beam limitation is employed.

Procedure

A series of three radiographs will be produced using the pelvis phantom centered to the midline of the x-ray table with a Bucky mechanism. Place a 14" by 17" (36 cm × 43 cm) slow speed (detail) film-screen recording system cassette in the Bucky tray, and center the film to the phantom. Using a 40" (102 cm) SID and a perpendicular angle of the central ray, select 80 kVp and the appropriate mAs to produce a 1.0 to 1.2 bone density.

Film 1: Proper Collimation for a Pelvis

1. Set up your procedure as indicated above. Open up your collimation to cover the entire film size. Mark the film with a lead number 1.
2. Expose and process the film.

Film 2: Proper Collimation for a Lumbosacral Articulation

1. Set up your procedure the same as for Film 1. Use the same size film, centering, and technical exposure factors.
2. Collimate your beam to cover an area of 6" (15 cm) wide by 8" (20 cm) long at the surface of the film. Mark the film with a lead number 2.
3. Expose and process the film.

Film 3: Proper Collimation for a Lumbosacral Articulation and Increased Exposure

1. Set up your procedure the same as for Film 2. Use the same size film, centering, technical exposure factors, and collimation to an area on the film of 6" by 8" (15 cm × 20 cm). Mark the film with a lead number 3.
2. Using the same 80 kVp, increase your mAs by 50% from the amount utilized for Film 1.
3. Expose and process the film.

Analysis

Compare the radiographic contrast, density, and visibility of the image among the three radiographs. You can now review the effect of significant beam limitation on the production of radiographic fog and its influence on the radiographic contrast, density, and visibility of the recorded image.

Conclusions

1. Identify and circle a midline bone structure on Film 3. Encircle the same structure on Films 1 and 2. Using a densitometer, measure the same midline structure density of all three radiographic images, and record their density in the spaces provided below:

   Film 1: _______ phantom density
   Film 2: _______ phantom density
   Film 3: _______ phantom density

2. Compare the radiographic image of Film 1 to Film 2 related to the overall visibility of the image (radiographic density and contrast). Is there a significant difference between the two images? Both images utilized the same exposure factors; therefore, if there is a difference, why has this occurred?

3. Compare the radiographic image of Film 1 to Film 3 related to the overall visibility of the image (radiographic density and contrast). Is the overall visibility of the images similar? How has the increase in exposure factors compensated for the loss of radiographic density due to the limitation of the beam?

4. You have performed a routine KUB (abdomen) on an adult, using a 14" by 17" (36 cm × 43 cm) film as a scout film for a combination upper gastrointestinal and cholecystogram procedure. Describe what will happen to the radiographic appearance of the recorded image and the appropriate adjustments in the selection of exposure factors necessary to maintain the radiographic density when additional films of the gallbladder are taken collimated to an image size of 6" by 8" (15 cm × 20 cm).

5. Indicate an additional radiographic procedure where the effect of significant beam limitation would require exposure factor adjustments in order to maintain the radiographic density.

Name ____ Date ________________

EXPERIMENT 62
RADIOGRAPHIC CONTRAST—Radiographic Fog: Lead Blockers

Objective

To demonstrate the effective use of lead blockers to reduce the influence of radiographic fog and to improve the radiographic contrast and visibility of the details of the recorded image in specific but limited applications.

Procedure

Using a pelvis phantom, place the phantom on its side to produce a lateral projection of the sacrum and coccyx. Center the sacrum and coccyx to the midline of the x-ray table, and secure the phantom in position with tape. Using a 10" by 12" (25 cm × 30 cm) slow speed (detail) film-screen recording system cassette in the Bucky tray, collimate the beam of radiation to the size of the film. Center the central ray to include the sacrum and coccyx, and center the film in the Bucky tray to the center of the central ray using a 40" (102 cm) SID.

Film 1: Collimated to Film Size

1. Set up your procedure as indicated above. Mark the film with a lead number 1.
2. Using 85 kVp, select sufficient mAs to produce a bone density of the sacrum of 1.0 to 1.2.
3. Expose and process your film.

Film 2: Collimated to Film Size and Using Lead Blockers

1. Set up the procedure the same as for Film 1 including the same technical exposure factors. Mark the film with a lead number 2.
2. Referring to Chapter 5, Figure 5-11, arrange a flat lead rubber blocker along the edge of the phantom's posterior margin at the level of the sacrum and coccyx and an additional lead blocker adjacent to the area of the buttock and thigh.
3. Using the same technical exposure factors as in Film 1, expose and process your film.

Analysis

Review the two radiographic images side by side, and compare the overall radiographic density of each as well as the visibility of the details and radiographic contrast of the recorded images.

Conclusions

1. Locate and circle the sacrococcygeal junction for each recorded image. Using a densitometer, determine and record the radiographic density of the selected area in the space provided below:

   Film 1: _______ sacrococcygeal density

   Film 2: _______ sacrococcygeal density

2. Describe the difference, if any, in the overall radiographic density between the two radiographic images. Why has this occurred?

3. Describe the difference, if any, in the visibility of the details and radiographic contrast between the two radiographic images.

4. Which radiographic image more closely approximates an image of radiographic quality? Explain.

5. How has the use of lead blockers contributed to the radiographic quality of the recorded image?

6. Identify at least three other projections or examinations that could benefit from the application of lead blockers to improve the radiographic contrast of the recorded image, and describe how you would use them to advantage.

Name ____ Date ________________

EXPERIMENT 63

RADIOGRAPHIC CONTRAST—Radiographic Fog: Scattered Radiation

Objective

To demonstrate the multidirectional relationship of scattered radiation, the radiographic fog it produces, and the reasons it must be controlled during the production of radiographic images.

Procedure

Arrange the pelvis phantom to the midline of the table as for a pelvis examination. Center your central ray appropriately to the phantom using a perpendicular beam and 40" (102 cm) SID. Open your collimator to cover adequately a 14" by 17" (36 cm × 43 cm) slow speed (detail) film-screen recording system cassette. Using 80 kVp, select the required mAs to produce a 1.0 to 1.2 bone density within the pelvis.

Measure a distance 12" (30 cm) from the top of the phantom toward the head of your x-ray table, and place your cassette on the tabletop in a vertical position with its front facing the phantom. Secure it in this position with a cassette holder or tape. A flat metal grid such as that employed for testing screen contact or a flat piece of wire mesh approximating the size of the film is taped to the cassette front (see the diagram below). Expose the phantom, and process the film placed adjacent to the phantom.

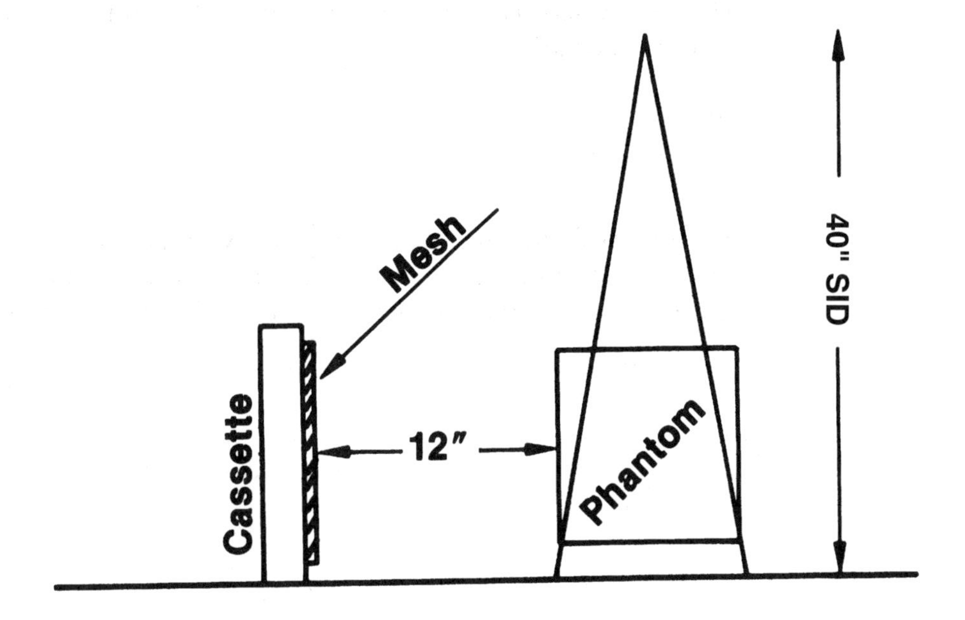

Analysis

Place the radiograph on an illuminator, and observe the amount of radiographic density and the pattern of scattered radiation produced by the phantom part.

Conclusions

1. Encircle a small area within five sections of the recorded image (center of the film, left and right film edges, top and bottom film edges). Using a densitometer, determine the density of the areas, and record this information in the spaces provided below:

   _______ Left film edge density

   _______ Right film edge density

   _______ Film center density

   _______ Top film edge density

   _______ Bottom film edge density

2. How was the radiographic density recorded on the film produced since the film was not in the direct beam of radiation, and the size of the beam of radiation was limited to the size of the film normally used for a pelvis examination? Is the radiographic density uniform throughout the entire recorded image?

3. Can you identify the pattern of the screen contact grid (wire mesh) on the radiograph? Is the pattern sharply recorded, or is there an unsharpness associated with the recorded image? Explain how the sharpness and unsharpness of the recorded image were produced.

4. Can scattered radiation produce a significant influence on radiographic contrast and image visibility? Is scattered radiation a significant contributor to radiographic fog? Explain your answers.

Name ______________________________ Date ______________

EXPERIMENT 64
RADIOGRAPHIC CONTRAST—Radiographic Fog: Nongrid versus Grid

Objective

To demonstrate the effectiveness of a radiographic grid in the clean-up of radiographic fog produced by scattered radiation.

Procedure

A series of three radiographs will be produced using both tabletop and Bucky techniques. All three will use a pelvis phantom aligned to the center of the table as if for a hip examination. Ten inch by twelve inch (25 cm × 30 cm) slow speed (detail) film-screen recording system cassettes are employed using a perpendicular angle of the central ray, collimated to the size of the film.

Film 1: Bucky Procedure

1. Set up your phantom as indicated above, place your cassette in your Bucky tray, and center your central ray for a hip examination. Align your film to your central ray and phantom. Mark your film with a lead number 1.
2. Using 80 kVp, select the required mAs to produce a 1.0 to 1.2 bone density within the acetabulum of the phantom.
3. Expose and process your film.

Film 2: Tabletop Procedure, Same Exposure

1. Set up your procedure the same as for Film 1. However, place your cassette directly upon the tabletop centered to the midline, and tape it in position so that it does not move when the phantom is placed upon it. Mark your film with the lead number 2.
2. Using the same technical exposure factors as for Film 1, expose and process your film.

Film 3: Tabletop Procedure, Reduced Exposure

1. Set up your procedure exactly as you did for Film 2. Mark your film with the lead number 3.

2. Determine the grid ratio of your Bucky mechanism. Maintaining the same 80 kVp, reduce your technical exposure factors (mAs) according to the grid ratio relationship identified in Chapter 5, Table 5-3, going from the mAs factor associated with the grid ratio you have used to a nongrid mAs.
3. Expose and process your film.

Analysis

You can compare the effect of a grid on radiographic density and the influence of scatter radiation on the visibility of the image. The overall radiographic density of Film 1 and Film 3 should be comparable. If not, adjust the technique of your tabletop procedure to equalize the overall density of the Bucky procedure, and repeat Film 3.

Conclusions

1. Make a circle around several areas of the recorded image (the acetabulum, the ilium wing, and a midline recorded image at the center of the film) on each of the radiographic images. Using a densitometer, determine those densities, and record them in the spaces provided below:

   Film 1: _______ acetabulum density _______ ilium wing density

   _______ midline image density

   Film 2: _______ acetabulum density _______ ilium wing density

   _______ midline image density

   Film 3: _______ acetabulum density _______ ilium wing density

   _______ midline image density

2. Compare the radiographic image of Film 1 with that of Film 2 (both using the same technical exposure factors). Describe the differences, if any, related to radiographic density, radiographic contrast, visibility of details, and radiographic fog. Discuss the factors or influences that may have contributed to any changes observed.

3. Compare the *overall* radiographic density of Film 1 to that of Film 3. Is it similar? Describe the radiographic scale of contrast between the two recorded images. Which film more closely approximates an image of radiographic quality? Explain your answer.

4. Should body parts measuring over 4" (10 cm) or requiring greater than 60 kVp be performed as tabletop procedures? Explain your answer.

5. Review the radiographic images produced by Film 1 and Film 3, and describe the principle of operation of the radiographic grid as a means of controlling scatter radiation and improving the visibility and radiographic contrast of the image.

Name ______________________________ Date ______________

EXPERIMENT 65
RADIOGRAPHIC CONTRAST—Parallel Grid: Source-Image Distance

Objective

To demonstrate the influence of different SIDs on the efficiency of a parallel grid as it relates to grid cutoff and the loss of the radiographic density of the image.

Procedure

A series of three radiographs will be produced using different SIDs. Select a 14" by 17" (36 cm × 43 cm) slow speed (detail) film-screen recording system parallel grid cassette or a parallel wafer-type grid taped to this size cassette. Set up for a tabletop procedure centering the cassette to the tabletop and taping it into position. Place a pelvis phantom upon the cassette centered as for an anteroposterior pelvis examination. Using a perpendicular angle, center your central ray to the middle of the film, and open the collimator to produce an image size equal to the size of the film. An 8:1 grid ratio can be used effectively for this experiment.

Film 1: 40" (102 cm) SID

1. Set up your procedure as indicated above using 40" (102 cm) SID. Mark your film with a lead number 1.
2. Using 80 kVp, select the appropriate mAs for the grid ratio employed to produce a 1.0 to 1.2 bone density within the acetabulum of the pelvis phantom.
3. Expose and process your film.

Film 2: 30" (76 cm) SID

1. Set up your procedure as you did for Film 1. Mark your film with a lead number 2.
2. Reduce your SID to 30" (76 cm), and open your collimator to cover the size of the film. **Note:** Depending on the equipment, it may be necessary to remove your collimator entirely in order to cover properly the desired image size at this reduced distance and not have image cutoff at the periphery of the film caused by the limitation of the collimator coverage at the reduced distance.
3. Using the same 80 kVp, adjust your mAs according to the inverse square law to maintain the radiographic density of the recorded image at the new distance.
4. Expose and process your film.

Film 3: 60" (153 cm) SID

1. Set up your procedure as you did for Film 1. Mark your film with a lead number 3.
2. Increase your SID to 60" (153 cm), and adjust your collimator to cover the size of the film. **Note:** Depending on the equipment, it may be necessary to move your materials to the floor and arrange your procedure there in order to achieve the required distance for Film 3.
3. Using the same 80 kVp, adjust your mAs according to the inverse square law to maintain the radiographic density of the recorded image at the new distance.
4. Expose and process your film.

Analysis

Place the radiographs side by side on illuminators, and compare their radiographic densities and overall image visibility at the middle of the images and at the periphery of the film near the top and bottom of the recorded image.

Conclusions

1. Using a densitometer, measure the radiographic density at the center of the film for the top, bottom, and middle of the recorded image in each radiograph. Record the density in the spaces provided below:

   Film 1: _______ density at the top of the recorded image
   _______ density at the middle of the recorded image
   _______ density at the bottom of the recorded image
   Film 2: _______ density at the top of the recorded image
   _______ density at the middle of the recorded image
   _______ density at the bottom of the recorded image
   Film 3: _______ density at the top of the recorded image
   _______ density at the middle of the recorded image
   _______ density at the bottom of the recorded image

2. Compare the radiographic densities at the middle of the recorded image for each of the three radiographs. Are they similar? Describe any differences in the image visibility between the three radiographs in the central portion of the recorded images.

3. Compare the radiographic densities at the top and bottom of the recorded images for Film 1 and Film 3. Are they similar? Is any grid cutoff evident? Describe the appearance of any grid cutoff and any differences in the image visibility between the two radiographic images in the peripheral portions of the recorded images.

4. Compare the radiographic densities at the top and bottom of the recorded images for Film 1 and Film 2. Are they similar? Is any grid cutoff evident? Describe the appearance of any grid cutoff and any differences in the image visibility between the two radiographic images in the peripheral portions of the recorded images.

5. Discuss the influence of the SID as it relates to grid cutoff when using a parallel grid. Identify the grid ratio used for this experiment. How would the grid cutoff be influenced by an increase/decrease in the grid ratio?

6. What precautions, if any, should be considered relative to the use of a parallel grid when changes in the SID are required?

7. Would the same precautions described in Question 6 for the examination of a large body part requiring a 14" by 17" (36 cm × 43 cm) film size be necessary when examining a small body part requiring an 8" by 10" (20 cm × 25 cm) film size? Explain.

8. In which radiographic image are the grid lines more obvious and visible? Explain.

Name ______________________________ Date ______________

EXPERIMENT 66

RADIOGRAPHIC CONTRAST—Parallel Grid: Angling Against the Direction of the Grid Lines

Objective

To demonstrate the influence of grid cutoff produced by the angulation of the central ray against the linear pattern of the grid lines when using a parallel grid. To demonstrate that a margin for error exists related to the application of radiographic grids.

Procedure

A series of three radiographs will be produced using different directions (angles) of the central ray. Select a 14" by 17" (36 cm × 43 cm) slow speed (detail) film-screen recording system parallel grid cassette or a parallel wafer-type grid taped to this size cassette. Set up for a tabletop procedure centering the cassette to the tabletop and taping it into position. Place a pelvis phantom upon the cassette centered as for an anteroposterior pelvis examination. Using 40" (102 cm) SID, collimate your beam to include the size of the film. Using 80 kVp, select the appropriate mAs for the grid ratio employed to produce a 1.0 to 1.2 bone density within the acetabulum of the pelvis phantom. An 8:1 grid ratio can be used effectively for this experiment.

Film 1: Central Ray Perpendicular

1. Set up your procedure as indicated above, marking your film with a lead number 1.
2. Using a perpendicular angle, center your central ray to the middle of the film, and open the collimator to produce an image size equal to the size of the film.
3. Expose and process your film.

Film 2: Central Ray Angled 5° Against the Grid Lines

1. Set up your procedure as indicated above, marking your film with a lead number 2.
2. Move your x-ray tube inferior to the phantom part, and angle back in a cephalic direction using a 5° angle of the central ray. Center your central ray to emerge through the middle of the film, and adjust the collimator to produce an image size equal to the size of the film.
3. Expose and process your film.

Film 3: Central Ray Angled 15° Against the Grid Lines

1. Set up your procedure as indicated above, marking your film with a lead number 3.
2. Move your x-ray tube inferior to the phantom part, and angle back in a cephalic direction using a 15° angle of the central ray. Center your central ray to emerge through the middle of the film, and adjust the collimator to produce an image size equal to the size of the film. Reduce your vertical tube height by 1.5" (4.8 cm) to compensate for the angle of the central ray.
3. Expose and process your film.

Analysis

Place the radiographs side by side in order of exposure on illuminators. Review the films for any changes in the overall radiographic density of the images. Observe for any patterns of recorded image loss due to grid cutoff.

Conclusions

1. Compare Film 1 with Film 2. Does either radiographic image demonstrate any obvious grid cutoff? If so, describe it. Are the overall radiographic densities of the two images similar? Compare the anatomical appearance and relationships of the recorded anatomy of Film 1 to those of Film 2, and describe the differences, if any, that are observed.

2. Compare Film 1 with Film 3. Does Film 3 demonstrate any grid cutoff? Describe the appearance of the grid cutoff, if any. Are the overall radiographic densities of the two images the same? Explain as needed. Compare the anatomical appearance and relationships of the recorded anatomy of Film 1 to those of Film 3, and describe the differences, if any, that are observed.

3. Is minimal angulation of the central ray against the grid lines a critical consideration in performing your radiographic procedure related to grid cutoff? Related to distortion? Is there a margin for error in the application of a radiographic grid? Explain.

4. In image visibility problems identified with grid cutoff, are the undesirable influences uniform throughout the entire recorded image? Explain.

5. Describe the limitations imposed on central ray angulation when employing a parallel grid. How would the grid ratio contribute to these limitations?

Name ______________________________ Date ______________

EXPERIMENT 67

RADIOGRAPHIC CONTRAST—Focused Grid: Source-Image Distance

Objective

To demonstrate the influence of different SIDs on the efficiency of a focused grid as it relates to grid cutoff and the loss of the radiographic density of the image.

Procedure

A series of three radiographs will be produced using different SIDs. Place a 14" by 17" (36 cm × 43 cm) slow speed (detail) film-screen recording system cassette in the Bucky tray. If possible, eliminate the motion of the Bucky mechanism so the grid lines will appear within the image. An alternative would be to use a wafer-type focused grid attached to a 14" by 17" (36 cm × 43 cm) cassette, and perform the experiment as a tabletop procedure. Depending on whether the experiment is performed as a tabletop or Bucky procedure, use the pelvis phantom centered as for an anteroposterior pelvis examination. Using a perpendicular angle, center your central ray to the middle of the film, and open the collimator to produce an image size equal to the size of the film. A 12:1 grid ratio can be used effectively for this experiment.

Film 1: 40" (102 cm) SID

1. Set up your procedure as indicated above using a 40" (102 cm) SID. Mark your film with a lead number 1.
2. Using 80 kVp, select the appropriate mAs for the grid ratio employed to produce a 1.0 to 1.2 bone density within the acetabulum of the pelvis phantom.
3. Expose and process your film.

Film 2: 30" (76 cm) SID

1. Set up your procedure as you did for Film 1. Mark your film with a lead number 2.
2. Reduce your SID to 30" (76 cm), and open your collimator to cover the size of the film. **Note:** Depending on the equipment, it may be necessary to remove your collimator entirely in order to cover properly the desired image size at this reduced distance and not have image cutoff at the periphery of the film caused by the limitation of the collimator coverage at the reduced distance.

3. Using the same 80 kVp, adjust your mAs according to the inverse square law to maintain the radiographic density of the recorded image at the new distance.
4. Expose and process your film.

Film 3: 60" (153 cm) SID

1. Set up your procedure as you did for Film 1. Mark your film with a lead number 3.
2. Increase your SID to 60" (153 cm) or to the greatest height that your equipment will allow you to use, and adjust your collimator to cover the size of the film.
3. Using the same 80 kVp, adjust your mAs according to the inverse square law to maintain the radiographic density of the recorded image at the new distance.
4. Expose and process your film.

Analysis

Place the radiographs side by side on illuminators, and compare their radiographic densities and overall image visibility at the middle of the images and at the periphery of the film near the top and bottom of the recorded image.

Conclusions

1. Using a densitometer, measure the radiographic density at the center of the film for the top, bottom, and middle of the recorded image in each radiograph. Record the density in the spaces provided below:

Film 1: _______ density at the top of the recorded image
_______ density at the middle of the recorded image
_______ density at the bottom of the recorded image

Film 2: _______ density at the top of the recorded image
_______ density at the middle of the recorded image
_______ density at the bottom of the recorded image

Film 3: _______ density at the top of the recorded image
_______ density at the middle of the recorded image
_______ density at the bottom of the recorded image

2. Compare the radiographic densities at the middle of the recorded image for each of the three radiographs. Are they similar? Describe any differences in the image visibility between the three radiographs in the central portion of the recorded images.

3. Compare the radiographic densities at the top and bottom of the recorded images for Film 1 and Film 3. Are they similar? Is any grid cutoff evident? Describe the appearance of any grid cutoff and any differences in the image visibility between the two radiographic images in the peripheral portions of the recorded images.

4. Compare the radiographic densities at the top and bottom of the recorded images for Film 1 and Film 2. Are they similar? Is any grid cutoff evident? Describe the appearance of any grid cutoff and any differences in the image visibility between the two radiographic images in the peripheral portions of the recorded images.

5. Discuss the influence of the SID as it relates to grid cutoff when using a focused grid. Identify the grid ratio used for this experiment. How would the grid cutoff be influenced by an increase/decrease in the grid ratio?

6. What precautions, if any, should be considered relative to the use of a focused grid when changes in the SID are required? How does the grid focusing distance enter into this consideration?

7. Would the same precautions described in Question 6 for the examination of a large body part requiring a 14" by 17" (36 cm × 43 cm) film size be necessary when examining a small body part requiring an 8" by 10" (20 cm × 25 cm) film size? Explain.

8. In which radiographic image are the grid lines more obvious and visible? Explain.

9. Compare the results of this experiment with the results of Experiment 65 using a parallel grid. With which type of grid would the SID be a more critical factor? Explain.

Name ______________________________ Date ______________

EXPERIMENT 68

RADIOGRAPHIC CONTRAST—Focused Grid: Angling Against the Direction of the Grid Lines

Objective

To demonstrate the influence of grid cutoff produced by the angulation of the central ray against the linear pattern of the grid lines when using a focused grid. To demonstrate that a margin for error exists related to the application of radiographic grids.

Procedure

A series of three radiographs will be produced using different directions (angles) of the central ray. Select a 14" by 17" (36 cm × 43 cm) slow speed (detail) film-screen recording system cassette, and place it in the Bucky tray lengthwise. (The length of the film should coincide with the longitudinal axis of the tabletop.) Set up for a Bucky procedure centering the pelvis phantom to the tabletop for an anteroposterior pelvis examination, but arrange the phantom to be centered with the longitudinal axis of the table rather than to the table width as you would normally position it. Using a 40" (102 cm) SID, collimate your beam to include the size of the film, remembering the position of the phantom and the film in the Bucky tray is opposite its normal placement for an anteroposterior pelvis examination. Using 80 kVp, select the appropriate mAs for the grid ratio employed to produce a 1.0 to 1.2 bone density within the acetabulum of the pelvis phantom. A 12:1 grid ratio can be used effectively for this experiment.

Film 1: Central Ray Perpendicular

1. Set up your procedure as indicated above, marking your film with a lead number 1.
2. Using a perpendicular angle, center your central ray to the middle of the film, and open the collimator to produce an image size equal to the size of the film. Align the position of your film in the Bucky tray to the central ray.
3. Expose and process your film.

Film 2: Central Ray Angled 5° Against the Grid Lines

1. Set up your procedure as indicated above, marking your film with a lead number 2.

2. The grid lines in a Bucky mechanism grid are aligned to go in the direction of the longitudinal axis of the tabletop. Depending on the mounting of your x-ray tube (tube stand or ceiling mounted), move your x-ray tube out from the center of the tabletop, and angle it back toward the midline using a 5° angle of the central ray. You will be angling in a cephalic direction to your phantom part. Center your central ray to emerge through the middle of the film, and adjust the collimator to produce an image size equal to the size of the film.
3. Expose and process your film.

Film 3: Central Ray Angled 15° Against the Grid Lines

1. Set up your procedure as indicated above, marking your film with a lead number 3.
2. Adjust to a 15° angle of the central ray. Center your central ray to emerge through the middle of the film, and adjust the collimator to produce an image size equal to the size of the film. Reduce your vertical tube height by 1.5" (4.8 cm) to compensate for the angle of the central ray.
3. Expose and process your film.

Analysis

Place the radiographs side by side in order of exposure on illuminators. Review the films for any changes in the overall radiographic density of the images. Observe for any patterns of recorded image loss due to grid cutoff.

Conclusions

1. Compare Film 1 with Film 2. Does either radiographic image demonstrate any obvious grid cutoff? If so, describe it. Are the overall radiographic densities of the two images similar? Compare the anatomical appearance and relationships of the recorded anatomy of Film 1 to those of Film 2, and describe the differences, if any, that are observed.

2. Compare Film 1 with Film 3. Does Film 3 demonstrate any grid cutoff? Describe the appearance of the grid cutoff, if any. Are the overall radiographic densities of the two images the same? Explain as needed. Compare the anatomical appearance and relationships of the recorded anatomy of Film 1 to those of Film 3, and describe the differences, if any, that are observed.

3. Is minimal angulation of the central ray against the grid lines a critical consideration in performing your radiographic procedure related to grid cutoff? Related to distortion? Is there a margin for error in the application of a radiographic grid? Explain.

4. In image visibility problems identified with grid cutoff, are the undesirable influences uniform throughout the entire recorded image? Explain.

5. Describe the limitations imposed on central ray angulation when employing a parallel grid. How would the grid ratio contribute to these limitations?

6. Compare the results of this experiment with those achieved in Experiment 66. Which type of grid appears more critical related to angulation against the direction of the grid lines? Comparing Film 3 of both experiments, which demonstrates greater grid cutoff? Consider any differences in the grid ratio between the two experiments in your answer.

Name ______________________________ Date ______________

EXPERIMENT 69
RADIOGRAPHIC CONTRAST—Focused Grid: Off-Centering

Objective

To demonstrate the influence of grid cutoff produced by the off-centering of the central ray from the middle of a focused grid and the loss of radiographic density and image visibility.

Procedure

Two radiographs will be produced using a 14" by 17" (36 cm × 43 cm) slow speed (detail) film-screen recording system cassette placed in the Bucky tray lengthwise. (The length of the film should coincide with the longitudinal axis of the tabletop.) Set up for a Bucky procedure centering the pelvis phantom to the tabletop for an anteroposterior pelvis examination, but arrange the phantom to be centered with the longitudinal axis of the table rather than to the table width as you would normally position it. Using 40" (102 cm) SID, collimate your beam to include the size of the film, remembering the position of the phantom and the film in the Bucky tray is opposite its normal placement for an anteroposterior pelvis examination. Using 80 kVp, select the appropriate mAs for the grid ratio employed to produce a 1.0 to 1.2 bone density within the acetabulum of the pelvis phantom. A 12:1 grid ratio can be used effectively for this experiment.

Film 1: Central Ray Centered to the Midline of the X-Ray Table

1. Set up your procedure as indicated above, marking your film with a lead number 1.
2. Using a perpendicular angle, center your central ray to the middle of the film, and open the collimator to produce an image size equal to the size of the film. Align the position of your film in the Bucky tray to the central ray.
3. Expose and process your film.

Film 2: Central Ray Off-Centered from the Midline of the X-Ray Table

1. Set up your procedure the same as for Film 1, marking your film with a lead number 2.
2. Release the x-ray tube centering locks, and manually move your tube off-center from the middle of the tabletop to the side a total of 4" (10 cm). Open your collimator in order to cover the film size and phantom, which is still centered to the midline of the table.

3. Using the same technical exposure factors for Film 1, expose and process your film.

Analysis

Place the radiographs side by side on illuminators, and compare their radiographic densities and the overall image visibility at the middle of the images and at their peripheries.

Conclusions

1. Using a densitometer, measure the radiographic density at the center of the film at the top, bottom, and middle of the recorded image in each radiograph. Record the density in the spaces provided below:

   Film 1: _______ density at the top of the recorded image

   _______ density at the middle of the recorded image

   _______ density at the bottom of the recorded image

   Film 2: _______ density at the top of the recorded image

   _______ density at the middle of the recorded image

   _______ density at the bottom of the recorded image

2. Compare the radiographic density and overall visibility at the center, top, and bottom of the recorded image in Film 1 with the same areas in Film 2. How has the off-centering of the central ray affected the recorded image? Has grid cutoff occurred? If so, is the grid cutoff the same at the top and bottom of the recorded image? Describe and explain any observed differences in radiographic quality.

3. Discuss the influence of central ray centering when employing a focused grid. Would this consideration be the same, increased, or decreased when using a parallel grid?

4. Would the same precautions described in Question 3 related to the examination of a large body part (requiring a 14" × 17" or 36 cm × 43 cm film size) be necessary when examining a small body part (requiring an 8" × 10" or 20 cm × 25 cm film size)? Explain your answer.

Name ______________________________ Date ______________

EXPERIMENT 70
RADIOGRAPHIC CONTRAST—Focused Grid: Tube Side Down

Objective

To demonstrate the objectionable pattern of grid cutoff and the loss of density and image visibility attributed to using a focused grid improperly, with the tube side placed toward the direction of the film.

Procedure

Place a 14" by 17" (36 cm × 43 cm) slow speed (detail) film-screen recording system cassette in the tray of a Bucky mechanism using a 12:1 ratio focused grid, and eliminate the motion of the grid, if possible. Alternatively, take the focused grid out of the Bucky mechanism, and center it on the top of the cassette for a tabletop procedure (tube side up). Place the pelvis phantom on the tabletop centered to the cassette as for an anteroposterior projection of the pelvis. Using a 40" (102 cm) SID and a perpendicular beam, center your central ray to the phantom/grid/cassette arrangement. Collimate your beam to produce an image size equal to the size of the film. Using 80 kVp, select the necessary mAs to produce a 1.0 to 1.2 bone density in the acetabulum of the pelvis phantom.

Film 1: Focused Grid, Tube Side Up

1. Set up your procedure as described above, and mark your film with the lead number 1.
2. Expose and process your film.

Film 2: Focused Grid, Tube Side Down

1. Reverse the position of your focused grid, placing the tube side down.
2. Set up your procedure the same as for Film 1, and mark your film with the lead number 2.
3. Using the same technical exposure factors as for Film 1, expose and process your film.

Analysis

Compare the radiographic images of the two films for grid cutoff and loss of radiographic density and visibility of the recorded image.

Conclusions

1. Using a densitometer, measure the radiographic density at the center of the film at the left edge, middle, and right edge of the recorded image in each radiograph. Record the density in the spaces provided below:

   Film 1: _______ density at the left edge of the recorded image

   _______ density at the middle of the recorded image

   _______ density at the right edge of the recorded image

   Film 2: _______ density at the left edge of the recorded image

   _______ density at the middle of the recorded image

   _______ density at the right edge of the recorded image

2. Compare the overall radiographic densities of Film 1 and Film 2. Has the reversal of the position of the focused grid had any influence on the radiographic density of the recorded image? Describe the difference, if any, observed between the recorded images of the two films.

3. Has there been any loss of image visibility associated with the grid cutoff caused by the improper use of the focused grid? If so, describe the appearance of this image loss.

4. Compare the grid cutoff produced by using the focused grid with the tube side down with the grid cutoff attributed to SID changes in Experiment 67, angling against the grid lines in Experiment 68, and off-centering against the focused grid in Experiment 69. Which factor appears to demonstrate the greater influence on the overall quality of the radiographic image? Explain your answer.

5. Would the reversal of a parallel grid produce a similar influence as the reversal of the tube side of a focused grid? Explain your answer.

Name ____ Date ________________

EXPERIMENT 71
RADIOGRAPHIC CONTRAST—Moving versus Stationary Grids

Objective

To demonstrate the objectionable image of recorded grid lines with a stationary focused grid compared to the image produced when the grid is in motion.

Procedure

Two radiographs will be produced using a 14" by 17" (36 cm × 43 cm) slow speed (detail) film-screen recording system cassette in the tray of a Bucky mechanism using a 12:1 ratio focused grid. Place the pelvis phantom on the tabletop centered to the midline as for an anteroposterior projection of the pelvis, and center your film in the Bucky tray to the phantom. Using a 40" (102 cm) SID and a perpendicular beam, center your central ray to the phantom/film arrangement. Collimate your beam to produce an image size equal to the size of the film.

Film 1: Moving Grid

1. Set up your procedure as indicated above, and mark your film with a lead number 1.
2. Using 80 kVp, select the required mAs to produce a 1.0 to 1.2 bone density of the acetabulum in the pelvis phantom.
3. Expose and process your film.

Film 2: Eliminate the Motion of the Bucky Mechanism

1. Set up your procedure the same as for Film 1. Use the same technical exposure factors, and mark the film with a lead number 2.
2. Eliminate the movement of your Bucky. Many pieces of equipment provide a toggle switch that turns off the Bucky movement. If this is not possible, remove the grid from the Bucky mechanism, lay it upon the surface of the cassette, and perform Film 2 as a tabletop procedure. **Note:** Be sure to increase your SID for the tabletop procedure in order to maintain the same SID used for Film 1.
3. Expose and process your film.

Analysis

Compare the overall radiographic density, contrast, and visibility of the radiographic image in each radiograph. Consider the influence of the moving versus the stationary grid on the radiographic quality of the image.

Conclusions

1. Are the lead strips of the grid pattern noticeable on either radiograph? Indicate which film demonstrates the grid lines, and describe why this has occurred. Does the appearance of grid lines improve or detract from the visibility of the recorded image? Explain your answer.

2. Explain the principle utilized to eliminate the appearance of the grid lines on a radiograph.

3. Has the recording of the grid lines produced any differences in the radiographic density, contrast, or visibility of the recorded details? Describe these differences, if any.

4. In the absorption of scattered radiation, would the function of a grid be more efficient if used as a stationary or a moving grid? Explain your answer.

Name ______________________________ Date ______________

EXPERIMENT 72

RADIOGRAPHIC CONTRAST—Using Kilovoltage To Change from Nongrid to Grid

Objective

To demonstrate the influence of the kilovoltage factor in the maintenance of the radiographic density of the image when changing from a nongrid procedure to a procedure employing a grid.

Procedure

A series of five radiographs using tabletop and Bucky procedures will be performed. Using 8" by 10" (20 cm × 25 cm) slow speed (detail) film-screen recording system cassettes at a 40" (102 cm) SID, the central ray is directed perpendicularly and centered to the part-film alignment for each radiograph, and the x-ray beam is collimated to produce an image size just smaller than the size of the film. The nongrid procedures will provide the standard for all of the subsequent grid procedures.

Nongrid Procedure

Film 1: 40 kVp

1. Arrange the hand phantom on your cassette as for a posteroanterior projection of the hand. Using 40 kVp, select the required mAs to produce a 1.0 to 1.2 bone density in the area of the wrist. Mark your film with a lead number 1.
2. Expose and process your film.

Film 2: 60 kVp

1. Arrange the knee phantom on your cassette as for an anteroposterior projection of the knee. Using 60 kVp, select the required mAs to produce a 1.0 to 1.2 bone density in the area of the distal femur. Mark your film with a lead number 2.
2. Expose and process your film.

Grid Procedure

Film 3: 60 kVp

1. Arrange the hand phantom on an 8:1 parallel grid cassette as for a posteroanterior projection of the hand similar to Film 1. Using the same mAs

as for Film 1, increase your kilovoltage by 20 so that it is 60 kVp. Mark your film with a lead number 3.

2. Expose and process your film.

Film 4: 80 kVp

1. Arrange the knee phantom on an 8:1 parallel grid cassette as for an anteroposterior projection of the knee similar to Film 2. Using the same mAs as for Film 2, increase your kilovoltage by 20 so that it is 80 kVp. Mark your film with a lead number 4.
2. Expose and process your film.

Film 5: 80 kVp, Bucky

1. Align the knee phantom to the midline of the table as for an anteroposterior projection of the knee similar to Film 2. Place your film into the Bucky tray using a 12:1 grid ratio. Using the same mAs as for Film 2, increase your kilovoltage by 20 so that it is 80 kVp. Mark your film with a lead number 5.
2. Expose and process your film.

Analysis

An increase of 20 kVp has been suggested by a number of references as a method of maintaining the radiographic density of an image when changing from a nongrid to a grid procedure. An analysis of this series of films will enable you to determine whether an increase of 20 kVp will be satisfactory related to a number of different circumstances, including the application of different kilovoltage levels and different grid ratios.

Conclusions

1. Using a densitometer, measure the radiographic density of the bones for each of the 5 films. Record the density in the spaces provided below:

Film 1: _______ phantom density

Film 2: _______ phantom density

Film 3: _______ phantom density

Film 4: _______ phantom density

Film 5: _______ phantom density

2. Review the overall radiographic density, radiographic contrast, and visibility of the recorded image in Film 1, performed as a nongrid procedure using 40 kVp, with the recorded image of Film 3, performed as an 8:1 parallel grid procedure with a 20-kVp increase. Are the overall appearances of the two images similar? Explain. Describe any differences in radiographic density and radiographic contrast between the two radiographic images. Which of the two images appears to represent the film of radiographic quality? Explain your answer.

3. Review the overall radiographic density, radiographic contrast, and visibility of the recorded image in Film 2, performed as a nongrid procedure using 60 kVp, with the recorded image of Film 4, performed as an 8:1 parallel grid procedure with a 20-kVp increase. Are the overall appearances of the two images similar? Explain. Describe any differences in radiographic density and radiographic contrast between the two radiographic images. Which of the two images appears to represent the film of radiographic quality? Explain your answer.

4. Does the kilovoltage change from 40 to 60 kVp for Films 1 and 3 appear to produce the same influence on the recorded image as the change from 60 to 80 kVp for Films 2 and 4? In what manner would the screen intensification factor influence the results of this adjustment?

5. Compare the recorded images of Films 2, 4, and 5. What happened to the radiographic density, radiographic contrast, and visibility of the recorded image when the grid ratio was increased from 8:1 to 12:1? In what manner would the grid ratio factor influence the results of the suggested kilovoltage adjustment?

6. Does it appear that a standardized 20-kVp increase would serve as a satisfactory adjustment to maintain the radiographic density of the image when changing from a nongrid to a grid procedure over the wide range of useful kilovoltage employed in diagnostic radiography? Explain your answer, giving reasons for your conclusions.

7. Does it appear as though a standardized 20-kVp increase would serve as a satisfactory adjustment to maintain the radiographic density of the image when changing from a nongrid to a grid procedure if different grid ratios are used? Explain your answer, giving reasons for your conclusions.

Name ______________________________ Date ______________

EXPERIMENT 73

RADIOGRAPHIC CONTRAST—Using mAs To Change from Nongrid to Grid

Objective

To demonstrate the influence of the mAs factor in the maintenance of the radiographic density of the image when changing from a nongrid procedure to a procedure employing a grid.

Procedure

A series of four radiographs using tabletop and Bucky procedures will be performed. Using 8" by 10" (20 cm × 25 cm) slow speed (detail) film-screen recording system cassettes at a 40" (102 cm) SID, the central ray is directed perpendicularly and centered to the part-film alignment for each radiograph, and the x-ray beam is collimated to produce an image size just smaller than the size of the film. The nongrid procedure will provide the standard for all of the subsequent grid procedures. A number of different grid ratios will be utilized for the grid procedures.

Nongrid Procedure

Film 1: 60 kVp

1. Arrange the knee phantom on your cassette as for an anteroposterior projection of the knee. Using 60 kVp, select the required mAs to produce a 1.0 to 1.2 bone density in the area of the distal femur. Mark your film with a lead number 2.
2. Expose and process your film.

Grid Procedure

Film 2: 60 kVp, 6:1 Grid Ratio, Parallel Grid

1. Arrange the knee phantom on a 6:1 parallel grid cassette as for an anteroposterior projection of the knee similar to Film 1. Using the same kilovoltage as for Film 1, increase your mAs according to the values identified in Chapter 5, Table 5-3, for the grid ratio used. Mark your film with a lead number 2.
2. Expose and process your film.

Film 3: 60 kVp, 8:1 Grid Ratio, Parallel Grid

1. Arrange the knee phantom on an 8:1 parallel grid cassette as for an anteroposterior projection of the knee similar to Film 2. Using the same kilovoltage as for Film 1, increase your mAs according to the values identified in Chapter 5, Table 5-3, for the grid ratio used. Mark your film with a lead number 3.
2. Expose and process your film.

Film 4: 60 kVp, 12:1 Grid Ratio, Parallel Grid

1. Arrange the knee phantom on a 12:1 parallel grid cassette as for an anteroposterior projection of the knee similar to Film 3. Using the same kilovoltage as for Film 1, increase your mAs according to the values identified in Chapter 5, Table 5-3. Mark your film with a lead number 4. **Note:** It may be necessary to use the Bucky mechanism with a focused grid in order to apply a 12:1 grid ratio.
2. Expose and process your film.

Analysis

An increase in the mAs according to the grid ratio employed has been suggested by a number of references as a method of maintaining the radiographic density of an image when changing from a nongrid to a grid procedure. An analysis of this series of films will enable you to determine whether mAs increases will be satisfactory for this purpose.

Conclusions

1. Using a densitometer, select a point within the distal end of the femur and a point outside and adjacent to the bone in each of the radiographs. Measure the radiographic density of the selected areas for each radiograph, and record the density in the spaces provided below:

   Film 1: _______ femur density; _______ adjacent density

   Film 2: _______ femur density; _______ adjacent density

   Film 3: _______ femur density; _______ adjacent density

   Film 4: _______ femur density; _______ adjacent density

2. Review the overall radiographic density, radiographic contrast, and visibility of the recorded image in Film 1 performed as a nongrid procedure with the recorded images of Films 2, 3, and 4 performed with different grid ratios. Are the overall appearances of the radiographic images similar? Explain. Describe any differences in radiographic density and radiographic contrast between the radiographic images. Are all of the recorded images of diagnostic quality? Explain your answer.

3. Which of the four radiographic images represents an image of radiographic quality? Explain your answer.

4. Discuss the function of mAs adjustments in the maintenance of radiographic density when changing from a nongrid to a grid procedure.

5. Discuss the maintenance of density using mAs adjustments when changing from nongrid to grid procedures as it relates to the overall radiation exposure of the patient.

6. Discuss the maintenance of density using mAs adjustments when changing from nongrid to grid procedures as it relates to possible motion unsharpness.

7. Comparing the use of kilovoltage in Experiment 72 to the use of mAs in this experiment as a method to maintain the radiographic density of an image when changing from a nongrid to a grid procedure, which method appears to be more accurate in achieving this goal? Explain. Which method appears to be less influenced by external factors? Explain.

Name ______________________________ Date ______________

EXPERIMENT 74
RADIOGRAPHIC CONTRAST—Intensifying Screens

Objective

To demonstrate the influence of intensifying screens on the radiographic appearance of the recorded image by comparing the scale of contrast produced with and without intensifying screens.

Procedure

Two series of exposures will be taken of an elbow phantom using direct exposure and intensifying screen techniques. Using a 40" (102 cm) SID, a 14" by 17" (36 cm × 43 cm) film size is employed for each series of exposures. The film for both the direct exposure and the intensifying screen procedures will be divided into four sections with lead rubber blocking dividers. For the initial exposure in both series of exposures, 50 kVp is employed.

Film 1: Intensifying Screen Procedure

Exposure 1

1. Divide the cassette into four equal sections, covering three of them with lead rubber blockers. Place the elbow phantom into the anteroposterior projection, and center it to the uncovered film section.
2. Using a perpendicular angle of the central ray, center the central ray over the phantom part, and collimate to a size slightly smaller than the film section. Mark the film section with a lead number 1.
3. Using 50 kVp, select the mAs necessary to produce a 1.0 to 1.2 bone density at the distal humerus.
4. Expose your film section.

Exposures 2, 3, and 4

1. Adjust the film division for each successive exposure, covering the exposed section and uncovering a new unexposed section. Mark the film sections successively with a lead number 2, 3, and 4.
2. Increase your kilovoltage by 2 kVp for each successive exposure: 52 kVp for Exposure 2, 54 kVp for Exposure 3, and 56 kVp for Exposure 4. All other technical factors remain the same.
3. Expose each film section in turn, and, when completed, process your film.

Film 2: Direct Exposure Procedure

Exposure 5

1. Using screen-type film in a nonscreen (cardboard) film holder cassette, divide the cassette into four equal sections, covering three of them with lead rubber blockers. Place the elbow phantom into the anteroposterior projection, and center it to the uncovered film section.
2. Using a perpendicular angle of the central ray, center the central ray over the phantom part, and collimate to a size slightly smaller than the film section. Mark the film section with a lead number 5.
3. Using 50 kVp, select the mAs necessary to produce a 1.0 to 1.2 bone density at the distal humerus. Increase your mAs by the intensification factor of screens for the kilovoltage selected. **Note:** It might be advantageous to take a separate exposure prior to performing the entire series of exposures to be sure that the radiographic density of the nonscreen procedure is similar to the radiographic density of the first exposure for the intensifying screen procedure.
4. Expose your film section.

Exposures 6, 7, and 8

1. Adjust the film division for each successive exposure, covering the exposed section and uncovering a new unexposed section. Mark the film sections successively with a lead number 6, 7, and 8.
2. Increase your kilovoltage by 2 kVp for each successive exposure: 52 kVp for Exposure 6, 54 kVp for Exposure 7, and 56 kVp for Exposure 8. The mAs used should be the same as for the first exposure of this series. All other technical factors remain the same.
3. Expose each film section in turn, and, when completed, process your film.

Analysis

Review the first exposure on each radiograph taken at 50 kVp. The mAs was adjusted to maintain the radiographic density between these two images. If their overall radiographic density is not similar, repeat the second series of nonscreen exposures until the appropriate mAs is employed so that Exposure 1 and Exposure 5 are comparable. Compare the radiographic density between the two films at the various kilovoltages employed. You can compare the overall differences in radiographic contrast between the two images produced at 50 kVp and the changes that occur when the kilovoltage is increased in each film.

Conclusions

1. Review the overall appearance of the radiographic density produced at 50 kVp in Exposure 1, and compare it with Exposure 5 produced at 50 kVp. Are the overall densities similar? Describe any density differences.

2. Analyze the overall radiographic contrast and visibility of the recorded image produced at 50 kVp in Exposure 1 compared with Exposure 5. Describe the differences in the appearances between the recorded images of the two elbows. Which image would be considered the image possessing greater radiographic quality? Why?

3. Which radiographic image appears to possess the higher scale of contrast? What properties of the procedure were involved in producing this higher contrast scale?

4. Analyze the differences in radiographic density and the visibility of the image as they relate to the radiographic contrast of the recorded image between Film 1 and Film 2 as the kilovoltage is increased from 50 to 52, 54, and 56 kVp. Do the differences in radiographic density and the visibility of the image due to increases in the kilovoltage appear greater with the nonscreen procedure or the procedure employing intensifying screens? Discuss the reasons for this observed phenomenon.

5. From your review of the radiographic appearance of both films, which type of procedure appears to possess a greater latitude of exposure? Why?

6. Does kilovoltage appear to have a greater influence on radiographic density and radiographic contrast with or without intensifying screens? Explain.

7. Discuss the principles of ALARA related to the selection of nonscreen versus intensifying screen procedures.

Index

A

B

C

D

E

F

G

H

I

K

L

M

N

O

P

Q

R

S

T

U

V

W

X

No Longer Property of
EWU Libraries